COMPUTERIZED ENGINE CONTROLS

FOURTH EDITION—1994 UPDATE

Dick H. King

Glendale Community College

DELMAR PUBLISHERS INC.™

I(T)P™

NOTICE TO THE READER

Publisher does not warrant or guarantee any of the products described herein or perform any independent analysis in connection with any of the product information contained herein. Publisher does not assume and expressly disclaims, any obligation to obtain and include information other than that provided to it by the manufacturer.

The reader is expressly warned to consider and adopt all safety precautions that might be indicated by the activities described herein and to avoid all potential hazards. By following the instructions contained herein, the reader willingly assumes all risks in connection with such instructions.

The publisher makes no representations or warranties of any kind, including but not limited to, the warranties of fitness for particular purpose or merchantability, nor are any such representations implied with respect to the material set forth herein, and the publisher takes no responsibility with respect to such material. The publisher shall not be liable for any special, consequential or exemplary damages resulting, in whole or in part, from the reader's use of, or reliance upon, this material.

Delmar Staff
Senior Administrative Editor: Joan Gill
Production Editor: Christopher Chien
Design Supervisor: Susan C. Matthews

Cover Photo
Manfred Kage/PETER ARNOLD INC.

For information, address Delmar Publishers Inc.
3 Columbia Circle, Box 15-015
Albany, New York 12212-5015

Delmar Publishers' Online Services
To access Delmar on the World Wide Web, point your browser to:
http://www.delmar.com/delmar.html
To access through Gopher: gopher://gopher.delmar.com
(Delmar Online is part of "thomson.com", an Internet site with information on more than 30 publishers of the International Thomson Publishing organization.)
For information on our products and services:
email: info@delmar.com
or call 800-347-7707

COPYRIGHT © 1994
BY DELMAR PUBLISHERS INC.

The trademark ITP is used under license.

All rights reserved. Certain portions of this work copyright © 1987, 1989, 1990, 1991 and 1993. No part of this work covered by the copyright hereon may be reproduced or used in any form or by any means—graphic, electronic, or mechanical, including photocopying, recording, taping, and information storage and retrieval systems—without written permission of the publisher.

Printed in the United States of America
Published simultaneously in Canada
by Nelson Canada,
a division of the Thomson Corporation

10 9 8 7 6 5 4 3   XXX 00 99 98 97 96 95

Library of Congress Cataloging-in-Publication Data:

King, Dick H.
    Computerized engine controls / Dick H. King.
      p. cm.
Includes index.
ISBN 0-8273-6853-4 (textbook)
1. Automobiles — Motors — Control systems. 2. Automobiles—Electronic equipment. I. Title
TI214.C64K56   1992
629.25 —dc20

90-44007

CONTENTS

Preface vi

Acknowledgments viii

Introduction

    Electrical Circuits versus Electronic Circuits/*1*   Electron Theory/*2*
    Magnetism/*4*   Conductors and Insulators/*12*   Circuits/*12*   Polarity/*17*
    Circuit Controls: Faults/*18*   Testing/*19*   Semiconductors/*25*   Integrated Circuits (ICs)/*36*

Chapter 1   Computers on Cars

    Objectives/*37*   Why Computers?/*37*   Emissions, Mileage, and Driveability/*44*
    Economy and Driveability/*48*   Attitude of the Technician/*50*   Review Questions/*51*   Glossary/*52*

Chapter 2   Common Components for Computerized Engine Control Systems

    Objectives/*55*   Common Features/*55*   Sensing Devices/*60*   Actuators/*73*
    Diagnosis and Service/*76*   Review Questions/*79*   Glossary/*80*

Chapter 3   General Motors' Computer Command Control

    Objectives/*83*   Electronic Control Module (ECM)/*84*   Operating Modes/*84*
    Inputs/*86*   Outputs/*93*   System Diagnosis and Service/*116*   Review Questions/*131*   Glossary/*132*

Chapter 4   General Motors' Electronic Fuel Injection

    Objectives/*135*   Electronic Control Module/*136*   Operating Modes/*137*
    Inputs/*138*   Outputs/*141*   System Diagnosis and Service/*151*
    Review Questions/*156*   Glossary/*158*

Contents

Chapter 5 General Motors' Port Fuel Injection

Objectives/*159*   Electronic Control Module/*160*   Operational Modes/*161*
Fuel Supply System/*162*   Injectors/*163*   Throttle Body/*165*
Non-ECM Emission Control/*166*   Inputs/*168*   Outputs/*177*
System Diagnosis and Service/*197*   Review Questions/*201*   Glossary/*202*

Chapter 6 Cadillac's Digital Fuel Injection

Objectives/*205*   Electronic Control Module (ECM)/*207*   Operating Modes/*209*
Inputs/*210*   Outputs/*213*   Body Computer Module (BCM)/*226*
System Diagnosis and Service/*235*   Review Questions/*243*   Glossary/*245*

Chapter 7 Ford's Electronic Engine Control IV

Objectives/*247*   Electronic Control Assembly (ECA)/*247*   Inputs/*254*
Outputs/*272*   Vehicle Speed Control (VSC)/*300*   System Diagnosis and
Service/*301*   Review Questions/*308*   Glossary/*310*

Chapter 8 Ford's EEC I, EEC II, and EEC III

Objectives/*311*   Electronic Engine Control (EEC) I/*311*   EEC II/*318*
EEC III/*321*   Review Questions/*323*   Glossary/*324*

Chapter 9 Ford's Microprocessor Control Unit

Objectives/*325*   MCU Module/*325*   Inputs/*327*   Outputs/*331*
System Diagnosis and Service/*334*   Review Questions/*335*   Glossary/*336*

Chapter 10 Chrysler's Single-point and Multipoint Fuel Injection

Objectives/*337*   Logic Module/Power Module/*338*   Inputs/*341*   Outputs/*347*
System Diagnosis and Service/*354*   Review Questions/*356*

Chapter 11 Chrysler's Oxygen Feedback System

    Objectives/*357*   Combustion Control Computer/*358*   Inputs/*358*
    Outputs/*362*   System Diagnosis and Service/*365*   Review Questions/*366*
    Glossary/*367*

Chapter 12 Robert Bosch Corporation's Motronic

    Objectives/*369*   Control Unit/*369*   Operating Modes/*370*   Injectors/*374*
    Inputs/*374*   Outputs/*377*   System Diagnosis and Service/*384*
    Review Questions/*385*   Glossary/*386*

Appendix: 1991 Automobiles/387
           1992 Automobiles/401
           1993 Automobiles/421
           1994 Automobiles/435

Index/449

PREFACE

The application of the microprocessor with its related components and circuits has made automotive technology both exciting and fast paced. Recent technological developments and those that follow will require entry level automotive service technicians to be well trained in the principles of automotive technology and to be career-long students. Those who respond positively to this requirement will find the task challenging but achievable and rewarding.

This text was written in response to a widely recognized need within the industry: to help student-technicians get a commanding grasp of how computerized engine control systems work and how to diagnose problems within them. The student-technician who studies this text will soon come to realize that no single component or circuit within any given computerized engine control system, other than the computer itself, is complicated.

Computerized Engine Controls is written with the assumption that the reader is familiar with the principles of traditional engine, electrical system, and fuel system operation.

While it is recognized that a computerized engine control system does in fact become an integral part of an engine's electrical and fuel system, it is much too significant and complex to be taught as just a unit in an engine performance textbook or class. For purposes of instruction, this topic should be taken out of context and examined as a stand-alone system. Once the student-technician fully understands the system's purpose, operation, and diagnostic approach, the diagnostic procedures within the service manual will put the system back in its proper perspective as an integral part of the engine's support system.

Computerized Engine Controls presents each popular, multifunction, computer control system in a separate chapter. Each system is fully covered, with enough specific information and detail to enable the reader to get a complete and clear picture of how the system works. This text is written with the premise that understanding how the system works and what it should be doing not only makes the diagnosis process easier, but also makes the diagnostic literature much easier to understand. Correctly interpreting diagnostic procedure directions is often a problem if the technician is not aware of what the procedure is trying to measure, what normal readings or responses should be, and what conditions will cause abnormal readings or responses.

Frequently, information contained in a manufacturer's service manual is presented in this text to help make a specific concept more clear and to help acquaint the reader with information found in the service manual. Other references are made to the service manual to impress upon the reader the importance of using the service manual in diagnostic procedures. This text is dedicated to helping the student-technician acquire the necessary knowledge to be able to diagnose and repair driveability problems with a computerized engine control system.

Objectives are provided at the beginning of each chapter to help the reader identify the major concepts to be presented. Review questions are provided at the end of each chapter, which help the reader assess his or her recall and comprehension concerning the most important concepts. The review questions are also written to reinforce the objectives.

Personal safety concerns peculiar to specific computerized engine control systems are highlighted where applicable. Terms peculiar to computerized engine control systems are highlighted when they are first introduced and their meaning is explained in the related text. A glossary of such terms is also provided at the end of each chapter. For those chapters where a significant number of abbreviations are used, a list of abbreviations is also provided at the end. Background information that is not of specific concern to understanding a given concept but might be of general interest to the reader is inserted in selected locations throughout the text.

SUGGESTIONS ON HOW TO USE TEXT

The Introduction should be read first. This chapter provides background on the operation of many components that are common to most systems. The remaining chapters, which are specific to individual systems, can be read in any order. Many students may not wish to study each specific chapter or their instructors may not choose to assign the study of each chapter. It is suggested, however, that at least three of the specific system chapters be selected for study following the completion of the Introduction. Remaining chapters can then be skimmed or serve as a reference for future use.

ACKNOWLEDGMENTS

I am grateful to several people who have helped with the many activities involved in the creation of this book: to Therese Murphy who generously gave of her personal time to type the proposal chapter; to Margery Rothschild for her hours of work and consistently quick turnaround time in processing illustrations; to Tim Turpin for the photographs used in the book; to Renault Catalano for his consultation concerning electronics, and to the following for their critical reviews and/or answers to questions:

Clem Childress
Fisher Technician Institute
Virginia Beach, Virginia

Jacob Darakjian
Leeward Community College
Pearl City, Hawaii

INTRODUCTION

A REVIEW OF ELECTRICITY AND ELECTRONICS

The earliest automobiles had little in the way of electrical systems, but as the automobile has become more sophisticated and more accessories have been added, *electrical* or *electronic* systems have become a major feature of today's car. Electronic control systems have been and will continue to be added to make the automobile comply with government standards and consumer demands. Automotive manufacturers tell us that before the 1990s have passed, most major automotive systems will be controlled by computers.

This increased use of electrical and electronic systems means two things for the automotive service technician: First, all service technicians need skills in electrical diagnosis and repair if they are going to be effective, regardless of their service specialty; second, technicians with such skills will command greater financial rewards.

There are several principles that electrical systems operate by, but they are all fairly simple; learning them is not difficult. As each principle is introduced to you through your reading or in class, ask questions and/or read until you understand it, review it frequently, and practice the exercises that your instructor assigns.

ELECTRICAL CIRCUITS VERSUS ELECTRONIC CIRCUITS

The differences between electrical circuits and electronic circuits are not always clear-cut or definite. This has led to some confusion about the use of those terms and how an *electronic* system is different from an *electrical* system. Perhaps the following comparisons will help.

Even though the use of solid-state components may be used as criteria to identify an electronic circuit, solid-state components such as a *power transistor* may also be used in an electrical circuit. A power transistor is a type of transistor designed to carry larger amounts of

Electrical Circuits	Electronic Circuits
Do work: motion, heat, light, or magnetism	Communicate information: voltage values or on-off signals
Use *electro/mechanical* devices such as switches, solenoids, or relays	Use *solid-state* devices with no moving parts *(semiconductors)*
Operate on relatively high amperage values	Operate on very low amperage values
Circuits contain relatively low resistance	Circuits contain very high resistance

Introduction A Review of Electricity and Electronics

amperage than are normally found in an electronic circuit.

ELECTRON THEORY

Molecules and Atoms

A study of electricity begins with the smallest pieces of matter. All substances—air, water, wood, steel, stone, and even the various substances that our bodies are made of—are made of the same bits of matter. Every substance is made of units called *molecules*. A molecule is a unit formed by combining two or more atoms and is the smallest unit that a given substance can be broken down to and still exhibit all of the characteristics of that substance. For example, a molecule of water is made up of two atoms of hydrogen and one atom of oxygen (H_2O: H is the chemical symbol for hydrogen and O is the chemical symbol for oxygen). If a molecule of water is broken down into its component atoms, it is no longer water.

As molecules are made up of atoms, atoms are in turn made up of:

- electrons, or negatively charged particles
- protons, or positively charged particles
- neutrons, or particles with no charge; these just add mass to the atom

The smallest and lightest atom is the hydrogen atom. It contains one proton and one electron (it is the only atom that does not have a neutron), Figure I-1. The next smallest and lightest atom is the helium atom. It has two protons, two neutrons, and two electrons, Figure I-2. Since the hydrogen atom is the smallest and lightest, and since it has one electron and one proton, it is given an atomic number of 1. Since helium is the next lightest with 2 electrons, protons, and neutrons it has an atomic number of 2. Every atom has been given an atomic number that indicates its relative size and weight (or its mass) and the number of

FIGURE I-1 Hydrogen atom

FIGURE I-2 Helium atom

electrons, protons, and neutrons it contains. An atom normally has the same number of electrons, protons, and neutrons.

Elements

Once the three different bits of matter are united to form an atom, two or more atoms combine to form a molecule. If all of the atoms in the molecule are the same, the molecule is called an *element*. Which element it is depends on how many protons, neutrons, and electrons the atoms contain. There are more than 100 different elements. Some examples of elements are hydrogen, oxygen, carbon, iron, lead, gold, silicon, and sodium. An element, then, is a pure substance whose molecules contain only one kind of atom. Substances such as water, which contains hydrogen and oxygen atoms, and steel, which contains iron and carbon atoms, are called *compounds*.

Atomic Structure and Electricity

Notice in Figures I-1 and I-2 that the protons and neutrons are grouped together in the center of each atom. This is called the *nucleus* of the atom. The electrons travel around the nucleus of the atom in an orbit similar to the way that the earth travels around the sun. But, because an atom usually has several electrons orbiting around its nucleus, the electrons form in layers, rather than all of them traveling in the same orbit, Figure I-3. Some, however, will share the same orbit, as seen in Figure I-3. For the purpose of this text, only the electrons in the last layer are of any real importance. This layer is often called the *outer shell* or *valence ring*.

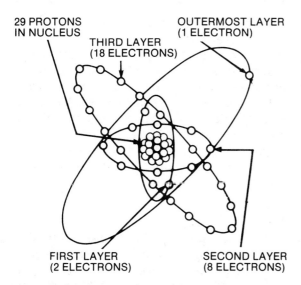

FIGURE I-3 Layers of electronics around copper atom

As mentioned, electrons are negatively charged and protons are positively charged. You have probably heard or know that like charges repel and unlike charges attract. And electrons are always moving; in fact, they are moving at the speed of light. These characteristics work together to explain many of the behaviors of an atom that makes current flow. *Current* is defined as having a mass of free electrons moving in the same direction.

Note: There are two types of current: direct current (DC) and alternating current (AC). Direct current always flows in one direction. Current from a battery is the best example. Most of the devices in an automobile use DC. Alternating current circuits continuously switch which end is positive and which end is negative so that current flow reverses direction repeatedly. The power that is available from commercial utility companies is AC and cycles 60 times per second. One cycle occurs when the current switches from forward to backward to forward again. The alternator (an alternating current generator) produces AC current, which is converted to DC before it leaves the alternator.

The fast-moving electron wants to move in a straight line, but its attraction to the proton nucleus makes it act like a ball tied to the end of a string that is being twirled around. The repulsive force between the electrons keeps them spread as far apart as their attraction to the nucleus will allow.

The fewer electrons there are in the outer shell of the atom and the more layers of electrons there are under the outer shell, the weaker is the bond between the outer electrons and the nucleus. If one of these outer electrons can somehow be broken free from its orbit, it will

Introduction A Review of Electricity and Electronics

travel to a neighboring atom and fall into the outer shell there, resulting in two unbalanced atoms. The first atom is missing an electron. It is now positively charged and is called a *positive ion*. The second atom has an extra electron. It is negatively charged and is called a *negative ion*. Ions are unstable. They want to either gain an electron or get rid of one so that they are balanced.

Potential

The first atom in the example has positive *potential*. It has more positive charge than negative charge because it has more protons than electrons. Suppose that this atom is at one end of a circuit, Figure I-4. Further suppose that there is an atom at the other end of the circuit that has an extra electron; it has negative potential. Because of the difference in potential at the two ends of the circuit, an electron at the negatively charged end will start moving toward the positively charged end. The greater the difference in potential (the greater the number of opposite charged ions) at each end of the circuit, the greater the number of electrons that will start to flow.

This situation can be created by putting something between the two ends of a circuit that will produce positive and negative ions. This is what a battery or generator does in a circuit, Figure I-4. If you connect the ends of the wire to a battery and put some kind of resistance in the wire (why the resistance is needed is explained later), the example will work perfectly. Remember that it is the difference in potential that makes current flow. Actually there are three factors that must be present for an electrical circuit to work.

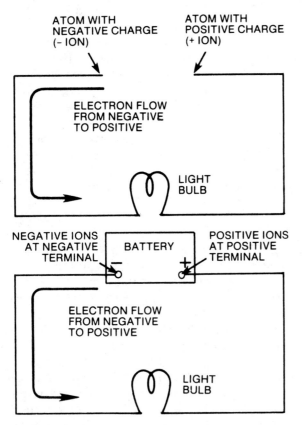

FIGURE I-4 Negative versus positive potential

MAGNETISM

Magnetism is closely tied to the generation and use of electricity. In fact, one of the prevailing theories is that magnetism is caused by the movement of electrons. Some materials strongly demonstrate the characteristics associated with magnetism and some do not. Those that strongly demonstrate the characteristics of magnetism such as iron are said to have high *permeability*. Those that do not such as glass, wood, and aluminum are said to have high *reluctance*.

Lines of Force

It is not known that there is such a thing as a *magnetic line of force*. What is known, however, is that magnetism exerts a force. Magnetic force is linear in nature and it can be managed to do many kinds of work. By assigning certain characteristics to these lines of force, the behavior of magnetism can be explained. Magnetic lines of force:

1. have a directional force (north to south outside the magnet).
2. want to take the shortest distance between two poles (just like a stretched rubber band between the two points from which it is held).
3. form a complete loop.
4. are more permeable to iron than air.
5. resist being close together (especially in air).
6. resist being cut.
7. will not cross each other.

Magnetic lines of force extending from a magnet make up what is commonly called a magnetic field and more correctly called magnetic *flux*, Figure I-5. If a magnet is not near an object made of permeable material, the lines of force will extend from the north pole through the air to the south pole (characteristic 1). The lines of force will continue through the body of the magnet to the north pole to form a complete loop (characteristic 3). Every magnet has a north and a south pole. The poles are the two points of a magnet where the magnetic strength is greatest. As the lines of force extend out of the north pole, they begin to spread out. Here you see opposition between characteristics 2 and 5. The lines of force want to take the shortest distance between the poles, but they spread out because of their tendency to repel each other (characteristic 5). The result is a magnetic field that occupies a relatively large area but has greater density near the body of the magnet.

Because the body of the magnet has high permeability, the lines of force are very concentrated in the body of the magnet (characteristic 4). This accounts for the poles of the magnet having the highest magnetic strength.

If there is an object with high permeability near the magnet, the magnetic lines of force

FIGURE I-5 Magnetic field

will distort from their normal pattern and go out of their way to pass through the object, Figure I-6. The tendency for the lines of force to pass through the permeable object is stronger than their tendency to take the shortest route. The lines of force will, however, try to move the object toward the nearest pole of the magnet.

Electromagnets

Early researchers discovered that when current passes through a conductor, a magnetic field forms around the conductor, Figure I-7. This principle makes possible the use of electromagnets, electric motors, generators, and most of the other components used in electrical circuits.

Introduction A Review of Electricity and Electronics

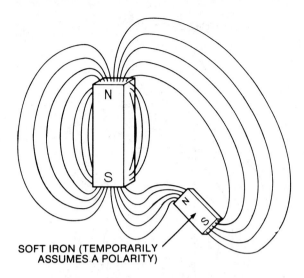

FIGURE I-6 Magnetic field distortion

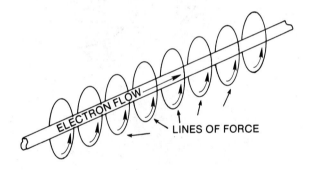

IF YOU IMAGINE PLACING YOUR LEFT HAND ON THE WIRE WITH YOUR THUMB POINTING IN THE DIRECTION OF ELECTRON FLOW, YOUR FINGERS WILL BE POINTING IN THE DIRECTION OF THE DIRECTIONAL FORCE OF THE MAGNETIC LINES OF FORCE. WHEN THINKING IN TERMS OF CONVENTIONAL CURRENT FLOW, THE SAME WOULD APPLY FOR THE RIGHT HAND.

FIGURE I-7 Lines of force formed around conductor

If a wire is coiled with the coils close together, most of the lines of force wrap around the entire coil rather than go between the coils of wire, Figure I-8. This is because if they do try to wrap around each loop in the coil, they must

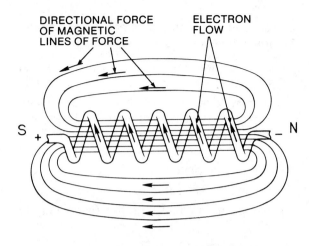

IMAGINE PLACING YOUR LEFT HAND, WITH THUMB EXTENDED, AROUND THE COIL WITH YOUR FINGERS POINTING IN THE DIRECTION OF THE ELECTRON FLOW THROUGH THE COILS. YOUR THUMB WILL POINT TO THE NORTH POLE OF THE MAGNETIC FIELD.

FIGURE I-8 Magnetic field around coil

cross each other, which they will not do (characteristic 7), Figure I-9.

If a high permeability core is placed in the center of the coil, the magnetic field becomes much stronger because the high permeability core replaces low permeability air in the center of the coil and more lines of force develop,

MAGNETIC LINES OF FORCE WILL NOT CROSS.

FIGURE I-9 Lines of force cannot cross

Introduction A Review of Electricity and Electronics

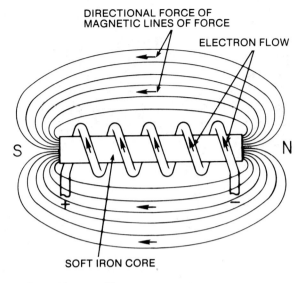

FIGURE I-10   Electromagnet

Figure I-10. If the core is placed toward one end of the coil, Figure I-11, the lines of force exert a strong force on it to move it toward the center so that they can follow a shorter path. If the core is movable, it will move to the center of the coil. A coil around an off-center, movable, permeable core is a *solenoid*. A spring is usually used to hold the core off center.

Motors

In an electric motor, current is passed through a conductor that is looped around the *armature* core, Figure I-12. The conductor loops are placed in grooves along the length of the core. The core is made of laminated discs of permeable, soft iron that are pressed on the armature shaft. The soft iron core causes the magnetic field that forms around the conduc-

FIGURE I-11   Solenoid

Introduction A Review of Electricity and Electronics

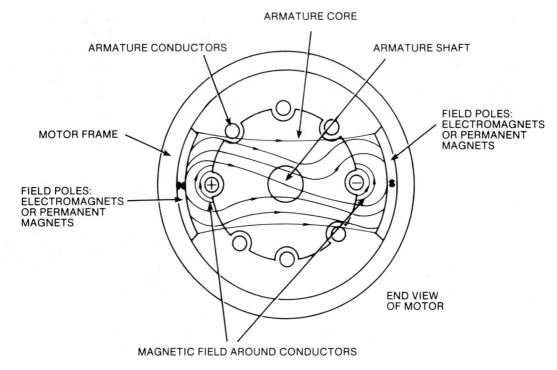

FIGURE I-12   Electric Motor

tor to be stronger due to its permeability. There are several conductor loops on the armature, but only the loop that is nearest the center of the field poles has current passing through it. The loops are positioned so that when one side of a loop is centered on one field pole, its other side is centered on the other field pole.

The field poles are either permanent magnets or pieces of soft iron that serve as the core of an electromagnet. If electromagnets are used, an additional conductor (not shown in Figure I-12) is wound around each field pole, and current is passed through these field coils to produce a magnetic field between the field poles. The motor frame that the poles are mounted on acts as the magnet body.

Looking at the armature conductor near the north field pole in Figure I-12, you see its magnetic field extending out of the armature core and that it has a clockwise force. The magnetic field between the field poles has a directional force from north pole to south pole. At the top of the armature conductor, the field it has produced has a directional force in the same direction as that of the lines of force between the field poles. The lines of force in this area are compatible, but combining these two fields in the same area produces a high-density field. Remember that magnetic lines of force do not like to be close together.

At the bottom of the armature conductor, the lines of force formed around it have a directional force that is in the opposite direction of those from the north field pole. The lines of force will not cross each other, so some from the field pole distort and go up and over the conductor into the already dense portion of the field above the conductor, and some just cease to exist. This produces a high-density field above the conductor and a low density below it.

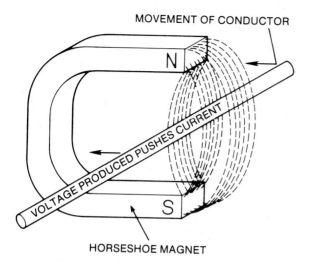

FIGURE I-13 Magnetic induction

The difference in density is similar to a difference in pressure. This produces a downward force on the conductor.

The other side of the armature loop, on the other side of the armature, is experiencing the same thing except that the current is now traveling the opposite way. The loop makes a U-shaped bend at the end of the armature. The magnetic field around this part of the conductor has a counterclockwise force. Here, the lines of force around the conductor are compatible with those between the field poles under the conductor but try to cross at the top. This produces an upward force on this side of the armature loop. The armature rotates counterclockwise. To change the direction in which the armature turns, change either the direction that current flows through the armature conductors or change the polarity of the field poles.

Magnetic Induction

Passing voltage through a wire causes a magnetic field to form around the wire. However, if lines of force can be formed around a conductor, a voltage is produced in the wire and current starts to flow. This is assuming, of course, that the wire is part of a complete circuit. Lines of force can be made to wrap around a conductor by passing a conductor through a magnetic field, Figure I-13. This phenomenon occurs because of characteristic 6. As the conductor passes through the magnetic field, it cuts each line of force. Because the lines of force resist being cut, they first wrap around the conductor much like a blade of grass would if struck by a stick, Figure I-14. This principle is used in generators to produce voltage and current flow. The principle will work regardless of whether:

- the conductor is moved through a stationary magnetic field, as in a direct current generator.
- a magnetic field is moved past stationary conductors, as in an alternating current generator.
- the lines of force in an electromagnetic field are made to move by having the circuit producing the magnetic field turned on and off, as in an ignition coil.

Note that in each case, movement of either the lines of force or the conductor is needed. A

FIGURE I-14 Cutting lines of force

magnetic field around a conductor where both are steady state will not produce voltage. The amount of voltage and current that is produced by magnetic induction depends on four factors:

1. The strength of the magnetic field (how many lines of force there are to cut). A tiny amount of voltage is induced in the wire by each line of force that is cut.
2. The number of conductors cutting the line of force. Winding the conductor into a coil and passing one side of the coil through the magnetic field cuts each line of force as many times as there are loops in the coil.
3. How fast and how many times the conductors pass through the magnetic field.
4. The angle between the lines of force and the conductor's approach to them.

Amperage

Amperage is a measure of the amount of current flowing in a circuit. One *ampere* equals 6,250,000,000,000,000,000 (six and one-quarter billion billion) electrons moving past a given point in a circuit per second. This is often expressed as one *coulomb*.

Voltage

A *volt* is a measure of the force or pressure that caused current to flow; it is often referred to as *voltage*. The difference in potential is voltage. The most common ways of producing voltage are chemically, as in a battery, or by *magnetic induction,* as in a generator. A more accurate but less used name is *electromotive force*.

Resistance

The fact that voltage is required to push current through a circuit suggests that the circuit offers *resistance*. In other words, you do not have to push something unless it resists moving. Resistance, which is measured in *ohms*, limits the amount of amperage that flows through a circuit, Figure I-15. If a circuit with-

LOW RESISTANCE ALLOWS MORE CURRENT TO FLOW.

HIGHER RESISTANCE ALLOWS LESS CURRENT TO FLOW.

FIGURE I-15 Resistance versus current flow

out enough resistance is connected across a reliable voltage source, wires or some other component in the circuit will be damaged by heat because too much current flows.

As mentioned, a bond exists between an electron and the protons in the nucleus of an atom. That bond must be broken for the electron to be freed so that it can move to another atom. Breaking that bond and moving the electron amount to doing work. Doing that work represents a form of resistance to current flow. This resistance varies from one conductive material to another, depending on the atomic structure of the material. For example, lead has more resistance than iron, and iron has more resistance than copper. It also varies with the temperature of the conductor. Loose or dirty connections in a circuit also offer resistance to current flow. Using current flow to do work to create heat, light, a magnetic field, or to move something also amounts to resistance to current flow.

Sometimes students confuse voltage and amperage while doing tests on electrical systems. Review these definitions and consider the influence that voltage and amperage have on a circuit. It might also help to remember that voltage can be present in a circuit without current flowing. However, current cannot flow unless voltage is present.

Voltage Loss

When voltage pushes current through a circuit, voltage is lost by being converted to some other energy form (heat, light, magnetism, or motion). This is usually referred to as *volt drop*. Every part of a circuit offers some amount of resistance—even the wires, although the resistance in the wires should be very low, Figure I-16. The volt drop in each part of the circuit is proportional to the resistance in that part of the circuit. **The total voltage dropped in a circuit must equal the source voltage.** In other words, all of the voltage applied to a circuit must be converted to another energy form within the circuit.

Ohm's Law

Ohm's law defines a relationship between amperage, voltage, and resistance. Ohm's law

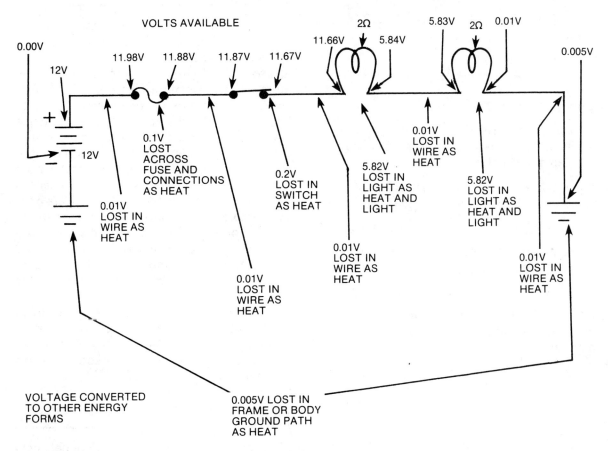

FIGURE I-16 Volt drop

Introduction A Review of Electricity and Electronics

says: It takes one volt to push one ampere through one ohm of resistance.

Ohm's law can be expressed in one of three simple mathematic equations:

$$E = I \times R \quad I = E/R \quad R = E/I$$

where: E = electromotive force or voltage
I = intensity or amperage
R = resistance

The simplest application of Ohm's law enables you to find the value of any one of the three factors: amperage, voltage, or resistance if the other two are known. For example, in the diagram, if the voltage is 12 volts and the resistance is 2 ohms, the current flow can be determined as follows:

I = E/R or I = 12 v/2 Ω = 6 amps

(The symbol Ω is often used as an abbreviation for ohms, and amps is often used as an abbreviation for amperes.)

If the resistance is 4 ohms and the current is 1.5 amps, the voltage applied can be found by:

E = I × R or E = 1.5 amps × 4 Ω = 6 v

If the voltage is 12 volts and the current is 3 amps, the resistance can be found by:

R = E/I or R = 12 v/3 amps = 4 Ω

Perhaps the easiest way to remember how to use these equations is to use the following diagram:

To find the value of the unknown factor, cover the unknown factor with your thumb and multiply or divide the other factors as their positions indicate.

There are many other applications of Ohm's law, some of which are quite complex. (A more complicated application is covered later in this chapter.) An automotive technician is rarely required to directly apply Ohm's law in order to find or repair an electrical problem on a vehicle. But knowing and understanding the relationship of the three factors just covered is a must for the technician who wants to effectively diagnose and repair electrical systems.

CONDUCTORS AND INSULATORS

Earlier, in discussing the electrons in the outer shell of an atom, it was said that the fewer electrons there are in the outer shell the easier it is to break them loose from the atom. If an atom has five or more electrons in its outer shell, the electrons become much harder to break away from the atom, and the substance made up of those atoms is a very poor conductor; so poor, it is classed as an insulator. Rubber, most plastics, glass, and ceramics are common examples of insulators. Substances with four electrons in their outer shell are poor conductors but can become good ones under certain conditions. Thus, they are called *semiconductors*. Silicon and germanium are good examples of semiconductors. (Semiconductors are covered in more detail later in this chapter.)

Conductors are substances made up of atoms with three or fewer electrons in their outer shell. These electrons are called *free electrons* because they can be freed to travel to another atom.

CIRCUITS

Almost all electrical circuits in vehicles will have some type of component to provide each of the following functions:

Introduction A Review of Electricity and Electronics

FIGURE I-17   Circuit components

Introduction A Review of Electricity and Electronics

- *Voltage Source,* such as a battery or generator, provides voltage to the power circuit.
- *Circuit Protection,* such as a fuse or circuit breaker, serves to open the circuit in the event of excessive current flow.
- *Circuit Control,* such as a switch or relay, provides the ability to open and close the circuit.
- *Load,* such as a motor or light, does the function the circuit was designed to do and also provides resistance to limit current.
- *Conduct Current,* usually wires or certain parts of the vehicle, such as the frame, body, or engine block.

These components are often represented by symbols as shown in Figure I-17.

Circuit Types

There are two distinct types of electrical circuits, plus a combination of the two.

Series Circuits. In a series circuit, there is only one path for current flow, and all of the current flows through every part of the circuit. Parts A and B of Figure I-17 show simple series circuits. Even though there is only one load in each of these circuits, they qualify as series circuits because there is only one path for current flow. Figure I-18 shows a better example of a series circuit. Not only is there just one path for current flow, there are also two loads in

FIGURE I-19 Series circuit

series with each other. When there are two or more loads in series, the current must pass through one before it can pass through the next. The characteristics of a series circuit include the following:

- Current flow is the same at all parts of the circuit.
- Resistance units are added together for total resistance.
- Current flow decreases as resistance units are added.
- A portion of source voltage is dropped at each resistance unit in the circuit.
- An open in any part of the circuit disrupts the entire circuit.

The following problems apply Ohm's law to a series circuit. Refer to Figure I-19, which shows a compound series circuit.

Problem 1. Assume that the resistance of R_1 is 2 ohms and that R_2 is 4 ohms; find the total current flow. In a series circuit, the resistance value of each unit of resistance can be added together because all of the current passes through each resistor.

$I = E/R$, or $I = E/R_1 + R_2$, or
$I = 12 v/2 \Omega + 4 \Omega$, or $I = 12 v/6 \Omega = 2$ amps

Problem 2. Assume the resistance values are unknown in Figure I-19, but that the total current flow is 3 amps. To find the total resistance:

$R_T = E/I$, or $R_T = 12 v/3 a = 4 \Omega$

Problem 3. Find the volt drop across R_1, using the same values from problem 1. Since you are

FIGURE I-18 Series circuit

FIGURE I-20   Parallel circuit

only concerned with the volt drop across R_1, use only the resistance of R_1.

$V_1 = R_1 \times I$, or $V_1 = 2\,\Omega = 2$ amps $= 4$ v

Parallel Circuits. In a parallel circuit, the conductors split into branches with a load in each branch, Figure I-17C and Figure I-20. Some current will flow through each branch with the most current flowing through the branch with the least resistance. Characteristics of a parallel circuit are:

- Current varies in each branch (unless resistance in each branch is equal).
- Total circuit resistance goes down as more branches are added and will always be less than the lowest resistance unit in the circuit.
- Total circuit current flow increases as more branches are added.
- Source voltage is dropped across each branch.
- An open in one branch does not affect other branches.

The following problems apply Ohm's law to a parallel circuit.

Problem 4. In the following diagram assume that R1 is 3 ohms and R2 is 4 ohms. Find the current flow through the R1 branch. Note that R1 is not affected by R2, so treat R1 like it is a series circuit.

$I = E/R$, or $I = 12$ v$/3\,\Omega = 4$ amps

Find the current flow through R2.

$I = E/R$, or $I = 12$ v$/4\,\Omega = 3$ amps

What is the total current flow in the circuit?

4 amps + 3 amps = 7 amps

Problem 5. Now that you know the current flow, you can easily find the total circuit resistance.

$R_t = E/I$, or $R_t = 12$ v$/7$ amps $= 1.71\,\Omega$

Note that to solve for total resistance you have simply solved each branch for current flow, added the flow of each branch, and divided the source voltage by total current. This method will work regardless of how many branches there are in the circuit.

There are other ways to find total resistance in a parallel circuit. If there are only two branches in the circuit, the product over the sum method can be used. In problem 4 there were two resistors in a parallel circuit with values of 3 ohms and 4 ohms. Using those values:

$$\frac{R1 \times R2}{R1 + R2} \text{ or } \frac{3\,\Omega \times 4\,\Omega}{3\,\Omega + 4\,\Omega} = \frac{12\,\Omega}{7\,\Omega} = 1.71\,\Omega$$

If there are more than two branches in a circuit, this method can still be used, but it has to be manipulated a bit. Assume there are four branches in a circuit. Select two branches and use the product over the sum method to solve them as just shown. Then take the other two and do the same for them. Now take the two answers and run them through the same formu-

Introduction A Review of Electricity and Electronics

Ia. The final result is total circuit resistance. There are still other methods, but you have learned all you need to know to solve problems you are likely to encounter. Many automotive electrical systems texts, electrical texts, and reference books present those methods if you want to pursue them.

Problem 6. Finding voltage drop in a parallel circuit is the easiest of all. Using the values for branch 1 in problem 4 again:

$$E = I \times R_1, \text{ or } E = 4 \text{ amps} \times 3\,\Omega = 12 \text{ v}$$

Branch 2 looks like this:

$$E = I \times R_2, \text{ or } E = 3 \text{ amps} \times 4\,\Omega = 12 \text{ v}$$

Voltage drop for the complete circuit:

$$E = I \times R_T, \text{ or } E = 7 \text{ amps} \times 1.71\,\Omega = 12 \text{ v}$$

Or you can just know that the volt drop across any branch of the whole circuit will be source voltage.

Series-Parallel Circuits. Some circuits have characteristics of a series as well as those of a parallel circuit. There are two basic types of series-parallel circuit. The most common is a parallel circuit with at least one resistance unit in series with all branches, Figure I-21. All of the current flowing through the circuit in Figure I-21 must pass through the indicator light before it divides to go through the two heat elements, which are in parallel with each other.

To solve for current, resistance, or volt drop values in a series-parallel circuit, you must identify how the resistance units relate to each other, then use whichever set of formulas (series circuit or parallel circuit) apply. For example:

Problem 7. In Figure I-21, assume a resistance of 4 ohms for the indicator light (R_3), 20 ohms for R_2, and 30 ohms for R_3. Find the total resistance for the circuit.

Branches R_1 and R_2 are in parallel with each other and in series with R_3. The battery sees the resistance of R_3 plus the resistance of the two parallel branches. Once you find the total resistance of the branch circuits, you can treat that as a single resistance value that is in series with R_3. Using the product over the sum method, first solve for the combined resistance of R_1 and R_2.

$$R = \frac{R_1 \times R_2}{R_1 + R_2} \text{ or } \frac{20\,\Omega \times 30\,\Omega}{20\,\Omega + 30\,\Omega} = \frac{600\,\Omega}{50\,\Omega} = 12\,\Omega$$

Now add the total resistance of R_1 and R_2 (12 ohms) to the resistance of R_3 (4 ohms).

$$R_T = R + R_3 \text{ or } 12\,\Omega + 4\,\Omega = 16\,\Omega$$

Problem 8. Find the total current flow for the circuit.

$$I_T = E/R_T, \text{ or } I_T = 12/16\,\Omega = 0.75 \text{ amp}$$

Before you can find the current flow through R_1 or R_2, you must find the voltage applied to R_1 and R_2. Remember that R_3 is in series with both R_1 and R_2. Also remember that in series circuit, a portion of the source voltage is dropped across each resistance unit. Therefore, because there is a volt drop across R_3, there is not a full 12 volts applied to R_1 and R_2.

Problem 9. Find the volt drop across R_3.

$$V_3 = I_T \times R_3, \text{ or } V_3 = 0.75 \text{ amp} \times 4\,\Omega = 3 \text{ v}$$

FIGURE I-21   Series-parallel circuit

Introduction A Review of Electricity and Electronics

FIGURE I-22 Series string in parallel

If 3 volts are dropped across R_3 from a source voltage of 12 volts, 9 volts are applied to R_1 and R_2.

Problem 10. Find the current flow through:

R_1: $I_1 = E/R_1$, or $I_1 = 9\text{ v}/20\text{ }\Omega = 0.45$ amp

R_2: $I_2 = E/R_2$, or $I_2 = 9\text{ v}/30\text{ }\Omega = 0.30$ amp

The second type of series-parallel circuit is a parallel circuit in which at least one of the branch circuits contains two or more loads in series, Figure I-22. This type of circuit is often referred to as *series string in parallel.* Finding the total current, resistance, or voltage drop for this circuit is easy. Identify the branch (or branches) that has loads in series and combine the resistances for each branch. Then treat the circuit like any other parallel circuit. When troubleshooting this kind of circuit in a car, failing to recognize the branch of a circuit you are testing as having loads in series could cause confusion.

Note: If you are not absolutely sure about the circuit you are testing consult the correct wiring diagram for the vehicle.

POLARITY

Just as a magnet has polarity—a north and a south pole that determine the directional force of the lines of force—an electrical circuit has a polarity. Instead of using north and south to identify the polarity, an electrical circuit's polarity is identified by positive and negative. An electrical circuit's polarity is determined by its power source. The best example is a battery, Figure I-23. The side of the circuit that connects to the positive side of the battery is positive and on most vehicles is the insulated side. The voltage on this side is usually near source voltage. Most of the voltage is dropped in the load where most of the work is done. The negative side of the circuit carries the current from the load to the negative side of the battery. There is usually very little resistance in the negative side of the circuit, and the voltage is near zero. Because this side of the circuit has the same potential as the vehicle's frame and sheet metal, this side of the circuit might not be insulated.

FIGURE I-23 Circuit polarity

Introduction A Review of Electricity and Electronics

CIRCUIT CONTROLS: FAULTS

Three kinds of faults can occur in an electrical or electronic circuit: opens, excessive resistance, and shorts.

Opens

By far the most common fault in electrical and electronic circuits is an open circuit, or open. An open circuit means that there is a point in the circuit where resistance is so high that current cannot flow. An open can be the result of a broken wire, loose or dirty connection, blown fuse, or faulty component in the load device. Experienced technicians know that wires rarely break except in applications where the wire experiences a lot of flexing. Most opens occur in connections, switches, and components. A switch in a circuit provides a way to conveniently open the circuit. Of course, when a circuit is opened deliberately, it is not a fault.

Excessive Resistance

A loose or dirty connection or a partially cut wire can cause excessive resistance in a circuit. Under these conditions, the circuit can still work but not as well as it could because the additional resistance reduces current flow through the circuit, Figure I-24. Excessive resistance can also result from a faulty repair or modification of a circuit in which a wire that is too long or too small in diameter has been installed. The location of excessive resistance in a circuit can be easily found with a series of voltage drop tests that are discussed under **Testing** in this chapter.

Shorts

Most faults in an electrical circuit are called shorts by the uninformed. A short is a fault in a circuit that causes current to bypass a portion or all of the circuit's resistance. The term *short* as used in electrical terminology means that the current is taking a shortcut rather than following the path it is supposed to take, Figure I-25. Remember Ohm's law: if resistance is reduced, current flow increases.

FIGURE I-25 Shorts

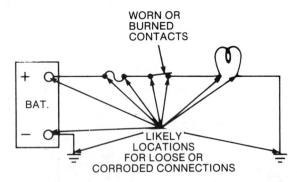

FIGURE I-24 Possible locations in a circuit for openings or high resistance

Short to Positive. In Figure I-25A, electrical contact has developed between two windings of a coil. This causes current to shortcut the loop or loops that exist between the points where the contact occurred, which lowers the coil's resistance. If only a relatively small number of the coil's loops are shorted, the increase in current flow might not be enough to further damage the circuit except possibly to blow a fuse if the circuit is fuse protected. The device might even continue to work but probably at reduced efficiency.

Short to Ground. If a short occurs up circuit from the load (between the load and the positive source), Figure I-25B, it provides a lower resistance path to ground for current flow. Because of the lower resistance, excessive current will flow, and the fuse element will melt, causing the circuit to be open. If the circuit does not have a fuse or some other type of circuit protection, the wire between the positive terminal and the short will overheat, possibly melting the wire or starting a fire.

If a short occurs down circuit (between the load and the negative terminal), Figure I-25C, the switch will not be able to provide the needed resistance to open the circuit. This is because the current has found a shortcut to ground that the switch cannot control.

If the circuit in question is controlled by an electronic switching circuit, as you will see in following chapters, this condition is quite threatening to the solid-state circuit because the additional current flow might damage it.

TESTING

Basic electrical test procedures are easy to learn. They are essential to effective electrical troubleshooting and can save considerable time in identifying a problem. Effective electrical testing requires the following:

- understanding the principles of electricity
- understanding the circuit being tested
- knowing how to use the test equipment
- knowing the three things above well enough to know what the test results mean

Testing Voltage

The simplest voltage test is an *available volts test*. It can be performed with a non-self-powered test light or a voltmeter. The test light usually comes with two leads: one with an alligator clip that hooks to a negative or ground source, and one with a sharp probe that can penetrate the insulation to contact a positive source, Figure I-26. By connecting the clip of a 12-volt test light to a known good ground (negative source) and contacting a positive source with the probe, the light will burn brightly if there are 12 volts available at the point being probed. If there are fewer than 12 volts available, the light will be less bright—as low as about 6 volts. A voltmeter used the same way indicates precisely how much voltage is available, Figure I-27.

Test Light versus Voltmeter

Most technicians prefer to use a test light for simple available volt tests. The test light is less expensive, requires less careful storage and handling, and does not have to be turned on and off; in short, it is more convenient to use. A voltmeter should be used whenever possible so that:

- you become more comfortable with its use. In some situations you will have to use a voltmeter. Do not make a diagnostic mistake because you make a mistake using or reading the meter.
- you develop a sense of what volt readings to expect on a properly operating circuit and what abnormal readings mean. A reading that is off by a few tenths of a volt can indicate a potential problem to a knowledgeable technician.

Introduction A Review of Electricity and Electronics

FIGURE I-26   Non-self-powered test light

FIGURE I-27   Available volts test

FIGURE I-28   Series circuit

The first thing to do when starting an electrical test procedure is to test the tester. With a battery-powered multimeter, test the meter's battery first. Most multimeters have a convenient built-in method to check the battery. A quick way to check a voltmeter or test light is to put it across the vehicle's battery. A fully charged 12-volt battery in good condition should read around 12.6 volts with no load across it. After verifying that the meter is working properly, select the correct scale on the meter. Most meters have at least two different scales for reading voltage: one for low values and one for high values. Select the scale that goes just a little beyond the voltage value you expect to see. For instance, if you are testing battery voltage on a 12-volt vehicle with the engine running, you should expect to see something between 12 and 15 volts. If your meter has a 3-volt scale and a 20-volt scale, select the 20-volt scale. Some meters are auto-ranging. That is, they automatically select the correct scale when connected to a circuit. **Do not condemn a component or circuit until you are sure the tester is working properly and you are using it correctly.**

Figure I-28 shows a circuit with several points indicated. Review the following chart and be sure that you can explain why each set of voltage values given indicates the faults listed.

Voltage Drop

Normally, in order for the load device in a circuit to work as well as it was intended, at least 85% of the source voltage should be dropped across it. This is assuming, of course, that the source voltage is what the circuit was designed for. When comparing the volt drop across the load to the source voltage, be sure to take the source voltage reading with the circuit in question turned on. This is because connecting the voltage source's terminals together, which is what the circuit does, lowers the difference in potential between the terminals. The lower the resistance in the circuit, the more it lowers the difference in potential.

Introduction A Review of Electricity and Electronics

Points			Fault	Fault Location
A	B	C		
12.4 v	12.4 v	12.4 v	Open	Down circuit from C
12.4 v	12.4 v	0	Open	At load
12.4 v	0	0	Open	Up circuit from C
12.4 v	12.2 v	.2 v	None	

Manufacturers usually allow for a maximum of about 15% of the source voltage to be dropped across long runs of wire, circuit protection devices, switches, and connections. If more than 15% of the source voltage is dropped across parts of the circuit other than the load, there is excessive resistance somewhere in the circuit, and current flow is too low. There are exceptions to this, however. Some examples are:

• the current limiting resistors used for the low and intermediate speeds of a blower motor control circuit, Figure I-29A.

FIGURE I-29 Current limiting resistor circuits

Introduction A Review of Electricity and Electronics

- the resistor used to limit current flow through the ignition coil of many ignition systems a few years ago, Figure I-29B.
- the current limiting resistor used in a temperature-sensing circuit, Figure I-29C.

You have to know or become familiar with the circuit you are testing.

To take a voltage drop reading, place the voltmeter across the portion of the circuit you want to test, Figure I-30. The meter will show the difference in voltage (volt drop) between the two points being probed. To make this test useful, the technician must know what portion of the circuit is between the probes and how much voltage should be dropped there. Remember that:

- all of the source voltage must be dropped across the circuit.
- the load in the circuit is where the work is being done and that is where you want to drop as much of the voltage as possible.
- the function of the rest of the circuit is to get the voltage and current from the source to the load and back again as efficiently as possible.
- the less voltage drop in the delivery portion of the circuit, the more efficiently that portion is working.
- the higher the volt drop reading, the more resistance there is in the portion of the circuit between the probes.

If the switch in Figure I-30 were open, the voltmeter would show exact source voltage. This is because the positive probe is connected to the positive battery terminal by way of the fuse and insulated circuit, and the negative probe is connected to the battery negative terminal by way of the motor and ground circuit. If the voltmeter were connected across any other part of the circuit while the switch was open (with the exception of the battery), the reading would be zero. This is worth remembering as a good diagnostic tool. If a volt drop reading is zero, there is no current flowing; if there is current flowing, there has to be some amount of volt drop. If the meter shows source voltage, you are across the open.

The volt drop is especially useful because it allows a technician to test the efficiency of a connection or component such as a switch that is not easy to get to by probing wires going to and from it where they are more accessible.

Ammeter Testing

An ammeter is used to measure the current flowing in a circuit. There are two types: the conventional type that must be put in series as part of the circuit, Figure I-31, and the induc-

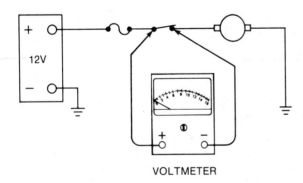

FIGURE I-30 Volt drop test

- - -> ELECTRON FLOW
———> CONVENTIONAL CURRENT FLOW

FIGURE I-31 Ammeter in circuit

Introduction A Review of Electricity and Electronics

FIGURE I-32 Inductive ammeter

tive type. The inductive type usually has leads that connect to a power source and one with an inductive clamp that clamps around a conductor in the circuit to be tested, Figure I-32. The clamp contains a Hall-effect device that uses the magnetic field around the conductor that it is clamped around to create an electrical signal. (Hall-effect devices are discussed in Chapter 2.) The strength of the signal depends on the strength of the magnetic field around the conductor, which in turn depends on the amount of current flowing through it. The ammeter displays a reading that corresponds to the strength of the signal. The inductive ammeter is more convenient to use because it eliminates having to open the circuit.

When using a conventional ammeter, it is important to select the correct scale. Passing 5 amps through an ammeter with a 3-amp scale selected can damage the meter. Some meters are fuse protected when on the lowest scale. Become familiar with an ammeter before attempting to use it.

Resistance Testing

Resistance is measured with an ohmmeter. An ohmmeter is self-powered and most have several scales to allow accurate testing of a wide range of resistance values. Most analog ohmmeters (those with a needle) measure from a fraction of 1 ohm to 100,000 ohms. Many digital ohmmeters measure as high as 20 megohms (20,000,000 ohms).

A common mistake when using an ohmmeter is to use a scale that is too low. For example, if you are going to measure the resistance of a circuit that has 1,500 ohms of resistance and you use a meter with the scale selector set on R × 1 or R × 10, you will get a reading of *infinity;* on an analog meter, the needle will not move from the left side of the scale, and on a digital meter you will get "OL" (open line) or some other indication that the circuit is open. This is because the resistance is higher than the selected scale can measure. To avoid this mistake, start with the highest scale. If you get a low reading (near the right side of the scale), select a lower scale. Continue to select a lower scale until you find the lowest scale that can read the resistance that is present. Some ohmmeters are auto-ranging. In other words, when the meter is turned on and its leads are contacting a circuit, it automatically selects the correct ohm scale. This type will show an ohm value and display a symbol (KΩ or MΩ) to show what scale is being used. Be sure to note the symbol so that you can interpret the resistance value correctly.

When reading an ohmmeter on which you have to manually select the scale, multiply the reading on the face of the meter by the value on the scale you have selected. Figure I-33 shows a meter reading of 3.5 ohms, but the scale selector switch is on the 100-ohm scale. So:

$$3.5 \times 100 = 350 \text{ ohms}$$

Most auto-ranging, digital ohmmeters would show the value shown above as:

350   Ω

3,500 ohms would be seen as:

3.5   KΩ (K = kilo = 1,000)

3,500,000 ohms would be seen as:

3.5   MΩ (M = mega = 1,000,000)

$3.5\Omega \times 100 = 350\Omega$

FIGURE I-33 Calculating ohmmeter readings

When using an ohmmeter it is important to remember that **the circuit or component being tested cannot have voltage in it from any source besides the battery in the ohmmeter.** If there is voltage from another source in a circuit when an ohmmeter is connected to it, the best thing that will happen is an inaccurate reading; the worst is a damaged meter.

When an ohmmeter is turned on, its battery tries to pass current through the meter coil and the two leads, but unless the leads are electrically connected together outside the meter, the circuit is open. If the two leads are touching or if they contact something that has *continuity,* the circuit is closed. If the leads are connected directly to each other, enough current should flow to cause the meter to swing full scale to the right, which is zero, Figure I-34. If the leads are across a conductor, component, or part of a circuit, the resistance of that unit will reduce the current flow through the meter's circuit, Figure I-35. The reduced current flow produces a weaker magnetic field, and the meter's needle will not be moved as far to the right. A digital ohmmeter has an electronic circuit that produces a digital readout in response to the amount of current flowing.

Most needle-type (analog) ohmmeters must be calibrated before each use. Calibration adjusts for voltage variation in the meter's battery due to the discharge that occurs during use. This is done by connecting the leads together and adjusting the zero knob until the needle reads zero, Figure I-34. Doing this es-

FIGURE I-34 Calibrating an ohmmeter

FIGURE I-35 Resistance testing

tablishes that the magnetic field produced by the current flowing through the meter's coil and external leads moves the needle to zero. Any additional resistance placed between the external leads will lower the current flow and will be displayed by the amount the needle lacks reaching the zero line. Many ohmmeters also must be recalibrated each time a different scale is selected. This is because selecting a different scale changes the meter's internal resistance.

Digital ohmmeters usually do not have a zero adjust knob, but the battery must be checked before each use to see that it meets or exceeds a minimum voltage that should be stated somewhere on the meter. If the battery is low, it must be replaced before the meter can be accurate.

Self-Powered Test Light

A self-powered test light is to an ohmmeter what a non-self-powered test light is to a voltmeter. It contains a battery, a light, and two leads, Figure I-36. If the leads make contact with each other, directly or by both touching some other conductive path, the light will come on. This light only tells you if you have continuity or not and is often called a continuity light.

SEMICONDUCTORS

As mentioned earlier in this chapter, a semiconductor is an element with four valence electrons. The two most used semiconductor materials for solid-state components are *silicon* and *germanium*. Of those two, silicon is used much more than germanium; therefore, most of this discussion will apply to silicon.

As said, an atom with three or fewer electrons in its outer shell easily gives them up. If an atom has more than four but less than eight electrons in its outer shell, it exhibits a tendency to acquire more until it has eight. If there are seven electrons in the outer shell, the tendency to acquire another one is stronger than if there are only six. Once there are eight, it becomes very stable; in other words, it is hard to get the atom to gain or lose an electron. In a semiconductor material, for example, a silicon crystal, the atoms share valence electrons in what are called *covalent bonds,* Figure I-37. Each atom positions itself so that it can share the valence electrons of neighboring atoms giving each atom, in effect, eight valence electrons. This lattice structure is characteristic of a crystal solid and provides two useful characteristics:

1. Impurities can be added to the semiconductor material to increase its conductivity; this is called *doping*.
2. It becomes a negative temperature coefficient, meaning that its resistance goes down as its temperature goes up. (This principle is put to use in Chapter 2.)

Doping

All of the valence electrons in a pure semiconductor material are in valence rings containing eight electrons as shown in Figure I-37.

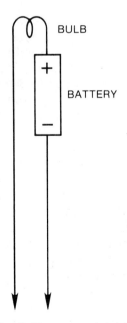

FIGURE I-36 Self-powered test light

Introduction A Review of Electricity and Electronics

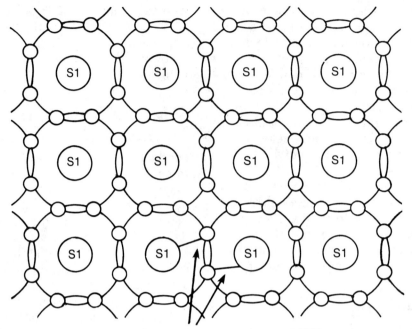

ATTRACTIVE BONDS OF PARENT AT ATOMS.

EACH SILICON ATOM SHARES AN ELECTRON WITH EACH NEIGHBORING ATOM.

FIGURE I-37 Covalent bonds

With this atomic structure there are no electrons that can be easily freed, and there are no holes to attract an electron even if some were available. The result is this material has a high resistance to current flow. Adding very small amounts of certain other elements can greatly reduce the semiconductor's resistance. Adding trace amounts (about 1 atom of the doping element for every 100 million semiconductor atoms) of an element with either five or three valence electrons can create a flaw in some of the covalent bonds.

Adding atoms with five valence electrons (referred to as *pentavalent* atoms) such as *arsenic*, *antimony*, or *phosphorus* achieves a crystal structure as shown in Figure I-38. In Figure I-38 phosphorus is the doping element and silicon is the base semiconductor element. Four of the phosphorus atom's five valence electrons are shared in the valence rings of neighboring silicon atoms, but the fifth is not included in the covalent bonding with neighboring atoms and is held in place only by its attraction to its parent phosphorus atom; a bond that can be easily broken. This doped semiconductor material is classified as an N-type material. Note that it does not have a negative charge because the material contains the same number of protons as electrons. It does have electrons that can be easily attracted to some other positive potential.

Adding atoms with three valence electrons (referred to as *trivalent* atoms) such as *aluminum*, *gallium*, *indium*, or *boron* achieves an atomic structure as seen in Figure I-39. In Figure I-39, boron is used as the doping material in a silicon crystal. The boron atom's three valence electrons are shared in the valence ring

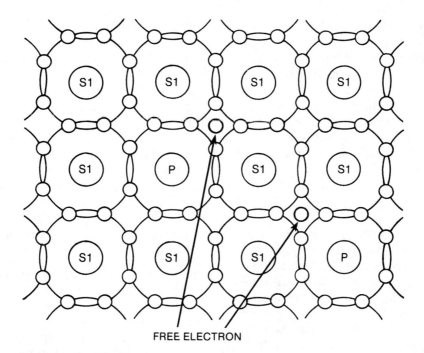

FIGURE I-38 Silicon doped with phosphorus

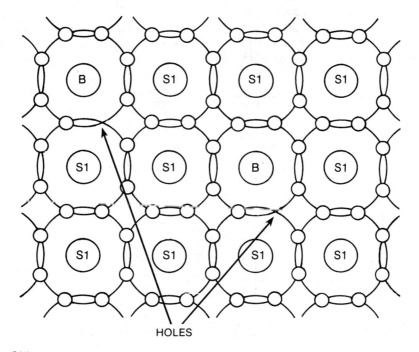

FIGURE I-39 Silicon doped with boron

Introduction A Review of Electricity and Electronics

of three of the neighboring silicon atoms, but the valence ring of the fourth neighboring silicon atom is left with a hole instead of a shared electron. Remember that a valence ring of seven electrons aggressively seeks an eighth electron. In fact, the attraction to any nearby free electron is stronger than the free electron's attraction to its companion proton in the nucleus of its parent atom. Thus this material is classified as a P-type.

Doping Semiconductor Crystals

In actuality, a PN junction is not produced by placing a P- and an N-type semiconductor back to back. Rather, a single semiconductor crystal is doped on one side with the pentavalent atom's opposite sides and on the other with trivalent atoms. The center of the crystal then becomes the junction. The doping is done by first bringing the semiconductor crystal to a molten temperature. In a liquid state, the covalent bonds are broken. The desired amount of doping material is then added. As the semiconductor crystal cools, the covalent bonds redevelop with the doping atoms included.

PN Junction

Figure I-40 shows a P-type and an N-type crystal separated from each other. Only the free electrons in the N-type and the holes in the P-type that result from how the doping atoms bond with the base semiconductor material is shown. If the P-type and the N-type are put in physical contact with each other, the free electrons near the *junction* in the N-type cross the junction and fill the first holes they come to in the seven-electron valence rings near the junction of the P-type, Figure I-41. The junction is the area that joins the P-type and N-type. This action quickly creates a zone around the junction in which:

- there are no more free electrons in that portion of the N-type.
- there are no more free holes in that portion of the P-type.
- the valence rings near the junction in both the N- and P-type have eight shared electrons so they are reluctant to gain or lose any more.

Another way to state the information in the preceding list is to say that there are no longer any current carriers in the zone around the junction. This zone is often referred to as the *depletion zone*. In addition:

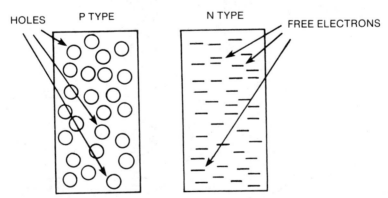

FOR SIMPLICITY, ONLY THE HOLES AND FREE ELECTRONS THAT RESULT FROM THE PENTAVALENT AND TRIVALENT ATOMS ARE SHOWN; THE STABLE ATOMS ARE LEFT OUT.

FIGURE I-40 P-type and N-type crystals

Introduction A Review of Electricity and Electronics

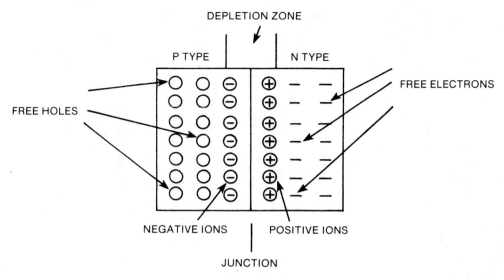

FOR SIMPLICITY, ONLY THE HOLES, FREE ELECTRONS, AND IONS ARE SHOWN; THE STABLE ATOMS ARE LEFT OUT.

FIGURE I-41 PN junction

- The phosphorus atoms near the junction in the N-type have each lost an electron, which makes them positive ions (the free electron that crossed the junction to fill a hole in the P-type abandoned a proton in the nucleus of the phosphorus atom).
- The boron atoms near the junction in the P-type have each gained an electron, which makes them negative ions (the electron that dropped into the valence ring around the boron atom is not matched by a proton in the nucleus of the boron atom).

The positive ions near the junction in the N-type are attracted to the free electrons that are farther from the junction, but they are more strongly repulsed by the negative ions just across the junction. Likewise, the negative ions in the P-type are attracted to the holes that are farther from the junction, but those holes are more strongly repulsed by the positive ions on the N-side of the junction. The extra electrons in the negative ion atoms are somewhat at-

Conventional Current Flow versus Electron Current Flow

Before electrons were known, Benjamin Franklin surmised that current was a flow of positive charges moving from positive to negative in a circuit. Franklin's belief became so accepted that even after electrons were discovered and scientists learned that current flow consists of electrons moving from negative to positive, the old idea was hard to give up. As a result, the "idea" of positive charges moving from positive to negative is still often used. It is referred to as *conventional current flow* or less formally as *hole flow*.

The conventional current flow theory has gotten a boost in recent years because it helps explain how semiconductors work: positively charged holes move from positive charges to negative charges as electrons move from negative charges to positive charges.

Introduction A Review of Electricity and Electronics

tracted to the nearby holes, but they are more strongly bound by the covalent bonding into which they have just dropped.

Keep in mind that a layer of negative ions along the junction in the P-type and a layer of positive ions along the junction of the N-type have been created. The opposing ionic charges on the two sides of the depletion zone create an electrical potential of about 0.6 volt (0.3 volt for germanium). This potential, often referred to as the *barrier potential,* cannot be measured directly, but its polarity prevents current from flowing across the junction unless it is overcome by a greater potential.

Diodes

What was just described in the preceding section is in fact the major part of a diode. A diode is basically a one-way electrical check valve; it will allow current to pass in only one direction. The best-known automotive application is probably the diodes used in the alternator to rectify the AC current produced to DC current. There are many other applications of diodes in today's vehicles such as in solid-state voltage regulators, electronic modules, and circuits where one-way control might be needed. Figure I-42 shows a rectifying diode in

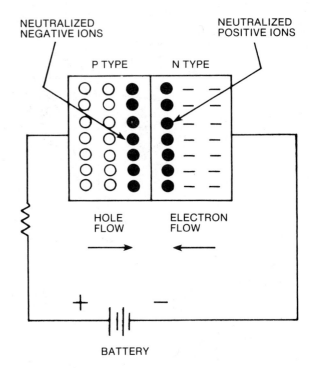

FIGURE I-43   Forward bias voltage applied to a diode

a circuit. If switch A is on and C is off, light B will get current, but light D will not because the diode will block it. If switch C is on and A is off both lights will get current.

Here is how the diode works. By applying voltage with correct polarity from an outside source such as a battery, current will flow across the PN junction, Figure I-43. Applying an external voltage of 0.6 volt or more with the positive to the P-side and the negative to the N-side is called *forward bias voltage.* The higher negative potential introduced by the forward bias voltage at the right side of the crystal repels the free electrons in the N-type. They move toward the junction, canceling the charge of the positive ions. At the same time, the higher positive potential introduced by the forward bias voltage on the left side of the crystal repels the free holes in the P-type. They move toward the junction, canceling the

FIGURE I-42   A rectifying diode application

Introduction A Review of Electricity and Electronics

FIGURE I-44   Reversed bias voltage applied to a diode

charge of the negative ions. With the barrier potential overcome, current easily flows across the junction, electrons toward the external positive potential, and holes toward the external negative potential. When the forward bias voltage is removed, barrier potential redevelops and the diode again presents high resistance to current flow.

Note: The amount of current that a diode or any other type of PN junction semiconductor can safely handle is determined by such things as its size, type of semiconductor and doping material used, heat dissipating ability, and surrounding temperature. If the circuit through which forward bias voltage is applied does not have enough resistance to limit current flow to what the semiconductor can tolerate, it will overheat, and the junction will be permanently damaged (open or shorted).

If a reverse bias voltage is applied, negative external potential to the P-side and positive external potential to the N-side, the positive potential attracts free electrons away from the junction, and the negative potential attracts holes away from the junction, Figure I-44. This causes the depletion zone to be even wider and the resistance across the junction to be even higher.

If the reverse bias voltage goes high enough (above 50 volts for most rectifier diodes [those designed to conduct enough current to do work, the most common type]), current will flow. It will rise quickly, and in most cases the diode will be damaged. This is called the *breakdown voltage*.

Zener Diode

A zener diode is one in which the crystals are more heavily doped. Because of this the depletion zone is much narrower, and its barrier voltage becomes very intense when a reverse bias voltage is applied. At a given level of intensity the barrier voltage pulls electrons out of normally stable valence rings, creating free electrons. When this occurs, current flows across the diode in a reverse direction without damaging the diode. The breakdown voltage at which this occurs can be controlled by the amount of doping material added in the manufacturing process. Zener diodes are often used in voltage regulating circuits.

Diode Symbols

The most commonly used symbol to represent a diode is an arrow with a bar at the point, Figure I-45A. The point always indicates the direction of current flow using conventional current. Electron flow would be opposite to the way the arrow is pointing. The arrow side of the symbol also indicates the P-side of the di-

Introduction A Review of Electricity and Electronics

FIGURE I-45 Diodes

ode, often referred to as the *anode*. The bar at the end of the arrow point represents the N-side and is often called the *cathode*. Figure I-45B shows a modified diode symbol that represents a zener diode.

On actual diodes, the diode symbol can be printed on the side to indicate the anode and cathode ends, or a colored band can be used instead of the symbol. In this case the colored band will be nearer to the cathode end, Figure I-45C and D. Figure I-45E shows a power diode. A power diode is one large enough to conduct larger amounts of current to power a working device. It will be housed in a metal case, which can serve as either the anode or the cathode connector and will also dissipate heat away from the semiconductor crystals inside. The polarity of a power diode can be indicated by markings, or in the case of specific part number applications such as in an alternator, may be sized or shaped so that they can be installed only one way.

Transistors

The transistor, probably more than any other single component, has made possible the world of modern electronics. Transistors most commonly used in automotive applications are called *bipolar transistors* because they use two polarities—electrons and holes. Bipolar transistors contain three doped semiconductor crystal layers called the *emitter,* the *base,* and the *collector,* Figure I-46. The base is always sandwiched between the emitter and the collector. The major difference between a diode and transistor is a transistor has two PN junctions instead of one. They can be arranged to have a P-type emitter and collector with an N-type base (a PNP transistor) or an N-type emitter and collector with a P-type base (an NPN transistor). Since most automotive applications are NPN transistors, they are discussed here.

The emitter is heavily doped with pentavalent atoms, and its function is to emit free

Introduction A Review of Electricity and Electronics

FIGURE I-46   Components of a bipolar transistor

electrons into the base. The base is lightly doped with trivalent atoms and is physically much thinner than the other sections. The collector is slightly less doped that the emitter but more than the base.

Recalling the discussion of the PN junction (review it if it is not clear), you know that a barrier potential forms at each junction, Figure I-47. If you forward bias the transistor by applying an external voltage of at least 0.6 volt from emitter to base with negative potential to the emitter and positive potential to the base, then apply an external voltage of at least 0.6 volt collector to emitter with positive potential to the collector and negative potential to the emitter, Figure I-48, you can get some unique results. A slightly higher voltage is usually used

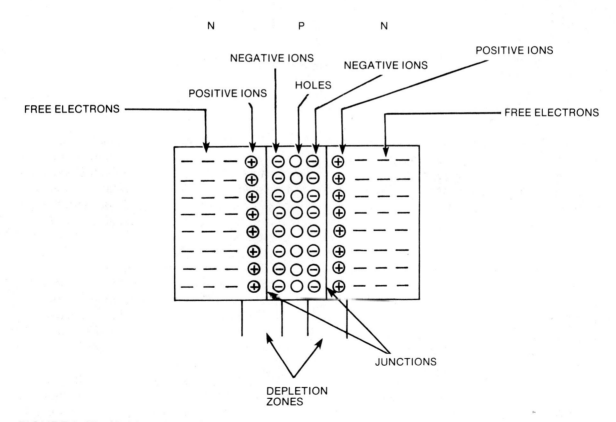

FIGURE I-47   Unbiased transistor

33

Introduction A Review of Electricity and Electronics

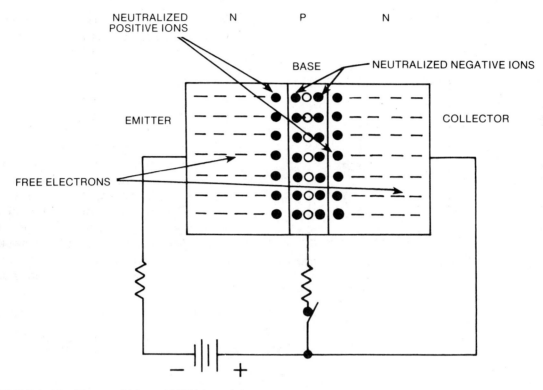

FIGURE I-48 Forward biased NPN transistor

between the collector and emitter, thus two separate voltage sources are shown.

Because it is lightly doped, the base does not have many holes for the free electrons to drop into. The holes that do exist are quickly filled. Because the base crystal is thin and has so few holes for free electrons to drop into, the majority of the free electrons coming from the emitter cross the base into the collector. The collector readily accepts them because the free electrons in the collector are attracted to the positive potential at the collector electrode leaving behind a lot of positive ions.

The positive potential that is applied to the base electrode is strong enough to attract some of the valence electrons there. As valence electrons near the base electrode flow from the base toward the emitter-base voltage source, nearby valence electrons move over to take their place.

Because primarily only valence electrons move from the base into the base circuit and because there are so few holes that will allow free electrons to drop into valence rings in the base (the base is very lightly doped), the current in the base-emitter circuit is quite small compared to the current in the collector-emitter circuit.

If the biasing voltage is removed from the base, the barrier potential is restored at the junctions and both base and collector currents stop flowing. The base current flow caused by the forwarding biasing voltage between the base to the emitter allows current to flow from the emitter to the collector.

Transistors are like switches that can be turned on by applying power to the base circuit. There are many types of transistors, but the most common fall into two categories: power and switching. Power transistors are larger because the junction areas must be larger to pass more current across them. Passing current across the junctions produces heat; therefore, a power transistor must be mounted on something that can draw heat from the transistor and dissipate it into the air. If not it will likely overheat and fail. A transistor is also much like a relay in that a relatively small current through the base circuit controls a larger current through the collector. The amount of current through the base is determined by the amount of voltage and resistance in the base circuit. In fact, carefully controlling the voltage applied to the base can control how much current flows through the collector. If enough voltage is applied to the base to just start reducing barrier potential, a relatively small amount of current will begin to flow from the emitter to the collector. This is called *partial saturation*. As base voltage is increased, collector current increases. When enough voltage is applied to the base, full saturation is achieved. There is usually a small voltage spread between minimum and full saturation, and once full saturation is achieved, increasing voltage at the base will not increase collector current. If too much voltage is applied, the transistor will break down. This principle is often used in power transistors. Most power transistors are NPN.

Switching transistors are much smaller. They are often used in information processing or control circuits, and conduct currents ranging from a few *milliamps* (thousandths of an ampere) down to a few *microamps* (millionths of an ampere). They are more often designed for extremely fast on and off cycling rates, and the base circuits that control them are designed to turn them on at full saturation or to turn them off.

Transistor Symbols

In electrical schematic illustrations transistors are usually represented by a symbol. Figure I-49 shows the symbols that are most often used. Of the two types of symbols, the circle containing the lines for emitter, base, and collector are most often used. The arrow is always used to indicate:

- the emitter side of the transistor
- the direction of conventional current flow
- whether the transistor is an NPN or PNP. If the arrow points outward, toward the circumference of the circle, it indicates an NPN transistor. If it points inward toward the base, it indicates a PNP transistor.

Effect of Temperature on a PN Junction

As the temperature of a semiconductor device goes up, the electrons in the valence ring move at a greater speed. This causes some to break out of the valence ring, creating more free electrons and holes. The increase in the number of free electrons and holes causes the depletion layer to become thinner, reducing the barrier potential. Barrier potential voltages of 0.7 for silicon and 0.3 for germanium semiconductors are true at room temperature only. At elevated temperatures, the lower barrier potential lowers the external bias voltage required to cause current to flow across the junction and raises the current flow.

This means that the expected operating temperature range of a semiconductor device must be considered when the circuit is being designed and that when in operation, the operating temperature must be kept within that temperature limit. If the operating temperature goes higher, unless some kind of compensating resistance is used, current values will go up, and the semiconductor might be damaged.

Introduction A Review of Electricity and Electronics

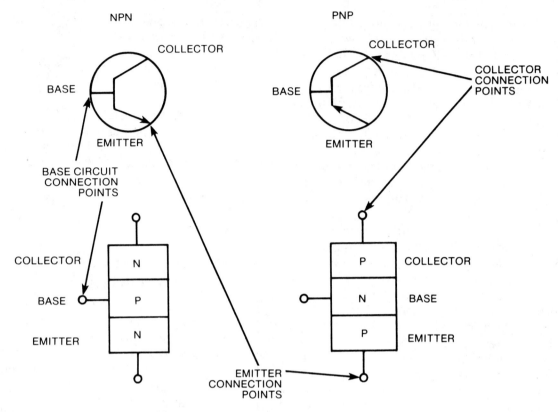

FIGURE I-49   Transistor symbols

INTEGRATED CIRCUITS (ICs)

Thanks to scientific research, manufacturers have the capability to produce microscopic transistors, diodes, and resistors. As a result, complete circuits are produced containing thousands of semiconductor devices and connecting conductor paths on a chip as small as two or three millimeters across. These *integrated circuits* operating with current values as low as a few milliamps or less can process information, make logic decisions, and issue commands to larger transistors. The larger transistors control circuits that operate on larger current values. Personal computers and the computers in today's vehicles became possible because of the development of ICs.

Because the components in ICs are so small, they cannot tolerate high voltages. Care must be taken to avoid creating high-voltage spikes such as those produced by disconnecting the battery while the ignition is on. Care must also be taken when handling a component with ICs in it to avoid exposing it to static discharges such as those you sometimes experience when touching something with ground potential after walking across a carpet.

1

COMPUTERS ON CARS

Objectives

After studying this chapter, you will be able to:
- name the major components that make up a microcomputer.
- list the cause or causes of CO, HC, and NO_x production.
- describe a stoichiometric air/fuel ratio.
- identify the air/fuel ratio needed to make a three-way catalytic converter work most effectively.
- explain why almost all cars currently being built for use in this country are equipped with a computerized engine control system.
- describe the difference between open-loop and closed-loop operation.
- explain the importance of performing a thorough prediagnostic inspection.

WHY COMPUTERS?

In 1963 positive crankcase ventilation systems were universally installed on domestic cars as original equipment. People in the service industry felt that dumping all of those crankcase gases in the induction system would plug up the carburetor and be harmful to the engine. Everyone in the service industry knew that all of the crankcase gases in the induction system would plug up the carburetor and ruin the engine. Engine life doubled. In 1968 exhaust emission devices were universally applied to domestic cars. Compression ratios began to go down, spark control devices denied vacuum advance during certain driving conditions, thermostat temperatures went up, air pumps were installed, heated air intake systems were used during warm-up, and air/fuel ratios began to become leaner. Through the seventies evaporation control systems, exhaust gas recirculation systems, and catalytic converters were added. Although some of the emissions systems actually tended to improve driveability and even fuel mileage, for the most part driveability and fuel mileage suffered in order to achieve a dramatic reduction in emissions.

In 1973 and 1974, when domestic car fuel economy was at its worst, we also experienced an oil embargo and an energy shortage. The federal government responded by establishing fuel mileage standards in addition to the already established emissions standards. By the late 1970s, the car manufacturers were hard pressed to meet the evermore stringent emissions and mileage standards; and the standards set for the eighties looked impossible, Figures 1-1 and 1-2. To make matters worse, the consumer was getting into the picture, too. Not only did the consumer's car get poor fuel mileage, it had poor *driveability* (idled rough, often hesitated or stumbled during acceleration if the engine was not fully warmed up, and had little power). What made the situation so difficult was that the three demands—lower emissions, better mileage, and better driveability—were largely in opposition to each other using the technology available at that time, Figure 1-3.

What was needed was a much more precise way to control engine functions, or *engine calibration*. The automotive industry had already looked to microprocessors. In 1968 Volkswagen introduced the first large-scale production, computer-controlled electronic fuel injection system, an early version of the Bosch D-Jetronic system. In 1975 Cadillac introduced a computer-controlled electronic fuel injection (EFI) system. In 1976 Chrysler introduced a computer-controlled electronic spark control system, the Lean-Burn system. In 1977

EMISSION REQUIREMENTS FOR PASSENGER CARS
(IN GRAMS PER MILE)

MODEL YEAR	HYDROCARBON (HC)		CARBON MONOXIDE (CO)		OXIDES OF NITROGEN (NO$_x$)	
	FEDERAL	CALIFORNIA	FEDERAL	CALIFORNIA	FEDERAL	CALIFORNIA
1978	1.5	0.41	15.0	9.0	2.0	1.5
1979	0.41	0.41	15.0	9.0	2.0	1.5
1980	0.41	0.39	7.0	9.0	2.0	1.0
1981	0.41	0.39	3.4	7.0	1.0	0.7
1982	0.41	0.39	3.4	7.0	1.0	0.4
1983	0.41	0.39	3.4	7.0	1.0	0.4
1984	0.41	0.39	3.4	7.0	1.0	0.4
1985	0.41	0.39	3.4	7.0	1.0	0.7
1986	0.41	0.39	3.4	7.0	1.0	0.7
1987	0.41	0.39	3.4	7.0	1.0	0.7
1988	0.41	0.39	3.4	7.0	1.0	0.7
1989	0.41	0.39	3.4	7.0	1.0	0.4
1990	0.41	0.39	3.4	7.0	1.0	0.4
1960	10.6*		8.4*		4.1*	

*Typical values before controls

FIGURE 1-1 Federal and California emissions standards compared to the uncontrolled vehicles of 1960.

FUEL ECONOMY STANDARDS

MODEL YEAR	MPG	TOTAL IMPROVEMENT OVER THE 1974 MODEL YEAR
1978	18.0	50%
1979	19.0	58%
1980	20.0	67%
1981	22.0	83%
1982	24.0	100%
1983	26.0	116%
1984	27.0	125%
1985	27.5	129%
1986	26.0*	116%
1987	26.0	116%
1988	26.0	116%
1989	26.5	120%
1990	27.5	129%

*Reduced from 27.5% by the federal government in 1986

FIGURE 1–2   Federally imposed fuel economy standards

Oldsmobile introduced a computer-controlled electronic spark control system that they called MISAR (microprocessed sensing automatic regulation). All of these systems had three things in common: they all controlled only one engine function, they all used an *analog* computer, and none of them started any landslide movement toward computer controls.

By the late seventies, the electronics industry had made great strides with *digital* microprocessors. These small computers with their comparatively low cost, compact size and weight, great speed, and application flexibility proved to be the answer for which the industry was looking. Probably the most amazing fea-

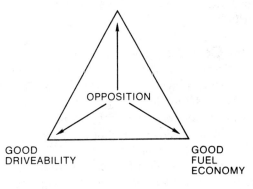

FIGURE 1–3   Opposition of exhaust emissions, driveability, and fuel economy to each other

Chapter 1 Computers on Cars

ture of the digital computer is its speed. To put it into perspective, one of these computers controlling functions on an eight-cylinder engine running at 3,000 RPM can send the spark timing command to fire a cylinder; reevaluate input information about engine speed, coolant temperature, engine load, barometric pressure, throttle position, air/fuel mixture, spark knock, and vehicle speed; make a new decision about air/fuel mixture, spark timing, whether or not to turn on the EGR valve, canister purge valve, and the torque converter clutch; then take a short nap before sending the commands, all before the next cylinder fires.

At last the automotive manufacturers had the technology to precisely monitor and control, to instantly and automatically adjust, while driving, enough of the engine's calibrations to make the vehicle comply with the government's demands for emissions and fuel economy and still satisfy the consumer's demands for better driveability. This brought about what is probably the first real revolution in the automotive industry's recent history; other changes have been evolutionary by comparison. After some experimental applications in 1978 and 1979 and some limited production in 1980, the whole domestic industry was using comprehensive computerized engine controls in 1981. At the same time, General Motors produced more computers than anyone else in the world and used half of the world's supply of computer parts.

How Computers Work

Contrary to some "informed" opinions and numerous suggestions from movies, computers cannot think for themselves; when properly programmed, however, they can carry out explicit instructions with blinding speed and almost flawless consistency.

Communication Signals. A computer uses voltage values as communication signals, thus voltage is often referred to as a *signal* or a *voltage signal*. There are two types of voltage signals: analog and digital, Figure 1-4. An *analog signal's voltage* is continuously variable within a given range and time is used for the voltage to change. An analog signal is generally used to convey information about a condition that changes gradually and continuously within an established range. Temperature-sensing devices usually give off an analog signal. *Digital signals* also vary but not continuously, and time is not needed for the change to occur. Turning a switch on and off creates a digital signal; voltage is either there or it is not. Digital signals are often referred to as square wave signals.

Binary Code. A computer converts a series of digital signals to a binary number made up of 1s and 0s; voltage above a given value converts to 1, and zero voltage converts to 0. Each 1 or 0 represents a *bit* of information. Eight bits equal a *byte* (sometimes referred to as a word). All communication between the microprocessor, the memories, and the interfaces is in binary code, with each information exchange being in the form of a byte. If a series of digital signals converts to a binary number of 01111010, the numerical value (in base$_{10}$, which we are most familiar with) can easily be derived, as shown in Figure 1-5A. A power is assigned to each place in the binary number,

FIGURE 1-4 Graphic illustration comparing analog and digital voltage signals. *Courtesy of GM Product Service Training.*

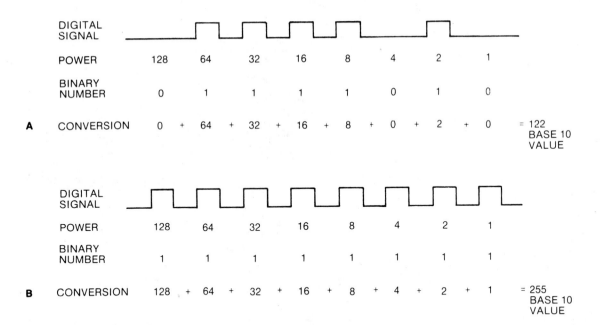

FIGURE 1-5   Binary numbers

starting from the right and working to the left. The rightmost place is given a power of 1, with the power doubling with each successive place to the left. Each digit in the binary number is multiplied by its respective power. The products are then added together to yield the base$_{10}$ numerical value. As you will see in Figure 1-5B, the largest value that an 8-bit computer can communicate is 255. Most automotive computers have 8-bit microprocessors, but as the computers take on more responsibility and have to communicate more information more rapidly, 16-bit units will become commonplace.

Interface. The microprocessor is the heart of a computer, but it needs several support functions, one of which is *interface*. A computer has an input and an output interface circuit, Figure 1-6. The interface has two functions: it protects the delicate electronics of the microprocessor from the higher voltages of the circuits attached to the computer, and it translates input and output signals. The input interface translates all analog input data to binary code; most sensors produce an analog signal. It is sometimes referred to as *A/D*, analog to digital. The output interface, *D/A*, translates digital signals to analog for any controlled functions that need an analog voltage.

Memories. The microprocessor of a small computer does the calculating and makes all of the decisions, or data processing, but it cannot store information. The computer is therefore equipped with information storage capability called *memory*. The computer actually has three memories: read-only memory (ROM), programmable read-only memory (PROM), and random access memory (RAM), Figure 1-7.

The ROM contains permanently stored information that instructs (programs) the microprocessor on what to do in response to input data. The microprocessor can read information from the ROM but cannot put information into it. The ROM unit is soldered into the computer and is not easily removed.

The PROM differs from the ROM in that it plugs into the computer and is more easily re-

Chapter 1 Computers on Cars

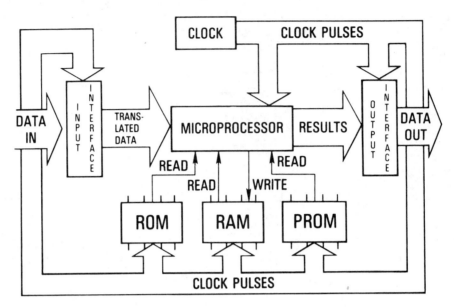

FIGURE 1-6 Interaction of the microprocessor and its support system. *Courtesy of GM Product Service Training.*

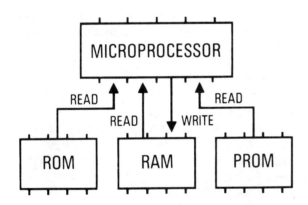

FIGURE 1-7 The three memories within a computer. *Courtesy of GM Product Service Training.*

moved and reprogrammed or replaced with one containing a revised program. It contains program information specific to different vehicle model calibrations.

Some of the information stored in both the ROM and the PROM is stored in the form of look-up tables similar to the tables one might find in the back of a chemistry or math textbook. These tables enable the computer to interpret and make decisions in response to sensor data. For example, a piece of information (a voltage value of 2.4 volts, about half throttle) is received from the throttle position sensor. This information plus information from the engine speed sensor is compared to a table for spark advance. This preprogrammed table tells the computer what the spark advance should be for that throttle position and engine speed. The computer will then modify this spark advance value, by consulting other tables, with information concerning engine temperature and atmospheric pressure. These tables are sometimes referred to as *maps.*

The RAM is where temporary information is stored. It can be both read from and written into. If, for example, data comes in from a sensor and will be used for several different decisions, such as manifold vacuum information, the microprocessor will record it in RAM until it is updated. In the example presented in the above paragraph, the information concerning throttle position and engine speed came from the RAM.

In most automotive computers, the computer tests at least some of the input and output signals to see that the circuit that sent the signal or received the command is working properly. The computer knows what the voltage will be if the circuit were open and what the voltage will be if it were shorted. It watches to see that the voltage remains between those voltage values. On some circuits, the oxygen sensor circuit for instance, the computer expects to see the voltage go up and down, crossing a certain value. If during its continuous testing it sees something wrong (a fault), it will record information in RAM concerning the fault. The RAM is also soldered in place and not easily removed.

There are two kinds of RAMs: volatile and nonvolatile. A volatile RAM (sometimes called *keep-alive memory*) is erased when disconnected from its power source. In automotive applications a volatile RAM is usually connected directly to the battery by way of a fuse or fusible link so that when the ignition is turned off, the RAM is still powered. A nonvolatile RAM does not lose its stored information if its power source is disconnected. Vehicles with digital display odometers usually store mileage information in a nonvolatile RAM.

The terms ROM, PROM and RAM are fairly standard throughout the computer industry; however, all car manufacturers do not use the same terms in reference to their computer's memory. For instance, instead of using the term PROM, Ford calls their equivalent unit an engine calibration assembly. Another variation of a PROM is an E-PROM. An E-PROM has a housing with a transparent top that will let light through. If ultraviolet light strikes the E-PROM, it reverts back to its unprogrammed state. To prevent de-programming or memory loss, a cover must protect it from light. The cover is often a piece of tape. An E-PROM is sometimes used where ease of changing stored information is important. General Motors uses it as a PROM on some applications because it has more memory capacity.

Another variation of RAM is KAM (Keep Alive Memory). Often used as a subsection of the RAM, KAM is usually powered by a fuse straight from the battery rather than by the ignition or an ignition-controlled circuit. Therefore, when the ignition is turned off, the KAM remains powered and its stored information is retained. Other types of memory that are unique to specific manufacturers will be introduced in appropriate chapters.

Clock Pulses. In order to maintain an orderly flow of information into, out of, and within the computer, a quartz crystal is used to produce a continuous, consistent time pulse, Figure 1-8. The pulse acts as a kind of rapid cadence by which information is transmitted. During the space or time interval between each pulse, one bit of binary code information is communicated from one part of the computer to another. This method provides an orderly flow of information.

The clock pulse is often referred by the term *baud rate*. The baud rate identifies the rate of the pulses. For example, a computer with a baud rate of 5000 could transmit 5000 bits of binary code information per second. Baud rates in automotive computers have gone up considerably since those used in the early eighties. The computers used by General Motors in the early eighties had a baud rate of 160 (Some of those computers were still used on some applications in the late eighties.), while their P-4 electronic control module introduced in the late eighties has a baud rate of 8192 with development underway to make them even faster. Ford introduced a computer in 1988 with a baud rate of 12500.

Data Links. When computers communicate with other electronic devices such as control panels, modules, some sensors, or other computers in the form of digital signals, they communicate through circuits called *data links*. Some data links transmit data in only one direction, although others transmit in both di-

Chapter 1 Computers on Cars

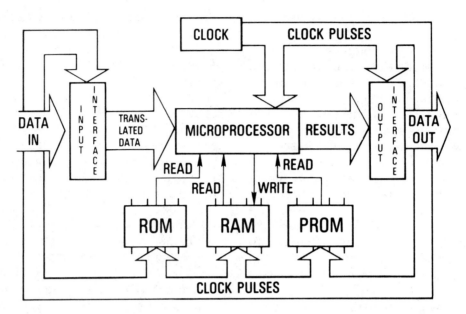

FIGURE 1-8 Clock pulses within a computer. *Courtesy of GM Product Service Training.*

rections. What makes a data link different from an ordinary circuit is the manner in which it is controlled. Some control must be used so that the devices at each end of the data link know when to transmit and when to receive. Several control methods can be used, but to discuss them here would go beyond the goals of this text and the interest of most automotive students and technicians.

EMISSIONS, MILEAGE, AND DRIVEABILITY

The computer of a computerized engine control system has a mission: to reduce emissions, improve mileage, and improve driveability. The priority varies, however, under some driving conditions. For example, during warm-up, driveability has a higher priority than mileage; and at full throttle, some systems give the performance aspect of driveability a higher priority than emissions or mileage.

There are three major emissions that share the greatest concern for control on an automobile: hydrocarbon (HC), carbon monoxide (CO), and oxides of nitrogen (NO_x).

1. *HC.* Gasoline is a hydrocarbon compound. When it is burned properly, the hydrogen and carbon atoms separate; each mixes with oxygen to form water (H_2O) and carbon dioxide (CO_2), Figure 1-9. If for any

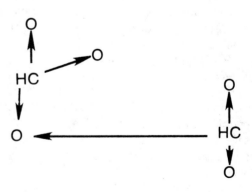

FIGURE 1-9 Chemical reaction during ideal combustion

reason the gasoline in the cylinder fails to burn, it is pumped into the exhaust system as a raw HC molecule. There can be several reasons for all or part of the HC in the cylinder not to burn:
- Any electrical malfunction, which prevents the spark plug from igniting the fuel mixture.
- A lack of mechanical integrity, which allows fuel to escape from the cylinder.
- An overly lean mixture, which hampers flame propagation.
- An exceedingly rich mixture, which uses up the oxygen before all of the HC has been even partially burned.
- A low cylinder surface temperature, which rapidly draws heat from those HC molecules in contact with it and thus prevents them from reaching or maintaining ignition temperature. A low cylinder surface temperature can be caused by a low coolant (low temperature or stuck open thermostat) or an overadvanced ignition timing.

2. *CO.* During combustion each carbon atom tries to take on two oxygen atoms (CO_2). If, however, there is a deficiency of oxygen in the combustion chamber, some carbon atoms are only able to combine with one oxygen atom and thus produce CO. An overly rich air/fuel ratio is the only cause of CO production.

3. *NO_x.* Air is made up of about 80% nitrogen and 20% oxygen. Under normal conditions nitrogen and oxygen do not chemically unite. When raised to a high enough temperature, however, they unite to form a nitrogen oxide compound, NO. In the presence of atmospheric oxygen, NO rapidly becomes nitrogen dioxide, NO_2. These compounds are grouped into a family of compounds referred to as oxides of nitrogen, NO_x. Oxide of nitrogen compounds start to form at 1,120° C (2,040° F), with production increasing as temperatures go higher.

Other Exhaust Emissions From Gasoline Engines

Aldehydes
Ammonia
Carboxylic acids
Inorganic solids
    lead (from leaded fuel)
    soot
Sulfur oxides (from fuel impurities)

Catalytic Converter

Controlling the air/fuel ratio and the spark timing, keeping the coolant temperature near 93° C (200° F) or more, and controlling intake air temperature and combustion temperature dramatically reduce exhaust emissions. Even these measures, however, do not eliminate them. The catalytic converter is the single most effective device for controlling exhaust emissions. Placed in the exhaust system, the *catalyst* agents cause low-temperature oxidation of HC and CO and reduction of NO_x, yielding H_2O, CO_2, O_2, and nitrogen. Placing the converter downstream in the exhaust system allows the exhaust gas temperature to have dropped significantly before the gas enters the converter and thus prevents the production of more NO_x in the oxidation process. The catalysts that are most often used, *platinum* and *palladium* for oxidation and *rhodium* for reduction, are coated over a porous material, usually aluminum oxide in pellet form or aluminum oxide in a honeycomb structure. Aluminum oxide's porosity provides a tremendous amount of surface area on which the catalyst material can be applied to allow maximum exposure of the catalyst to the gases.

There have been three different types of catalytic converters used on automobiles: a single-bed, two-way, introduced on cars in the mid-seventies; a single-bed, three-way; and a dual-bed, three-way. The three-way converters were introduced on cars in the late seventies.

Chapter 1 Computers on Cars

FIGURE 1-10 Single-bed, three-way catalytic converter. *Courtesy of GM Product Service Training.*

FIGURE 1-11 Dual-bed, three-way catalytic converter. *Courtesy of GM Product Service Training.*

A single-bed converter has all of the catalytic materials in one chamber, Figure 1-10. A dual-bed converter has a reducing chamber for breaking down NOₓ and an oxidizing chamber for oxidizing HC and CO. The exhaust passes through the reducing chamber first, Figure 1-11. Additional oxygen is introduced into the oxidizing chamber to make it more efficient. A two-way converter oxidizes HC and CO; a three-way converter oxidizes HC and CO and reduces NO_x.

The rhodium-reducing catalyst of a catalytic converter causes the NO_x molecules to be broken back down into nitrogen and oxygen. It only does so, however, if there is deficiency of oxygen in the exhaust stream. A rich air/fuel ratio is therefore needed; a rich mixture yields only a small amount of leftover oxygen. On the other hand, the platinum- and palladium-oxidizing catalysts need abundant available oxygen to oxidize HC and CO. A lean air/fuel mixture is needed to yield leftover oxygen, Figure 1-12. A *stoichiometric air/fuel ratio* (14.7 to 1 at sea level) is essential to make the three-way converter work effectively, Figure 1-13. If the fuel mixture is leaner than 14.7 to 1, the reducing side of the converter is not effective. If the mixture is richer than 14.7 to 1, the additional CO and the oxygen liberated by the reducing catalyst, plus additional oxygen pumped into the oxidizing chamber in the case of a dual-bed converter, causes the converter to overheat as a result of the abundant oxidation process.

FIGURE 1-13 Converter efficiency. *Courtesy of GM Product Service Training.*

One of the major reasons for creating a computerized engine control system was to have the ability to maintain a 14.7 to 1 air/fuel ratio and therefore enhance the effectiveness of the three-way catalytic converter.

Most manufacturers have experienced some degree of problem with catalytic converters becoming restricted and have developed methods of testing for this condition. Depending on the severity of the restriction, the symptoms vary from slight loss of power after several minutes of driving to the engine not starting.

Other Emission-Control Devices

Other emission-control devices currently being used on most light-duty vehicles and that can be included as part of the computerized engine control system or can be in addition to it are discussed here.

EGR Valve. This valve introduces exhaust gas into the intake manifold to reduce combustion temperature. It is included as a part of and is controlled by the control system on most

3 WAY CONVERTER

NOₓ RICHER ← (14.7:1 / 14.7:1) → CO & HC LEANER

FIGURE 1-12 Three-way catalytic converter. *Courtesy of GM Product Service Training.*

> ### Stoichiometric Air/Fuel Ratios
>
> A stoichiometric air/fuel ratio provides just enough air and fuel for the most complete combustion resulting in the least possible total amount of leftover combustible materials: oxygen, carbon, and hydrogen. This maximizes the production of CO_2 and H_2O and minimizes the potential for producing CO, HC, and NO_x. It also provides good driveability and economy.
>
> The most power is obtained from an air/fuel ratio of about 12.5 to 1, although the best economy is obtained at a ratio of about 16 to 1. Air/fuel ratios this far from stoichiometric, however, are not compatible with the three-way catalytic converter. The leaner mixtures also increase engine temperature as a result of slower burn rate.

vehicles but is independent of the control system on other vehicles. It is not used at all on an increasing number of engines.

Intake Charge Heating. Either electric or exhaust heat is used to heat the air/fuel mixture as it enters the intake manifold. Intake charge heating also is included as one of the controlled functions on some applications and not on others. It is not used at all on some systems.

Fuel Evaporation Control. This system prevents HC vapors from escaping from the fuel tank and carburetor float bowl. It is included as a controlled function on some control systems and not on others.

Air Injection System (Pump or Pulse Air). This system pumps air into the exhaust ports to oxidize HC and CO coming out of the exhaust valve. Intake charge heating is controlled by the computer on some applications, while on others it is controlled by conventional methods or not used at all.

Heated Air Inlet. This system heats the air before it enters the air cleaner during engine warm-up. It is not controlled by the engine control system and is not used on most port fuel injection systems.

PCV System. This system recirculates crankcase gases through the combustion chambers to burn them. It is a separate system on all engine applications.

ECONOMY AND DRIVEABILITY

Computerized engine control systems contribute to fuel economy and driveability in several ways.

Fuel Economy Controls

Air/Fuel Mixture. Maintaining a 14.7 to 1 air/fuel ratio promotes fuel economy by eliminating the fuel waste that occurs with richer fuel mixtures. Richer mixture usage can easily occur without being noticed in an uncontrolled engine. It also provides an efficient mixture for combustion and thus enhances economy and driveability. Systems that use electronic fuel injection contribute to economy even further with more efficient fuel control such as

- shutting off fuel during deceleration.
- replacing the mechanical choke with a more articulate, electronically controlled cold engine enrichment.
- replacing the accelerator pump and vacuum-controlled full-power enrichment functions of a carburetor with an electronically controlled acceleration enrichment.

In many cases these features also help to improve driveability.

Electronic Ignition Timing. Ignition timing is well known to be a major factor in both fuel economy and driveability. It is also a critical factor in the control of exhaust emissions. With computer-controlled ignition timing, timing does not have to be dependent on just engine speed and load. It can be adjusted to meet the greatest need during any driving condition. For example, following a cold start, it can be advanced for driveability. During light load operation with a partially warmed-up engine, it can be retarded slightly to hasten engine warm-up and reduce exhaust emissions. During acceleration or wide-open throttle operation, it can be adjusted for maximum torque.

Idle Speed Control. Electronic idle speed control can more nearly maintain the best compromise between a speed at which the engine is prone to stall and one that wastes fuel and produces an annoying lurch as the automatic transmission is pulled into gear. It can keep idle speed from yo-yoing up and down as the air-conditioning clutch cycles on and off and can be designed to prevent dieseling. It can also adjust idle speed in an effort to compensate for low-charging system voltage, low engine temperature, and engine overheating.

Torque Converter Clutch Control. Many engine control computer systems control a clutch in the torque converter. This function eliminates converter slippage and heat production and thus contributes to fuel economy. On some manual transmission applications, the computer operates a *shift light,* which alerts the driver when to upshift in order to obtain maximum economy.

EGR Valve Control. The EGR valve has not been the best thing to happen to fuel economy or driveability. Many earlier EGR systems had the EGR valve open anytime the engine was off idle and short of full throttle, any control of it is an improvement. Most engine control computer systems control when the EGR valve is turned on and leave it off for more of the time when combustion temperatures are not high enough to produce significant amounts of NO_x. Some systems even control how much it opens *(modulate it).*

Air-conditioning Control. Some computerized engine control systems feature an air-conditioning clutch cutout function that disengages the air-conditioning clutch during heavy throttle operation. This function contributes somewhat to driveability.

Turbocharger Boost Control. To increase performance, each manufacturer is offering at least one turbocharged engine option. In most cases, the computer controls the amount of boost that the turbocharger can develop. It is worth noting that each of the manufacturers are using an air-to-air intercooler on selected turbo applications. This lowers manifold air temperature by 120° to 150°F and allows for even more aggressive turbo boost and spark advance.

Closed Loop, Open Loop

Closed Loop. Made possible by the development and use of the oxygen sensor, a feature of all comprehensive computerized engine control systems is the ability to operate the engine in a closed-loop mode. The term *closed loop* describes an intimate, triangular relationship between the oxygen sensor, the computer, and the fuel-metering control device. The oxygen sensor tells the computer what the air/fuel mixture is. The computer sends a command to the fuel-metering control device to adjust the air/fuel ratio toward a stoichiometric air/fuel ratio (14.7 to 1 at sea level). The adjustment usually causes the air/fuel ratio to cross stoichiometric, and it starts to move away from 14.7 in the opposite direction. The oxygen sensor sees this and reports it to the computer. The computer again issues a command to adjust back toward 14.7 to 1. With the speed at which this cycle occurs, the air/fuel ratio never gets very far from 14.7 to 1 in either direction. This cycle repeats continuously.

Closed loop is the most efficient operating mode, and the computer is programmed to keep the system in closed loop as much as possible. There are, however, criteria that must be met before any of the systems can go into closed loop.

- The oxygen sensor must reach operating temperature (around 315°C [600°F]).

- The engine coolant temperature must reach a criterion temperature (varies somewhat but tends to be around 65° C (150° F).
- A predetermined amount of time must elapse from the time the engine was started (this varies from a few seconds to one or two minutes).

There are other operating conditions, such as hard acceleration or, in some cases, prolonged idle, that force the system out of closed loop.

Open Loop. *Open loop* is used during periods of time when a stoichiometric air/fuel ratio is not appropriate such as during engine warm-up or at wide-open throttle (WOT). During this operational mode, the computer uses input information concerning coolant temperature, engine load, barometric pressure, and engine speed to determine what the air/fuel ratio should be. Once the necessary information is processed, the computer sends the appropriate command to the mixture control device. The command does not change until one of the inputs changes. In this mode the computer does not use oxygen sensor input and therefore does not know if the command it sent actually achieved the most appropriate air/fuel ratio for the prevailing operating conditions. As a result of this weakness of the open-loop mode, the computer puts the system into closed loop as soon as possible and keeps it there for as much of the time as possible. Certain failures within the system prevent it from going into closed loop or cause it to drop back out of closed loop.

ATTITUDE OF THE TECHNICIAN

Historically many automotive technicians had a negative attitude about the changes that the manufacturers continuously put into cars. Changes mean always having to learn something new, having to cope with a new procedure, and so on. There is another way to look at it, though: every time new knowledge or a new skill is required, it represents a new opportunity to get ahead of the pack. Most of us are in this business to make money; the more you know that other people have not bothered to learn yet, the more money you can make. Successful, highly paid automotive technicians earn the money they get because of what they know. Their knowledge enables them to produce more with the same effort.

Computerized automotive control systems represent the newest and probably the most significant and most complicated development ever to occur in the history of automotive service. (It is complicated, not because it is more difficult, but because it involves more processes.) We have seen only the beginning. It is likely that before long, every major component and function on the car will be electronically controlled. Those who appreciate the capability of these systems and learn everything they can about them will be able to service them and do very well while other people scratch their heads and complain.

Approaching Diagnosis

Diagnosing problems on complicated electronic systems takes a little different approach than many automotive technicians are used to. The flat-rate pay system has encouraged many of us to take shortcuts whenever possible; however, shortcuts on these systems get you into trouble. Use the service manual and follow procedures carefully. It is very useful to spend some time familiarizing yourself with the diagnostic guides and charts that manufacturers present in the service manual or manuals you will be using.

Prediagnostic Inspection. Although most computerized engine control systems feature some degree of self-diagnostic capability, none of them monitors spark plugs, spark plug wires, valves, vacuum hoses, PCV valves, and other engine support and emission-control components that are not a part of the computer system. A seemingly unrelated part such as a spark plug wire can have a direct impact on the performance of the system. For example, a shorted spark plug wire prevents its cylinder from firing. The unburned oxygen coming out of that cylinder, if it is on the same side as the oxygen sensor on a V-type engine, causes the oxygen sensor to mistakenly see a lean condition. The computer responds by enriching the air/fuel mixture. Replacing the oxygen sensor, half or all of the other sensors, the computer, the headlights, and the front bumper does not solve the problem. Replacing the shorted spark plug wire and possibly cleaning the spark plug does solve the problem.

The computer knows the engine operating parameters (throttle position, atmospheric pressure, engine speed, temperature, load, etc.), and with this information it can calculate exactly what the spark timing should be. It continuously makes calculations and sends its spark timing commands. It does not, however, know what base timing is; and the command it sends out is added to base timing. If base timing is incorrect, a spark plug will fire at the wrong time in spite of the computer.

Having a closed thermostat stick or a restricted radiator causes the engine to overheat, possibly causing the computer to think that the coolant temperature-sensing device or its connecting circuit is shorted and setting a fault code in its diagnostic memory. The coolant-sensing device, however, is only reporting what it sees; the fault is in the cooling system.

Do not overlook the possibility that someone has replaced a closed 195° thermostat with one that opens at 160° or that the charcoal canister is saturated with gasoline or that it is being purged when it should not be. Do not overlook the possibility of loose or shorted electrical wires or of cracked or misrouted vacuum hoses. Any such problem, although it may not be directly related to the computer system, can cause a driveability problem and can in some cases cause the computer to mistakenly report a problem in one of its circuits. The service manual for each vehicle with an engine control computer system should provide instructions for making a prediagnostic inspection, because the system's diagnostic procedures assume that all such unrelated, or perhaps it would be better to say semirelated, systems are functioning properly.

All of the engine components, especially engine support components, and some non-engine components such as the transmission are more related in their influence on each other than they have ever been before computer control systems. It is extremely important to remember that in order for the computer control system to work properly, all other engine-related systems must be operating to manufacturer's specifications. If routine maintenance, such as changing oil, PCV valve, spark plugs, spark plug wires, is due, it is advisable to do it before beginning diagnostic procedures.

Excessive HC Production

Overadvancing ignition timing by 6° can cause HC production to go up by as much as 25% and NO_x production to increase by as much as 20% to 30%. A thermostat that opens at 160° F can cause HC production to go up as much as 100 to 200 parts per million.

REVIEW QUESTIONS

1. Name two types of voltage signal.
2. The term *square wave* refers to what?
3. Most sensors create what type of signal?

Chapter 1 Computers on Cars

4. How many different characters are in the binary code?
5. Data coming into the automotive-type microcomputer is first processed by _____ .
6. What does the abbreviation ROM stand for and what is its function?
7. Name at least two ways that a PROM is different from a ROM.
8. Name at least two ways that a RAM is different from a ROM.
9. Name two types of RAMs.
10. What is the function of the clock pulse in a computer?
11. Name at least three things that cause excessive HC from the exhaust.
12. What causes high CO production?
13. What causes high NO_x production?
14. What air/fuel ratio is necessary for the three-way catalytic converter to work effectively?
15. Name two different types of three-way catalytic converters.
16. Describe the condition(s) that must exist before the computer system can function properly and before a diagnostic procedure should be begun.

ASE-type Questions. (Actual ASE test questions will probably not be this product specific.)

17. Technician A says that computerized engine control systems are intended to reduce exhaust emissions. Technician B says that computerized engine control systems are designed to improve fuel mileage and driveability. Who is correct?
    a. A only
    b. B only
    c. both A and B
    d. neither A nor B
18. A car with a driveability complaint is being checked. During the inspection it is found that base timing is off by 4° and one spark plug is faulty. Also the engine control computer's self-diagnostic test shows an oxygen sensor code. Which of the three faults identified are most likely related?
    a. the timing and the oxygen sensor code
    b. the oxygen sensor code and the spark plug
    c. all three
    d. none of them
19. Technician A says that a fault code indicating a coolant temperature sensor problem could be the result of a faulty sensor or one of its circuit connections. Technician B says that it could be the result of a faulty thermostat. Who is correct?
    a. A only
    b. B only
    c. both A and B
    d. neither A nor B
20. Technician A says that for a car to use a three-way catalytic converter, it must also have some way to test and control its air/fuel mixture. Technician B says that if a car has a three-way catalytic converter, the converter can be either a single-bed or a dual-bed design. Who is correct?
    a. A only
    b. B only
    c. either A or B
    d. neither A nor B

GLOSSARY

A/D A converter circuit that translates an analog signal into digital quantities.

ANALOG A signal that continuously varies within a given range and with time being used for the change to occur.

BAUD RATE In a stream of binary signals, one baud is one bit per second. The term "baud" is in honor of the Frenchman Baudot, who is responsible for much of the early development in unit data transmission.

BIT A single piece of information in a binary code system; somewhat comparable to a single letter within a word or a single digit within a number.

BYTE Eight pieces of binary code information strung together to make a complete unit of information.

CATALYST An agent that causes a chemical reaction between other agents without being changed itself.

CLOSED LOOP Describes an intimate, signal-response, triangular relationship between the O_2 sensor, the ECM, and the M/C solenoid.

D/A A converter circuit that translates digital signals to analog voltage values.

DIGITAL Using values (numbers) expressed as digits with specific meanings to represent the variables in a set of operations.

DRIVEABILITY Those factors, including ease of starting, idle quality, acceleration without hesitation, and so on, that affect the ease of driving and reliability.

ENGINE CALIBRATION Adjustments or settings that produce desired results such as spark timing, air/fuel mixture, EGR control, and so on.

INTERFACE Circuitry that converts both input and output information to analog or digital signals as needed and filters external circuit voltage to protect computer circuits.

MICROPROCESSOR A processor contained on an integrated circuit, processor (central processing unit) that makes the arithmetic and logic decisions in a microcomputer.

MODULATE To find a position between two extremes; between on and off.

OPEN LOOP An operational mode in which the air/fuel mixture is calculated based on coolant temperature, engine speed, and engine load without benefit of the oxygen sensor.

PALLADIUM (Pd) A silver-white, ductile, metallic element; number 46 on the Periodic Table.

PLATINUM (Pt) A heavy, grayish white, ductile, metallic element; number 78 on the Periodic Table.

PROM Programmable read-only memory.

RAM Random access memory.

RHODIUM (Rh) A white, hard, ductile, metallic element; a member of the noble metals family; number 45 on the Periodic Table.

ROM Read-only memory.

STOICHIOMETRIC In automotive terminology it refers to an air/fuel ratio in which all combustible materials are used with no deficiencies or excesses; 14.7 parts air to 1 part fuel, by weight, at sea level.

2

COMMON COMPONENTS FOR COMPUTERIZED ENGINE CONTROL SYSTEMS

Objectives

After studying this chapter, you will be able to:

- state the advantages and disadvantages of a multipoint fuel injection system.
- define the terms *pulse width* and *duty cycle*.
- describe how an oxygen sensor works.
- describe how a thermistor is used as a temperature-measuring sensor.
- describe how a potentiometer is used as a position-measuring sensor.
- describe the operation of a Hall-effect switch.
- describe a sensor that is commonly used to measure air pressure.
- describe how a simple switch is used as a sensor.
- define the term *actuator*.

Among the different car manufacturers, vehicle models, and years, there are many different computerized engine control systems. Although there are significant differences between the various systems, when compared they are actually more alike than different. Some of the more common components and circuits will be discussed in this chapter to avoid needless duplication when discussing the specific systems. As you read the following chapters, you may find it useful to refer to this chapter to clarify how a particular component or circuit works.

COMMON FEATURES

Computers

The comprehensive automotive computer is a small, highly reliable, solid-state digital computer protected inside a metal box. It receives information in the form of voltage signals from several sensors and other input sources. With this information, which the computer reads several thousand times per second, it can make decisions about engine calibration functions such as air/fuel mixture, spark tim-

Chapter 2 Common Components

ing, EGR application, and so forth. These decisions appear as commands sent to the *actuators*—the solenoids, relays, and motors that carry out the output commands of the computer. Commands usually amount to turning on or off an actuator. In most cases the ignition switch provides voltage to the actuators either directly or through a relay. The computer controls the actuator by using one of its internal solid-state switches (transistors) to ground the actuator's circuit. In this way the transistor only has to deal with the low-voltage side of the circuit (the voltage is dropped and most of the heat is produced in the actuator, outside the computer).

WARNING: No attempt should be made to open the computer's metal housing. The housing protects the computer from static electricity. Opening it or removing any circuit boards from it in any other than carefully controlled laboratory conditions will likely result in damage to some of its components.

FIGURE 2-1 General Motors electronic control module and PROM. *Courtesy of GM Product Service Training.*

Location. On most systems the computer is located inside the passenger compartment. This protects it from the harsh, high-temperature environment in the engine compartment.

Engine Calibration. Because of the wide range of vehicle sizes and weights, engine and transmission options, axle ratios, and so forth, and because many of the decisions the computer makes must be adjusted for those variables, the manufacturers use some type of an engine calibration unit, Figure 2-1. The calibration unit is a chip that contains information that has to be specific to each vehicle. For example, the vehicle's weight affects the load on the engine, so to optimize ignition timing, ignition timing calibration must be programmed for that vehicle's weight. In a given model year, a manufacturer might have more than a hundred different vehicle models counting different engine and transmission options, but might use less than a dozen different computers. This is made possible by using an engine calibration unit.

The engine calibraiton unit may be referred to as the PROM, or some other term may be used. Some manufacturers make the calibration unit removable, and some make it a permanent part of the computer.

Five-volt Reference

With few exceptions all of the systems use 5 volts to operate their sensors. Within the electronics industry, 5 volts has been almost universally adopted as a standard for information transmitting circuits. This voltage value is high enough to provide reliable transmission and low enough not to damage the tiny circuits on the chips in the computer.

Fuel Injection

There are two types of fuel injection systems used with comprehensive computerized engine control systems: single point and multipoint. They each use an intermittent or timed spray to control fuel quantity.

Constant Spray and Timed Spray Injection

There are two basic methods of controlling the quantity of fuel introduced by a fuel injection system: the constant spray injection and the timed spray injection. Historically these have been port injection systems (placing an injector in the center of the intake manifold, above the throttle blades, is a comparatively new development). As the term *constant spray* implies, in this system each injector sprays a continuous stream of fuel into each intake port; and the quantity of fuel introduced is controlled by a variable fuel pressure. Fuel pressure is controlled by an airflow meter. The Bosch K-Jetronic is the most common example of this type of system.

In timed spray systems, fuel pressure is constant and the injectors are turned on and off. Fuel quantity is controlled by how long the injectors are turned on, and the injectors are controlled by an electronic module (a single-function computer system). The Bosch D-Jetronic and L-Jetronic systems are the best known examples of this type of system.

Single-point Injection. Single-point injection is often referred to as *throttle body injection*. Single point means that fuel is introduced into the engine from one location. This system uses an intake manifold similar to what would be used with a carbureted engine, but the carburetor is replaced with a throttle body unit, Figure 2-2. The throttle body unit contains one or two solenoid-operated injectors that spray fuel directly over the throttle blade (or blades). Fuel under pressure is supplied to the injector. The throttle blade is controlled by the throttle

Chapter 2 Common Components

FIGURE 2-2 TBI unit. *Courtesy of GM Product Service Training.*

linkage just as in a carburetor. The computer controls voltage pulses to the solenoid-operated injector, which opens and sprays fuel into the throttle bore. The amount of fuel introduced is controlled by the length of time the solenoid is energized. This is referred to as *pulse width*. The amount of air introduced is controlled by the opening of the throttle blade.

The EFI system is characterized, especially on smaller engines, by excellent throttle response and good driveability. Experience has shown, however, that the system is best suited for engines with small cross-sectional area manifold runners that at low speeds will keep the fuel mixture moving at a higher velocity. This reduces the tendency for the heavier fuel particles to fall out of the airstream.

Multipoint Injection. Multipoint injection is often referred to as *port injection* and means that fuel is introduced into the engine from more than one location. This system uses an injector at each intake port. Fuel is sprayed directly into the port, just on the manifold side of the intake valve.

The multipoint injection system provides the most advanced form of fuel control yet developed. It offers the following advantages:

- Spraying precisely the same amount of fuel directly into the intake port of each cylinder eliminates the unequal fuel distribution so inherent when already mixed air and fuel are passed through an intake manifold.
- Because there is no concern about fuel condensing as it passes through the intake manifold, there is less need to heat the air or the manifold.
- Because there is no concern about fuel molecules falling out of the airstream while moving through the manifold at low speeds, the cross-sectional area of the manifold runners can be larger and thus offer better cylinder-filling ability (volumetric efficiency, VE) at higher speeds.
- Most of the manifold-wetting process is avoided; some wetting still occurs in the port areas. If fuel is introduced into the intake manifold, some will remain on the manifold floor and walls especially during cold engine operation and acceleration. Fuel metering has to allow for this fuel in order to avoid an overlean in the cylinders. It has to be accounted for again during high-vacuum conditions because it will then begin to evaporate and go into the cylinders.

In general port fuel injection provides better engine performance and excellent driveability while maintaining or lowering exhaust emission levels and increasing fuel economy.

Fuel Pressure Regulators. All fuel injection systems that are part of a comprehensive computerized engine control system use pulse width as the primary means of controlling fuel metering. A pressure regulator is used to provide a constant pressure to the injector. An electric fuel pump supplies under pressure to the regulator.

The fuel pressure regulator is located on the fuel rail or manifold and allows excess fuel to return to the tank. It has a diaphragm with

Chapter 2 Common Components

Fuel Injection History

Fuel injection for gasoline engines dates at least as far back as the middle 1930s. It has captured the interest of automotive engineers and performance enthusiasts for years by promising to be a superior method of fuel induction. Only recently, however, has its technology evolved far enough (see **Measuring Air Mass** in this chapter) to enable fuel injection to surpass the carburetor's ability to deliver performance, driveability, economy, and emissions reduction under all driving conditions.

---

fuel pressure on one side and atmospheric pressure on the other side, Figure 2-3. A diaphragm spring pushes the diaphragm against fuel pressure. As the diaphragm moves in the direction the spring is pushing it, a flat disc in the center of the diaphragm closes off the fuel return passage. This allows less fuel to return to the tank and causes fuel pressure to go up, closer to fuel pump pressure. The increased fuel pressure moves the diaphragm against the spring. This allows more fuel to escape through the return line and thus causes fuel pressure to drop. The diaphragm will always find a balanced position that will keep fuel pressure at the desired value.

On some single-point injection systems, the spring side of the diaphragm is exposed to atmospheric pressure. Changes in atmospheric pressure produce a slight change in fuel pressure. Some multipoint injection systems connect manifold vacuum to the spring side of

FIGURE 2-3 Fuel pressure regulator. *Courtesy of Cadillac Motor Car Division.*

59

Chapter 2 Common Components

the diaphragm. Changes in atmospheric pressure or in manifold pressure change the total pressure on the spring side of the diaphragm and therefore the fuel pressure. The intent, however, is not to increase or decrease fuel delivery but to maintain a constant pressure drop across the injector. For example, when manifold pressure goes up, the injector has to spray fuel into a higher pressure environment. In order for the correct amount of fuel to be delivered during the pulse width, fuel pressure must be increased in proportion to the manifold pressure increase.

SENSING DEVICES

Exhaust Oxygen Sensors

One of the major components of all comprehensive computerized engine control systems is the oxygen sensor, Figure 2-4. The amount of free oxygen in the exhaust stream is a direct result of the air/fuel ratio. A rich mixture yields little free oxygen; most of it is consumed during combustion. A lean mixture, on the other hand, yields considerable free oxygen because all of the oxygen is not consumed during combustion. With an oxygen sensor in or near the exhaust manifold, measuring the amount of oxygen in the exhaust reveals the air/fuel ratio burned in the cylinder.

The heart of the oxygen sensor is a hollow ceramic body *(zirconium dioxide)* closed at one end, Figure 2-5. The inner and outer surfaces of the ceramic body are coated with separate, thin, gas-permeable films of platinum. The platinum coatings serve as electrodes. The outer electrode surface is covered with a thin, porous, ceramic layer to protect against contamination from combustion residue. This body is placed in a metal shell similar to a spark plug shell except that the shell has a louvered nose that encloses the closed end of the ceramic body. When the shell is screwed into the exhaust pipe or manifold, the louvered end extends into the exhaust passage. The outer electrode surface contacts the shell; the inner electrode connects to a wire that goes to the computer. Ambient oxygen is allowed to flow into the hollow ceramic body and to contact the inner electrode.

The oxygen sensor has an operating temperature range of 300°C to 850°C (572°F to 1562°F). When the zirconium dioxide reaches 300°C it becomes oxygen-ion conductive. If the oxygen in the exhaust stream is less than that in the ambient air, a voltage is generated. The greater the difference, the greater the voltage, Figure 2-6. Voltage generated by the oxygen sensor will normally range from a minimum of 0.1 volt (100 millivolts) to a maximum of 0.9 volt (900 millivolts). The computer has a preprogrammed value, called a *set-point,* that it wants to see from the oxygen sensor during closed-loop operation. The set-point is usually between 0.45 volt and 0.5 volt and equates to the desired air/fuel ratio. Voltage values below the set-point are interpreted as lean, and values above it are interpreted as rich.

By the mid to late 1980s, most manufacturers had begun using heated oxygen sensors on some engine applications. The operation of the sensor itself is no different; however, a heat element is placed inside it to keep it hot enough. This is largely due to a tendency to move the sensor a little lower in the exhaust system so that the exhaust sampled by the sensor is more of an average of what all of the cylinders are producing. Placing the oxygen sensor lower in the exhaust pipe means that it

FIGURE 2-4 Oxygen sensor

Chapter 2 Common Components

FIGURE 2-5 Exhaust gas oxygen sensor. *Courtesy of Ford Motor Co.*

FIGURE 2-6 Relationship of oxygen sensor voltage to air/fuel ratio (approximate)

is exposed to less heat due to the rapid loss of heat through the pipe wall. Two additional wires are attached to the oxygen sensor to power the heat element. In most cases, one powers the heat element when the ignition is turned on and the other goes to a fixed ground. The circuit is not part of the computer control system.

Oxygen sensors are subject to contamination under certain conditions. Contaminating either electrode surface (the outside electrode is the most vulnerable) will shield it from oxygen and adversely affect its performance. Prolonged exposure to rich fuel mixture exhaust can cause it to be carbon fouled. This can sometimes be burned off by operating in a lean condition for two or three minutes. The oxygen sensor will also be fouled by exposure to:

- exhaust from leaded fuel.
- the vapors from some silicone-based gasket-sealing compounds.
- the residues of coolant leaking into the combustion chamber.

Fouling by these materials will probably require that the oxygen sensor be replaced.

61

Chapter 2 Common Components

FIGURE 2-7 Coolant temperature sensor. *Courtesy of GM Product Service Training.*

Thermistors

A *thermistor* is a resistor made from a semiconductor material. Its electrical resistance changes greatly and predictably as its temperature changes. At −40° C (−40° F) a typical thermistor can have a resistance of 100,000 ohms. At 100° C (210° F) the same thermistor can likely have a resistance between 100 and 200 ohms. Even small changes in temperature can be observed by monitoring the thermistor's resistance. This characteristic makes it an excellent means of measuring the temperature of such things as water and air.

There are negative temperature coefficient (NTC) and positive temperature coefficient (PTC) thermistors. The resistance of the NTC thermistor goes down as its temperature goes up, although the resistance of the PTC type goes up as its temperature goes up. Most thermistors, including those used as temperature sensors on automotive computer systems, are the NTC type. A good example is an engine coolant temperature sensor, common to just about all computerized engine control systems. The coolant temperature sensor consists of a thermistor in the nose of a metal housing, Figure 2-7. The housing screws into the engine, usually in the head, with its nose extending into the water jacket so that the thermistor element will be the same temperature as that of the coolant. The computer sends a regulated voltage signal (reference voltage, usually 5 volts) through a fixed resistance and then on to the sensor, Figure 2-8. A small amount of current flows through the thermistor (usually shown as a resistor symbol with an arrow diag-

FIGURE 2-8 Temperature-sensing circuit

Chapter 2 Common Components

FIGURE 2-9 Potentiometer as throttle position sensor

onally across it) and returns to ground. This is a voltage divider circuit (current flows through the first resistance unit then has an alternate path in parallel with the second resistance unit) and is commonly used as a temperature-sensing circuit. Because the resistance of the thermistor changes with temperature, the voltage drop across it changes also. The computer monitors the voltage drop across the thermistor and, using preprogrammed values, converts the voltage drop to a corresponding temperature value.

Potentiometers

A potentiometer is another application of a voltage divider circuit. It has a movable center contact or *wiper* that taps or senses the voltage between two resistance units, Figure 2-9. The potentiometer is usually used to measure either linear or rotary motion. If we were going to use a potentiometer to measure throttle position, we would probably attach it to one end of the throttle shaft. As the throttle is opened, the rotation of the throttle shaft moves the wiper, which slides along the wire-wound resistor.

63

The computer sends a reference voltage to one terminal of the potentiometer, point A. If the wiper is positioned near point A (wide-open throttle on most applications), there will be a low voltage drop between points A and B (low resistance) and a high voltage drop will exist between points B and C. When the wiper is positioned near point C (idle position on most applications), there is a high voltage drop between points A and B and a low drop between points B and C. The computer monitors the voltage drop between points B and C (C should be the same as D) and interprets a low voltage, 0.5 for example, as idle position. A high voltage, around 4.5, will be interpreted as wide-open throttle (WOT). Voltages between these values will be interpreted as a proportionate throttle position.

The insert in Figure 2-9 shows a slight variation in the throttle position sensor circuit. Some applications use this variation to enhance diagnostic capability. The high resistance connection between the reference voltage circuit and point B causes near reference voltage to appear at the sensor signal terminal if an open circuit occurs in the harness that connects the sensor to the computer.

Pressure Sensors

Pressure sensors are commonly used to monitor intake manifold pressure and/or barometric pressure. Intake manifold pressure is a direct response to throttle position and engine speed, with throttle position being the most significant factor. The greater the throttle opening, the greater the manifold pressure becomes (lower vacuum). At WOT manifold pressure is nearly 100% of atmospheric pressure. Intake manifold pressure can therefore be translated as engine load and is a critical factor in determining many engine calibrations. Barometric pressure also impacts on manifold pressure and in most systems is considered when calculating engine calibrations such as air/fuel mixture and ignition timing.

Intake manifold pressure (usually a negative pressure or *vacuum*) and/or barometric pressure are, on most systems, measured with a silicon diaphragm that acts as a resistor, Figure 2-10. The resistor/diaphragm (about 3 millimeters wide) separates two chambers. As the pressure on the two sides of the resistor/diaphragm varies, it flexes. The flexing causes the resistance of this semiconductor material to change its electrical resistance. The computer applies a reference voltage to one side of the diaphragm. As the current crosses the resistor/diaphragm, the amount of voltage drop that occurs depends on how much the diaphragm is flexed. The signal is passed through a filtering circuit before it is sent to the computer as a DC analog signal. The computer monitors the voltage drop and, by using a look-up chart of stored pressure values, determines exactly what the pressure is to which the diaphragm is responding.

There are two slightly different pressure sensor designs that use the *piezoresistive* silicon diaphragm: the absolute pressure and the pressure differential sensors, Figure 2-11. In one design the chamber under the diaphragm is sealed and contains a fixed reference pressure. The upper chamber is exposed to either intake manifold pressure or to atmospheric pressure. If the upper chamber is connected to the intake manifold, the sensor functions as a *manifold absolute pressure sensor*. Absolute, as used in MAP, refers to a sensor that compares a varying pressure to a fixed pressure. The output signal (return voltage) from this sensor increases as pressure on the variable side of the diaphragm increases (wider throttle opening). If the upper chamber is exposed to atmospheric pressure, the sensor functions as a *barometric pressure sensor*.

The differential pressure sensor combines the functions of both the MAP and the BARO sensors. Instead of using a fixed pressure, one side of the resistor/diaphragm is exposed to barometric pressure and the other side is connected to intake manifold pressure. The output signal is the result of subtracting manifold vacuum from barometric pressure. Opposite to the MAP sensor, however, as manifold pressure

Chapter 2 Common Components

FIGURE 2-10 Silicon diaphragm pressure sensor

decreases (higher vacuum), output voltage increases. The pressure differential sensor had limited use on some early General Motors Computer Command Control systems.

Other types of pressure sensors are discussed in Chapters 7 and 8.

FIGURE 2-11 Two types of silicon diaphragm pressure sensors. *Courtesy of GM Product Service Training.*

Speed Density Formula. In order to know how much fuel to meter into the cylinder, the microprocessor must know how much air, as measured by weight, is in the cylinder. Since the microprocessor has no way to actually weigh the air going into a cylinder, some other method must be used to determine this value. Most systems use a mathematical calculation called the *speed density formula.*

$$\frac{EP \times EGR \times VE \times MAP}{AT} =$$
air density in cylinder

Where: MAP = manifold absolute pressure
VE = volumetric efficiency
EP = engine parameters
EGR = EGR flow
AT = air temperature

To make this calculation, all of the factors that influence how much air gets into the cylinders must be considered. Throttle position and engine speed affect air intake and engine temperature (measured as coolant tempera-

Chapter 2 Common Components

ture) affects how much heat the air gains as it passes through the induction system and thus affects the air's density in the cylinder. These factors are grouped together and called *engine parameters*. The microprocessor gets these pieces of information from some of its sensors that are common to all systems.

The amount of air getting into the cylinders is reduced by the amount of exhaust gas that the EGR valve meters into the induction system. Some systems use a sensor to determine EGR flow, and some use estimates (based on engine parameter) of EGR flow that are stored in computer memory.

You know the diameter, length, and shape of the intake manifold runners; the valve size, lift, and timing; the combustion chamber design; and the cylinder size and compression ratio all affect how much air can get into the cylinders. This group of factors is called *volumetric efficiency*. A volumetric efficiency value for any given set of engine parameters is stored in the computer's memory.

The air's temperature affects its density and, therefore, how much air can get into the cylinders. Most systems use an air temperature sensor to measure air temperature. Some have estimates of air temperature stored in memory. The estimates of air temperature are based on a fairly predictable relationship between coolant temperature and air temperature.

You also know that barometric pressure pushes air past the throttle valve and that manifold pressure pushes that air past the intake valve into the cylinder. The last known values that must be determined are how much air the engine will inhale during any given set of operating conditions and barometric pressure and manifold pressure. The MAP sensor, as you will see in later chapters, almost always provides both pressure values. A MAP sensor identifies those systems that use the speed density formula.

Measuring Air Mass

Port fuel injection or multipoint injection has become the predominate fuel-metering system. Although a multipoint injection system offers several distinct advantages over other types of fuel-metering systems (discussed earlier in this chapter), it has one shortcoming. A multipoint injection system provides less opportunity for the fuel to evaporate than a system where the fuel is introduced into the center of the intake manifold does, and only the fuel that is evaporated before combustion occurs is useful. In the past this problem was probably most apparent as a lean stumble when accelerating from idle and was the result of the high manifold pressure that comes from the sudden increase in throttle opening. With the throttle open, atmospheric pressure forces air into the manifold faster than the engine can use it. The turbulence produced during the intake and compression strokes (and the increased oxygen density on turbocharged applications) helps to atomize the fuel and hasten its evaporation. The condition is only completely overcome, however, by spraying in additional fuel to compensate for the failure of the heavier hydrocarbon molecules to evaporate. The capability of the digital microprocessor provides the ability to calculate just the right amount of fuel to avoid a lean stumble without sacrificing economy or emissions.

For an engine at a given operating temperature and an identified set of atmospheric conditions (air temperature, pressure, and humidity), there is one precise amount of fuel that should be injected into the intake port for each manifold pressure value within the operating range, that will provide an evaporated 14.7 to 1 air/fuel ratio in the combustion chamber with the lowest possible leftover unevaporated HC.

Mass Airflow (MAF) Sensor. The MAF sensor used with the speed density formula has been pretty effective in enabling the micropro-

Chapter 2 Common Components

cessor to accurately determine the mass of the air that goes into the cylinder. Keep in mind that the microprocessor quantifies both the air and the fuel by weight when calculating air/fuel ratios. The vane meter has also been fairly effective in providing the microprocessor with the needed information to calculate the air's mass. The vane meter is discussed in the **Inputs** section of Chapter 7. Each of these two methods have, however, been unable to measure one factor that affects the air's mass—humidity.

The MAF sensor, introduced in the mid-1980s, provides information to the microprocessor that accounts for the airflow's rate, temperature, and humidity and can actually determine the airflow mass. It does this by measuring the air's ability to cool. Air flowing over an object that is at a higher temperature than the air carries heat away from the object. The amount of heat carried away depends on several factors: the difference in temperature between the air and the object, the object's ability to conduct heat, the object's surface area, the air's mass, and its flow rate.

The MAF sensor is placed in the duct that connects the air cleaner to the throttle body so that all of the air entering the induction system must pass through it, Figure 2-12. In the MAF

FIGURE 2-12   MAF sensor location. *Courtesy of Ford Motor Co.*

67

Chapter 2 Common Components

FIGURE 2-13 MAP sensor. *Courtesy of Ford Motor Co.*

sensor's main passageway is a smaller passage that a fixed percentage of the air will pass through. In the smaller passage is a thermistor and a heated wire, Figure 2-13. The thermistor measures the incoming air temperature. The heated wire is maintained at a predetermined

FIGURE 2-14 MAP sensor circuit

temperature above the air's temperature by a small electronic module mounted on the outside of the sensor's body.

The heated wire is actually one of the resistors of a balanced bridge circuit. Figure 2-14 represents a simplified MAF sensor circuit. The module supplies battery voltage to the balanced bridge. Resistors R1 and R3 form a series circuit in parallel to resistors R2 and R4. The voltage at the junction between R1 and R3 is equal to the voltage drop across R3. Because the values of R1 and R3 are fixed, the voltage at this junction is constant; it will vary but only as battery voltage varies. Resistors R1 and R2 have equal values. With no airflow across R4, its resistance is equal to R3, and the voltage at the junction between R2 and R4 is equal to the voltage at the R1-R3 junction.

As airflow across R4 increases, its temperature drops, and because it is a positive temperature coefficient resistor, its resistance goes down with its temperature. As the air's mass increases, R4's resistance goes down even more. As the resistance of R4 goes down, a smaller portion of the voltage applied to the circuit formed by R2-R4 is dropped across R4. Therefore, a larger portion of the voltage is dropped across R2. So, as the resistance of R4 goes down, the voltage at the R2-R4 junction goes down. The computer reads the voltage difference between the R1-R3 junction and the R2-R4 junction as the MAF sensor value. By looking at a table in the computer's memory, the microprocessor converts the voltage value to a mass airflow rate.

Hall-effect Switches

A *Hall-effect switch* is frequently used to sense engine speed and crankshaft position. It provides a signal each time a piston reaches top dead center, and its signal serves as the primary input on which ignition timing is calculated. Because of the frequency with which this signal occurs and the critical need for accuracy, the Hall-effect switch is a popular choice because it and its related circuitry provide a

FIGURE 2-15 Hall-effect switch. *Courtesy of Ford Motor Co.*

digital signal. On some applications a separate Hall-effect switch is used to monitor camshaft position as well. Systems that do not use a Hall-effect switch, use either the magnetic pick-up unit in the distributor as the crankshaft speed and position information source or its equivalent outside the distributor.

A Hall-effect switch consists mainly of a permanent magnet, a *gallium arsenate crystal* with its related circuitry, and a shutter wheel, Figure 2-15. The permanent magnet is mounted so that a small space is between it and the gallium arsenate crystal. The shutter wheel, rotated by a shaft, alternately passes its vanes through the narrow space between the magnet and the crystal, Figure 2-16. When a vane is between the magnet and the crystal, the vane intercepts the magnetic field and thus shields the crystal from it. When the vane moves out, the gallium arsenide crystal is invaded by the magnetic field. A steady current is being passed through the crystal from end to end. When the magnetic lines of force from the permanent magnet invade the crystal, the current flow across the crystal is distorted. This results in a weak voltage potential being produced at the crystal's top and bottom surfaces, negative at

Chapter 2 Common Components

one surface and positive at the other. As the shutter wheel turns, the crystal provides a weak high/low voltage signal.

This signal is usually modified before it is sent to the computer. For example, as shown in Figure 2-16, the signal is sent to an amplifier, which strengthens and inverts it. The inverted signal (high when the signal coming from the Hall-effect switch is low) goes to a *Schmitt trigger* device, which converts the analog signal to a digital signal. The digital signal is fed to the base of a switching transistor. The transistor is switched on and off in response to the signals generated by the Hall-effect switch assembly.

When the transistor is switched on, current flows through it from another circuit to ground. The other circuit can come from the ignition switch or from the computer, as shown in Figure 2-16. In either case it has a resistor between its voltage source and the transistor. A voltage-sensing circuit in the computer connects to the circuit between the resistor and the transistor. When the transistor is on and the circuit is

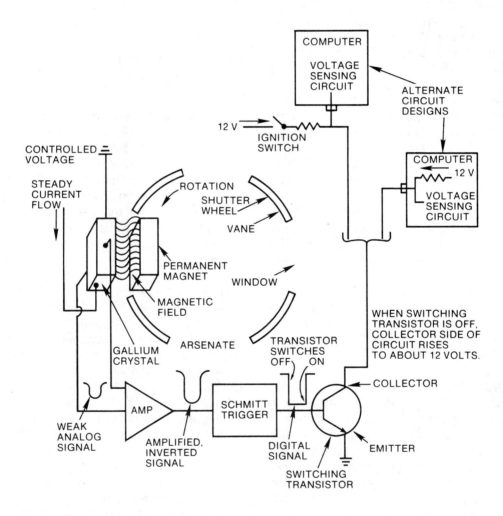

FIGURE 2-16 Hall-effect switch circuit

complete to ground, a volt drop occurs across the resistor. The voltage signal to the voltage-sensing circuit is less than 1 volt. When the transistor is switched off, there is no drop across the resistor; and the voltage signal to the voltage-sensing circuit is near 12 volts. The computer monitors the voltage level and can determine by the frequency at which it rises and falls what engine speed is. Each time it rises, the computer knows that a piston is approaching top dead center.

Detonation Sensors

To optimize performance and fuel economy, ignition timing must be adjusted so that combustion will occur during a specific number of crankshaft degrees beginning at top dead center of the power stroke. If it occurs any later, less pressure will be produced in the cylinder. If it occurs any sooner, detonation will occur. Due to variations in fuel quality, cylinder-cooling efficiency, machining tolerances and their effect on compression ratios, and so forth, a preprogrammed spark advance schedule can result in spark knock under certain driving conditions. In order to be able to provide an aggressive spark advance and avoid spark knock, many systems use a detonation sensor, often referred to as a *knock sensor*. It alerts the computer when spark knock occurs so that the timing can be retarded.

The knock sensor, Figure 2-17, screws into some section of the engine where it will be subjected to the high-frequency vibration caused by the spark knock. Most knock sensors contain a piezoelectric crystal. *Piezoelectric* refers to a characteristic of certain materials whereby they produce an electrical signal in response to physical stress or experience stress in response to an electrical signal. The most commonly used piezoelectric material produces an oscillating voltage signal of about 0.3 volt or more in response to pressure or vibration. The oscillations occur at the same frequency as the spark knock (5,000 to 6,000 hertz [Hz], or cycles per second). Although

FIGURE 2-17   Knock sensor

most vibrations within the engine and many elsewhere in the vehicle cause the knock sensor to produce a signal, the computer only responds to signals of the correct frequency.

Switches

Most systems have several simple switches that are used to sense and communicate information to the computer. The information includes whether or not the transmission is in gear, the air-conditioning is on, the brakes are applied, and so on. The same type of circuit is usually used for each switch regardless of what condition the switch is measuring.

As an example, let's take a switch used to tell the computer whether or not the automatic transmission is in gear, Figure 2-18. The ignition switch feeds 12 volts through 1 k ohms of resistance to the P/N switch. Let's assume that when the transmission is in park or neutral, the switch is closed and thus allows current to go to ground. During this condition the only significant resistance between the voltage source and ground is the 1 k-ohm resistor, and all but a

Chapter 2 Common Components

FIGURE 2-18 Park/neutral switch

tiny fraction of the 12 volts is dropped across it. The voltage signal to the computer is near zero. When the transmission is shifted into gear, the switch opens. Now the only path to ground is through the 1 k-ohm resistor and then the 10 k-ohm resistor in the computer. As the current goes through the 1 k-resistor, about 1 volt is dropped and the remaining 11 volts is dropped across the 10 k-ohm resistor. The computer recognizes that when voltage is present at the sensing circuit, the transmission is in gear.

E-cells

Some selected systems use an E-cell to indicate to the computer when a specified amount of engine operation has occurred. The

E-cell actually measures a specific amount of ignition on time, which equates to an estimated number of vehicle miles. Its purpose is to alert the computer when the predetermined mileage has occurred so that the computer can adjust specific calibrations to compensate for engine wear.

The E-cell contains a silver cathode and a gold anode. Ignition voltage is applied to the E-cell through a resistor. The passage of current through the cell controls a signal to the computer. As current passes through the cell, a chemical reaction takes place; the silver on the cathode is attracted to the gold anode. After a certain amount of on time, the silver cathode is completely deplated and the circuit opens. How long this takes is controlled by the amount of current allowed to flow through the E-cell (how much resistance is in the circuit).

ACTUATORS

Solenoids

The most common actuator is a valve powered by a solenoid. The valve usually controls vacuum, but it can also be a valve designed to control fuel vapor, airflow, oil or water flow, and so forth. A solenoid consists of a coil of wire with a movable iron core, Figure 2-19. When the solid-state switching circuit in the computer grounds the solenoid circuit, the coil becomes energized. The magnetic field developed by the current passing through the coil

FIGURE 2-19 Typical solenoid-controlled vacuum valve

Chapter 2 Common Components

windings pulls the iron core so that it occupies the entire length of the coil. This movement of the core opens the valve attached to its other end.

Duty-cycled and Pulse Width Solenoids. In some cases the switching circuit that drives the solenoid turns on and off rapidly. If the solenoid circuit is turned on and off (cycled) ten times per second, then the solenoid is *duty cycled.* To complete a cycle, it must go from off to on to off again. If it is turned on for 20% of each tenth of a second and off for 80%, it is said to be on a 20% duty cycle. In most cases where a solenoid is duty cycled, the computer has the ability to change the duty cycle to achieve a desired result. The best example of this is the mixture control device on many computer-controlled carburetors, where the duty cycle will frequently change to control the air/fuel mixture.

If a solenoid is turned on and off rapidly but there is no set number of cycles per second at which it is cycled, the on time is referred to as *pulse width.* For example, let's assume that the computer has decided that the EGR valve should be open but that it should be open to only 60% of its capacity. The computer can energize a solenoid valve such as the one in Figure 2-19 to allow vacuum to the EGR valve. To control or *modulate* the amount of vacuum to the EGR valve, the computer issues pulse width commands to a solenoid that when energized opens a port that allows atmospheric pressure to bleed into the EGR vacuum line, Figure 2-20. Because the computer wants the EGR valve open to 60% of its capacity, it selects

FIGURE 2-20 Typical duty-cycled vent valve

pulse widths from a look-up chart to appropriately weaken the vacuum signal to the EGR valve. Some literature refers to pulse width as *dithering*.

Relays

A *relay* is a remote control switch that allows a light-duty switch such as an ignition switch, a blower motor switch, a starter switch, or a driver switch in a computer to control a device that draws a relatively heavy current load. The most common example in automotive computer application is a fuel pump relay, Figure 2-21. When the computer wants the fuel pump on, it turns on its driver switch (transistor). The transistor applies 12 volts to the relay coil, which develops a magnetic field in the core. The magnetic field pulls the armature down and closes the normally open contacts. The contacts complete the circuit from the battery to the fuel pump. On many systems the fuel pump relay is one of the few examples where the computer supplies the voltage to operate a device instead of the ground connection.

Electric Motors

When electric motors are used as actuators, there are two unique types used.

Stepper Motors. A stepper motor, as used in these applications, is a small electric motor powered by voltage pulses. It contains a permanent magnet armature and usually either two or four field coils, Figure 2-22. A voltage pulse applied to one coil or pair of coils causes the motor to turn a specific number of degrees. The same voltage pulse applied to the other coil or pair of coils makes the motor turn an equal number of degrees in the opposite direction.

The armature shaft usually has a spiral on one end and this spiral connects to whatever the motor is supposed to control. As the motor turns one way, the controlled device, a *pintle valve* for instance, is extended. As it turns the other way, the valve is retracted. The computer can apply a series of pulses to the motor's coil windings in order to move the controlled device to whatever location is desired. The computer can also know exactly what position the

FIGURE 2-21 Typical fuel pump relay circuit

Chapter 2 Common Components

FIGURE 2-22 Typical stepper motor

valve is in by keeping count of the pulses applied. The stepper motor has been used to control air/fuel mixture and idle speed.

Permanent Magnet Field Motors. Some systems use this type of motor to control idle speed, Figure 2-23. It is a simple, reversible, DC motor with a wire-wound armature and a permanent magnet field. The polarity of the voltage applied to the armature winding determines the direction in which the motor spins. The armature shaft drives a tiny gear drive assembly that either extends or retracts a plunger. The plunger contacts the throttle linkage; and when it is extended, the throttle blade opening is increased. When it is retracted, the throttle blade opening is decreased. The computer has the ability to apply a continuous voltage to the armature until the desired idle speed is reached.

DIAGNOSIS AND SERVICE

Special Tools

A good service manual that applies specifically to the vehicle being serviced is essential.

FIGURE 2-23 Permanent magnet field reversible DC motor

It should contain all of the necessary charts and diagnostic procedures and should list the tools necessary for diagnostic operations.

Most systems have a data link used in the factory to test the system's operation as it leaves the assembly line. Several companies have designed and are selling test equipment that can connect to the data link and get service code information from the computer. In most cases other information such as sensor voltage values and the state of position of some actuators can also be obtained. This information is either shown as a digital readout on the test instrument or is printed on a piece of paper. This type of tool is very helpful and in some cases is essential.

High-impedance Digital Volt-Ohmmeter (DVOM). When diagnosing most systems, a DVOM is required for two reasons.

1. Many of the readings to be taken require a higher degree of accuracy than is obtained by reading a needle position on a scale.
2. Many of the circuits that will be tested have high resistance and very low voltage and amperage values. Applying a standard meter with its relatively low internal resistance to such a circuit can easily constitute a short to the circuit. In other words the meter's circuit offers an easier electrical path than the circuit or portion of the circuit being tested. A *high-impedance* meter has a minimum of 10 million ohms of internal resistance; a standard meter has only about 10 thousand.

Service Tips

Oxygen Sensor. When handling or servicing the oxygen sensor, several precautions should be observed.

- A boot is often used to protect the end of the sensor where the wire connects. It is designed to allow air to flow between it and the sensor body, Figure 2-24. The vent in the boot and the passage leading

FIGURE 2-24 Exhaust gas oxygen sensor. *Courtesy of Ford Motor Co.*

into the chamber in the oxygen-sensing element must be open and free of grease or other contaminants.

- Those with soft boots must not have the boot pushed onto the sensor body so far as to cause it to overheat and melt.
- No attempt should be made to clean the sensor with any type of solvent.
- Do not short the sensor lead to ground, and do not put an ohmmeter across it. Any such attempt can permanently damage the sensor. The single exception is that a digital high-impedance (10 megohm minimum) voltmeter can be placed across the sensor lead to ground (the sensor body).
- If an oxygen sensor is reinstalled, its threads should be inspected and if necessary cleaned with an 18 mm spark plug thread chaser. Before installation its threads should be coated with an antiseize compound, preferably liquid graphite and glass beads. Replacement sensors may already have the compound applied.

PERSONAL SAFETY: It is often recommended that the oxygen sensor be removed while the engine is hot to reduce the possibility of thread damage. If doing so wear leather gloves to prevent burns on the hand.

Vacuum Leaks. Vacuum leaks can have far-reaching effects. Examine Figure 2-25. Assume that the diaphragm ruptured in the primary vacuum break (or it could be a vacuum hose cracked or left off anywhere in that circuit). The results would be as follows:

- The primary vacuum break would not work.

FIGURE 2-25 Typical vacuum circuit. *Courtesy of GM Product Service Training.*

- The secondary vacuum break would not work (rich condition, possibly stalling during warm-up).
- The heated air system would not work (poor cold driveability).
- The vacuum sensor would report "no vacuum" to the computer, which would in turn issue erroneous spark and fuel mixture commands.
- A vacuum sensor code would probably be set.

Vacuum leaks downstream from the airflow sensor on a multipoint injection system cause poor idle if the engine idles at all.

Charcoal Canister. Charcoal canister saturation can often be the result of fuel tank overfilling. Slowly filling the tank right to the top of the filler neck causes the antioverfill chamber in the tank to fill and thus leaves no room for fuel expansion. When the fuel does expand, it is pushed up and into the charcoal canister. Canister purging then dumps so much hydrocarbon into the intake manifold that the system may not be able to compensate, even in closed loop. Many driveability problems can be linked to charcoal canister problems; don't overlook it in diagnostic procedures.

Avoid Damaging the Computer. When servicing a computerized engine control system vehicle, observe the following in order to avoid producing a voltage spike that could damage the computer:

- Avoid starting the vehicle with the aid of a battery charger.
- Turn the ignition off before connecting jumper cables.
- Turn the ignition off before disconnecting the battery or computer connectors.
- Replace any system actuator with less than specified resistance or that has a shorted clamping diode. A *clamping diode* is a diode placed across an electrical winding to suppress voltage spikes that are produced when the winding is turned on and off.
- Be sure that the winding of any electrical accessory added to the vehicle has sufficient voltage spike containment capability. A resistor of 200- to 300-watt capacity can be used instead of a diode.
- Use caution when testing the charging system, and never disconnect the battery cables with the engine running.

REVIEW QUESTIONS

1. What does the term *duty cycle* mean?
2. How is duty cycle different from pulse width?
3. How does the oxygen sensor know if the air/fuel mixture is rich?
4. What type of thermistor has higher resistance at lower temperatures?
5. Diagram a voltage divider circuit.
6. Diagram a potentiometer circuit.
7. Describe how a Hall-effect switch turns on and off a switching transistor.
8. Describe how the Hall-effect switch's switching transistor signals the computer.
9. What information does the Hall-effect switch provide for the computer?
10. Name a resistance device commonly used to measure air pressure.
11. Why is 5 volts most commonly used as a reference voltage for sensor operation?
12. Describe how a simple switch is used as a sensing device.
13. What is an actuator?
14. Describe the operation of a solenoid.
15. Describe the operation of a relay.
16. Describe how to reverse the direction of rotation of a permanent magnet field motor.
17. For what is a permanent magnet field motor usually used?

ASE-type Questions. (Actual ASE test questions will probably not be as product specific.)

18. Technician A says that a rich exhaust stream contains little free oxygen because most of the available oxygen was combined with other elements during combustion. Technician B says that little free oxygen is present in a rich exhaust stream because there was not enough oxygen in the air/fuel mixture from the beginning. Who is correct?
    a. A only
    b. B only
    c. both A and B
    d. neither A nor B

19. The voltage reading from the oxygen sensor of a car with a computerized engine control system is consistently around 0.7 volt. This indicates that the system is operating in a _____ condition.
    a. rich
    b. lean
    c. normal
    d. abnormal

20. Technician A says that a potentiometer is usually used to monitor throttle position. Technician B says that a thermistor is usually used to monitor throttle position. Who is correct?
    a. A only
    b. B only
    c. both A and B
    d. neither A nor B

21. Technician A says that an oxygen sensor can be fouled by fuel additives such as lead and antioxidation agents. Technician B says that an oxygen sensor can be fouled by use of a gasket compound containing silicon and by fuels containing alcohol. Who is correct?
    a. A only
    b. B only
    c. both A and B
    d. neither A nor B

GLOSSARY

ACTUATOR Any mechanism, such as a solenoid, relay, or motor, that when activated causes a change in the performance of a given system or circuit.

CLAMPING DIODE A diode placed across a coil or winding to trap the voltage spikes produced when the device is turned on or off.

DUTY CYCLE The portion of time during each cycle when an electrical device is turned on, when the device is cycled at a fixed number of cycles per unit of time.

GALLIUM ARSENATE CRYSTAL A semiconductor material that changes its conductivity when exposed to a magnetic field.

HALL-EFFECT SWITCH A magnetic switching device often used to signal crankshaft position and speed to the computer.

Hz (HERTZ) Cycles per second.

IMPEDANCE The total circuit resistance including resistance and reactance.

PIEZOELECTRIC A crystal that produces a voltage signal when subjected to physical stress such as pressure or vibration.

PIEZORESISTIVE A semiconductor material whose electrical resistance changes in response to changes in physical stress.

PINTLE VALVE A valve with a stem, some part of which is cone shaped. The cone-shaped portion of the stem fits against a seat to close the passage through the valve.

POTENTIOMETER A three-terminal variable resistor that acts as a voltage divider.

PULSE WIDTH The portion of time during a cycle when an electrical device is turned on, when the rate at which the cycles occur varies.

SCHMITT TRIGGER A device that trims an analog signal and converts it to a digital signal.

THERMISTOR A temperature-sensing device containing a coil of wire whose resistance changes dramatically as the temperature changes. Most thermistors have a negative temperature coefficient (the resistance goes up as the temperature goes down).

VOLUMETRIC EFFICIENCY (VE) A measure of cylinder filling. VE is expressed as the percentage of atmospheric pressure to which the cylinder is filled at the completion of the compression stroke.

ZIRCONIUM DIOXIDE (ZrO_2) A white crystalline compound that becomes oxygen-ion conductive at about 315°C (600°F).

3
GENERAL MOTORS' COMPUTER COMMAND CONTROL

Objectives

After studying this chapter, you will be able to:

- describe torque converter clutch operation.
- explain how a dwellmeter is used to evaluate air/fuel mixture control.
- describe the operation of the mixture solenoid in a computer command control system.
- describe how idle speed is controlled in a computer command control system.
- state when the air management system directs air to the exhaust manifold, catalytic converter, atmosphere.
- describe what a trouble code is and how it is obtained from the computer.

The computer command control (CCC) system was General Motors' first widely used comprehensive computerized engine control system. It was first introduced in mid-1980. Beginning with the 1981 model year, all of General Motors' carbureted passenger cars have made use of the CCC system. It should be noted that the CCC system varies slightly from one engine application to another and from one model year to another. In 1982 four different CCC systems were used. Most 1982 engines used the full function system, which was a slightly updated version of the 1981 system. Beginning in 1982 Oldsmobile 5-liter engines were equipped with a limited control system; the T car (Chevrolet Chevette and Pontiac

From EFC to CCC

General Motors introduced its first electronic closed-loop fuel control system in 1978. This single function system, called electronic fuel control (EFC), was limited to selected four-cylinder engines. In 1979 GM introduced its first comprehensive engine control system on selected applications: computer-controlled catalytic converter (C-4). The system was improved for 1981 production and was renamed the computer command control. Someone at GM decided they did not want their engine control system named after a Ford transmission.

T-1000) used a minimum function system. The Chevrolet and GMC S truck (S-10 and S-15, introduced in 1982) with the California emissions package and a four-cylinder gas engine used an imported electronic control module (ECM) that was slightly different from the other CCC systems. There are also slight variations within each of these systems.

Although all of this can seem a bit bewildering, it need not be. All versions of the CCC system are much more alike than different. Once the basic system and function are understood and a good service manual is at hand for specific directions and specifications, you will find the CCC system very manageable.

ELECTRONIC CONTROL MODULE (ECM)

The engine control computer used in the CCC system and in all other General Motors systems is called an electronic control module, Figure 3-1. The inputs it receives and the functions it controls are shown in Figure 3-2. The ECM is located in the passenger compartment, usually near the glove compartment or behind the passenger kick panel. The PROM (programmable read-only memory) is located under an access cover, Figure 3-3. It is responsible for fine-tuning engine calibrations for each specific vehicle and when plugged in becomes a functioning part of the ECM.

> Note: In earlier ECMs, the PROM can easily be installed backwards, which will result in electrical damage to it. Consult a service manual or the **System Diagnosis and Service** section of this chapter.

Keep Alive Memory (KAM)

The ECM monitors its most important sensor and selected actuator circuits. If a fault such as an open circuit occurs during normal operation, a fault code will be stored in the KAM. The code can be retrieved at a later time to assist the technician in locating the problem. This feature is especially useful because it enables the ECM to report a fault that occurred recently but is no longer present, as well as those that are still present. The KAM receives battery power at all times so that the code can be retained in memory.

OPERATING MODES

The two basic modes of operation are closed loop and open loop.

Closed Loop

Before the system can go into closed loop, the coolant must reach a temperature of approximately 65° C (150° F), the oxygen sensor must reach at least 300° C (570° F), and a predetermined amount of time must elapse following engine start-up. This time is programmed into the PROM and varies with engine application. It can be as little as a few seconds on some engines or as much as a couple of minutes on others.

FIGURE 3-1 ECM. *Courtesy of GM Product Service Training.*

Chapter 3 General Motors' Computer Command

Inputs	Codes	ECM	Codes	Outputs
Coolant Temperature Sensor	14, 15		23	Fuel Mixture
Vacuum Sensor	34		42	Electronic Spark Timing
Barometric Pressure Sensor (if used)	32	ROM and PROM process		Electronic Spark Retard (if used)
Throttle Position Sensor	21	information		Idle Speed
Distributor Reference (crank position—engine speed)	12, 41	(inputs) and issue commands		AIR Management EGR
Oxygen Sensor	13, 44, 45	(outputs).		Canister Purge
Vehicle Speed Sensor	24			Torque Converter Clutch (or shift light on
Ignition On				manual transmission)
Air Conditioner On/Off				
Park/Neutral Switch		RAM monitors		Air Conditioner
System Voltage		indicated circuits,		Early Fuel Evaporation
Transmission Gear Position Switch(es)		sets codes, and reads codes out		Diagnosis (check engine light)
Spark Knock Sensor (if used)	43	when put in		(test terminal)
EGR Vacuum Indicator Switch (only on 1984 and later models)	53	diagnostics.		(serial data)
Idle Speed Control Switch	35			

FIGURE 3-2 Overview of computer command control. Inputs and outputs with code numbers are monitored for faults. Those without are not.

FIGURE 3-3 ECM and PROM. *Courtesy of GM Product Service Training.*

Chapter 3   General Motors' Computer Command

The system drops out of closed loop if the vehicle is operated at or near wide-open throttle (WOT), if the coolant temperature drops below the criterion value, or if certain system component failures occur.

Note: Nineteen eighty-two and later T cars (Cheverolet Chevette and Pontiac T-1000) used a minimum function CCC system. These systems use a coolant temperature switch instead of a coolant temperature sensor. As the engine reaches operating temperature, the switch closes and thus puts the system into closed loop (assuming the O_2 sensor is hot and the proper time has expired). If the engine then cools down and the switch opens, the system stays in closed loop.

Open Loop

This is an operating mode that includes several suboperational modes.

Start-up Enrichment. This provides a rich command to the mixture control solenoid for a short time after each engine start. The duration depends on engine temperature, as measured by the coolant temperature sensor, when the engine is started. If the engine is started hot, this mode will be shorter than if it is started cold.

Blended Enrichment. This mode occurs during engine warm-up (engine coolant and/or the oxygen sensor are not up to proper operating temperature). In this mode the ECM controls the air/fuel ratio by using information from sensors concerning coolant temperature, throttle position, manifold pressure, barometric pressure, and engine speed. As engine temperature comes up, the air/fuel ratio is adjusted leaner by the ECM. In this open-loop mode, the ECM attempts to adjust the air/fuel ratio using fixed, programmed instructions in its read-only memory. If the air/fuel ratio is not correct because of a carburetor malfunction, vacuum leak, and so forth, the ECM will not recognize the problem and will not respond to it. In other words in open loop the ECM does what it is programmed to do, but the engine still does not run correctly unless other engine components function properly.

Power Enrichment. This mode occurs when the vehicle is operated at or near wide-open throttle. The ECM sends a steady power enrichment signal to the mixture control solenoid. This provides the rich mixture required by an engine when manifold vacuum is near zero.

Shutdown. This mode turns off the M/C solenoid, and thus creates a full, rich operating condition, when the reference pulse to the ECM indicates 200 RPM or less and when battery voltage to the ECM is less than 9 volts.

Limp In. This mode allows the engine to continue operating in spite of most major failures, such as an ECM failure, that can occur within the system. This is a *fail soft* mode that provides no fuel control (full rich or full metered) and no spark advance.

INPUTS

Coolant Temperature Sensor (CTS)

With the exception of some T car applications (Chevrolet Chevette and Pontiac T-1000), all CCC systems use a CTS, Figure 3-4, as described in Chapter 2. It is most often located near the thermostat housing and is the single most important sensor in the system. Its input impacts on just about every decision the computer makes. (This statement can generally be applied to all systems.)

Some 1982 and later T cars are equipped with a minimum function system and use a coolant temperature switch instead of a CTS. The switch is open while the coolant is below a specified temperature and closes as the coolant approaches normal operating temperature. In this system the ECM only knows whether the coolant is below or above a specific temperature.

Chapter 3 General Motors' Computer Command

FIGURE 3-4 Coolant temperature sensor. *Courtesy of GM Product Service Training.*

FIGURE 3-5 MAP sensor. *Courtesy of Buick Motor Division.*

Note: The MAP and BARO sensors used on 1980 and early 1981 vehicles were round metal units, both mounted on the same bracket, under the hood.

Pressure Sensors

There are three different types of pressure sensors used on various CCC applications. They are all of the piezoresistive silicon diaphragm type and are nearly identical in appearance, Figure 3-5. Each can, however, be identified by a colored plastic insert at its electrical harness connector.

Manifold Absolute Pressure (MAP) Sensor. The MAP sensor has either a red or orange (orange for turbo engines) connector. The MAP sensor is often used in combination with a BARO sensor.

Barometric Pressure (BARO) Sensor. The BARO sensor has a blue connector. On some applications it is located under the hood and near the MAP sensor. On others it is inside the passenger compartment and under the instrument panel.

Pressure Differential (VAC—for vacuum) Sensor. The VAC sensor has a gray or black connector and is used by itself.

Oxygen (O_2) Sensor

The O_2 sensor used on 1981 and earlier models had two wires and used a vented silicone boat to cover and protect its open end, Figure 3-6. The second wire served as a backup ground; the sensor shell is already grounded to the manifold. Beginning in 1982 the ground

FIGURE 3-6 Oxygen sensor, early style (two wire). *Courtesy of GM Product Service Training.*

Chapter 3 General Motors' Computer Command

wire was not used, and a metal boot replaced the silicone boot. The silicone boot melted if it was pushed too far onto the O_2 sensor toward the exhaust manifold.

Note: There is an O_2 sensor ground wire that runs from the ECM to the engine block (not connected to the O_2 sensor).

Throttle Position Sensor (TPS)

The TPS is a potentiometer mounted in the bowl section of the carburetor, Figure 3-7. When the throttle is depressed, the accelerator pump lever forces the TPS plunger down. This action moves the potentiometer's wiper. The TPS is the only sensor in the CCC system that can be adjusted. Because engine vacuum and

FIGURE 3-7 Throttle position sensor. *Courtesy of GM Product Service Training.*

Chapter 3 General Motors' Computer Command

FIGURE 3-8 Electronic spark control circuit. *Courtesy of GM Product Service Training.*

> **Service Interval for O₂ Sensors**
>
> The O_2 sensors used on GM vehicles prior to 1981 had a service interval of 15,000 or 30,000 miles. No service interval is required for those used on 1981 and later models.

throttle position are so closely related, the ECM expects to see the readings from these two sensors closely synchronized.

Distributor Reference Pulse (REF)

The REF can also be abbreviated as DIST or REF pulse. This signal tells the ECM what engine speed is and what position the crankshaft is in. The REF is obtained by tapping into the pickup coil circuit inside the HEI module, Figure 3-8.

The pickup coil signal is generated as the distributor shaft and timer core turn, moving the points of the timer core past the points of the magnetic pickup unit. Each time the magnetic field strengthens or weakens as a result of the points aligning and misaligning, a voltage signal is produced in the pickup coil. A signal converter in the HEI module changes the analog signal produced by the pickup coil to a digital signal. On HEI applications prior to CCC, the pickup coil signal was used to signal the HEI module when to turn off the main switching transistor, which turned off the primary ignition circuit. On CCC applications the only time that the pickup coil signal is used by the module is during starting and some system failure conditions. During normal operation the pickup coil signal is used only as a REF signal to the ECM. This is explained further in the **Outputs** section of this chapter.

From 1982 to 1985, a Hall-effect switch, Figure 3-9, was mounted between the pickup coil and the rotor on the Chevrolet, 3.8 liter (229 cid.), odd-firing V6. It was used to produce greater timing accuracy. The odd-firing engine is especially susceptible to signal converter error. Notice in Figure 3-10 that the pickup is still there. It, however, is only used to start the engine. Notice also that terminal R of the HEI module is not used (on applications without a

89

Chapter 3 General Motors' Computer Command

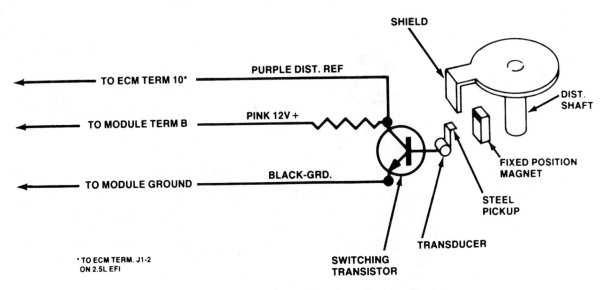

FIGURE 3-9 Hall-effect switch. *Courtesy of GM Product Service Training.*

Hall-effect switch, terminal R feeds the REF signal to the ECM) and that the REF signal comes to the ECM directly from the Hall-effect switch. Once the engine has started, the solid-state switching device, illustrated in Figure 3-10 as a relay, has disconnected the pickup coil from the main switching transistor and connected in its place the EST lead from the HEI terminal E. The ECM is now in charge of spark timing.

FIGURE 3-10 Electronic spark control system with Hall-effect switch. *Courtesy of GM Product Service Training.*

Chapter 3 General Motors' Computer Command

FIGURE 3-11 Late-style knock sensor

FIGURE 3-12 Vehicle speed sensor. *Courtesy of GM Product Service Training.*

Detonation Sensor

The detonation sensor, Figure 3-11, (knock sensor) is part of the electronic spark control (ESC) system that has been made a subsystem of CCC on some applications. A similar spark control system, also called ESC, has been a stand-alone system for several years on non-CCC vehicles.

On many General Motors engines, the knock sensor is screwed into the block in the crankcase area to reduce its sensitivity. On those applications it was found that if the knock sensor were mounted in the upper part of the engine, a clicking valve lifter could cause the sensor to produce its "knock occurring" signal.

Vehicle Speed Sensor (VSS)

Two types of VSS are used on CCC applications. The earlier type consists of a *light-emitting diode* (LED) and a phototransistor. Both are housed in a plastic connector that plugs into the back of the speedometer housing, near the speedometer cable, Figure 3-12. The LED is powered by the ignition switch on 1982 and later models and by the ECM on earlier models. When the ignition is turned on, the LED directs its infrared light beam, invisible to the human eye, toward the back of the speedometer cup, which is painted black. The drive magnet, which spins with the speedometer cable, has a reflective surface. As the drive magnet moves into the light from the LED, the light is reflected back to the phototransistor. Every time the light strikes the phototransistor, a voltage is produced. These voltage pulses are fed to a buffer switch.

The signal from the VSS is an analog signal that varies in amplitude as well as frequency as vehicle speed changes. The buffer switch modifies the raw VSS signal to 2,002 digital pulses per mile. This information is used by the ECM to determine when to apply the torque converter clutch.

Some later models that feature an electronically operated speedometer and no speedometer cable use a pulse generator device as a VSS. This device mounts on the transmission where the speedometer cable drive gear would be if it were used. This transmission-driven device rotates a magnet near a coil. As each pole of the magnet swings by the coil, a weak volt-

Chapter 3 General Motors' Computer Command

age signal, similar to a magnetic pickup unit in a distributor, is produced in the coil. This signal produces an analog voltage signal that is proportional to the speed of the car. These signals are fed to a buffer, which converts them to 4,004 digital pulses per mile. These signals are sent to the ECM, which can calculate vehicle speed from the signal frequency.

Ignition Switch

The ignition switch is one of the two power supplies to the ECM. When the ignition is turned on, the ECM initializes (starts its programmed routine) and is ready to function. The ignition switch also powers most of the actuators that the ECM controls. The other power supply comes from the battery by way of a fuse. It powers the diagnostic memory. These power supplies are also the inputs by which the ECM monitors system voltage.

Park/Neutral (P/N) Switch

The P/N switch consists of another pair of contacts being added to the neutral safety switch that has been on cars with automatic transmissions for years. It provides a signal to the ECM indicating whether the transmission is in gear or not. This information is used to help control engine idle speed. On some applications the ECM will not issue any spark advance or EGR commands unless the transmission is in gear.

Air-conditioning (A/C) Switch

On CCC cars equipped with air-conditioning, the A/C switch is connected by a wire to the ECM. When the A/C is turned on or off, the ECM is informed and uses this information in its control of idle speed.

Idle Speed Control (ISC) Switch

At closed throttle the throttle lever, being pulled by the return spring, presses against the ISC plunger. The pressure exerted on the ISC plunger causes a set of contact points at the base of the plunger to close, Figure 3-13. This

FIGURE 3-13 Idle speed control motor. *Courtesy of GM Product Service Training.*

signals the ECM to be in charge of idle speed. Opening the throttle allows the contacts to open. This signals the ECM not to control idle speed, Figure 3-14.

FIGURE 3-14 Idle speed control motor connector. *Courtesy of GM Product Service Training.*

FIGURE 3-15   EGR diagnostic switch circuit. *Courtesy of GM Product Service Training.*

Transmission Switches

Most transmissions used in CCC-equipped vehicles have one or more hydraulically operated electric switches screwed into the valve body. These switches provide the ECM with signals that tell it what gear the transmission is in. This information enables it to more effectively control torque converter clutch operation.

EGR Vacuum Diagnostic Control Switch

Beginning in 1984 most CCC systems included a vacuum-operated switch that tied into the vacuum hose between the EGR valve and the EGR control solenoid, Figure 3-15. When vacuum is applied to the EGR valve, the switch closes. If the ECM detects a closed EGR diagnostic switch during starting, idle, or any other time it has not commanded the EGR to be applied, it will turn on the check engine light and set a fault code in its diagnostic memory. If it sees an open switch during any time it has commanded the EGR to be applied, it will turn on the check engine light and set the same code.

OUTPUTS

Mixture Control (M/C) Solenoid

In a CCC system, the M/C solenoid effectively replaces the full power system that would otherwise be used in the carburetor of a non-CCC vehicle. It, however, has much more articulate control of air/fuel ratio over a much wider range of engine operating conditions than the conventional full power system does, Figures 3-16, 3-17, and 3-18. The M/C solenoid is powered by the ignition switch. It is duty cycled by a solid-state grounding switch in the ECM, Figure 3-19. When the ECM grounds the M/C solenoid circuit, the solenoid is turned on

Chapter 3 General Motors' Computer Command

FIGURE 3-16 M/C solenoid, Rochester Dualjet or Quadrajet. *Courtesy of GM Product Service Training.*

and drives the solenoid plunger down. When the plunger is driven down, it drives a metering rod(s) down into the main metering jet(s) to restrict fuel flow. in some cases the end of the solenoid plunger itself provides the restriction. This produces a lean air/fuel mixture. When the ECM opens the circuit, the solenoid is turned off and the restriction is removed. This allows more fuel to flow into the main metering system and thus produces a rich mixture.

The ECM is programmed to cycle the M/C solenoid ten times per second. It cannot vary from ten cycles per second except in shutdown mode, when it is turned off. The ECM can, however, vary the duty cycle anywhere from 90% to 10% (hold the solenoid down for 90% of the cycle or 10% of the cycle, Figure 3-20).

Measuring the Duty Cycle. The relative time that the M/C solenoid is turned on or off can be measured with a dwellmeter, Figure

FIGURE 3-17 M/C solenoid, Rochester Varajet. *Courtesy of GM Product Service Training.*

Chapter 3 General Motors' Computer Command

FIGURE 3-18 M/C solenoid, Holley 6510 C. *Courtesy of GM Product Service Training.*

FIGURE 3-19 M/C solenoid circuit

FIGURE 3-20 M/C solenoid duty cycle. *Courtesy of GM Product Service Training.*

Chapter 3 General Motors' Computer Command

FIGURE 3-21 Dwell test lead. *Courtesy of GM Product Service Training.*

3-21. The wire that runs from the M/C solenoid to the ECM has a special connector (dwell test connector) for this purpose. A dwellmeter, when connected to the dwell test connector and to ground and set on the six-cylinder (60°) scale, indicates in degrees the relative amount of time that the M/C solenoid is turned on, Figure 3-22.

Dwell Readings. During start-up enrichment or WOT operation, a dwell reading of 6° is expected. During engine warm-up (open loop) an initial reading around 10° is considered normal, depending on how cold the engine is. As the engine warms up, the dwell reading should steadily increase. In open loop the ECM selects an air/fuel ratio based on engine temperature, engine load, barometric pressure, throttle position, and engine speed. This produces the best results in terms of exhaust emissions, fuel mileage, and driveability. The fuel mixture, as indicated by the dwell reading, remains constant until some of the input information changes or until the system goes into closed loop. When the coolant reaches a tem-

FIGURE 3-22 Dwell readings. *Courtesy of GM Product Service Training.*

perature of about 150° F, the O_2 sensor reaches about 600° F, and the required time has elapsed, the ECM puts the system into closed loop. At that moment the dwellmeter needle begins wagging. It usually wags, or varies, over a range of 5° to 10°. The dwell reading is now taken by selecting the center or average of the range over which it is sweeping.

The dwell is varying because of the way the ECM responds to the O_2 sensor to control the air/fuel ratio. When the O_2 sensor sees a lean condition, lean being anything leaner than 14.7 to 1, it reports "lean" to the ECM. The ECM responds with a rich command to the M/C solenoid. This produces a rich condition, which the O_2 sensor immediately reports. This in turn evokes a lean command. In this way the air/fuel ratio constantly bounces back and forth across 14.7 to 1. It most often remains within a range of about 14.5 to 14.9 to 1.

At the same time that the M/C solenoid is cycling or stroking the main metering system, it is also stroking an idle air bleed valve in the idle speed-low speed system. The idle air bleed valve, Figure 3-23, is designed as a variable restriction to the amount of air that enters the idle speed-low speed circuit. When the M/C solenoid is down, the idle air bleed valve plunger drops and allows maximum air into the idle speed-low speed circuit (lean mixture). When the M/C solenoid plunger is up, the idle air bleed valve plunger is driven up and allows minimum air into the idle speed-low speed circuit (rich mixture). With this arrangement the M/C solenoid is controlling the air/fuel ratio in whatever circuit is being used.

Electronic Spark Timing (EST)

The EST function of the ECM eliminates the vacuum and centrifugal advance mechanisms of earlier distributors, Figure 3-24. In order to achieve this, the HEI module has been modified to work with the ECM. The ECM now controls timing by signalling the HEI module when to turn the primary ignition circuit off. The HEI module only responds to these spark

FIGURE 3-23 Rochester Dualjet or Quadrajet. *Courtesy of GM Product Service Training.*

Chapter 3 General Motors' Computer Command

FIGURE 3-24 HEI circuit with mechanical and vacuum advance. *Courtesy of GM Product Service Training.*

timing commands from the ECM once the engine has started.

Note: In the following discussion and in the accompanying illustrations, the HEI module will be treated as though it has a mechanical bypass relay with a double set of contact points. This is done to make its operation easier to explain. Solid-state components are actually used to achieve the functions discussed.

During cranking, the bypass relay is in the de-energized or module mode. In this mode the main switching transistor of the HEI module is connected to the pickup coil, Figure 3-25. The pickup coil signal, after it is converted from an analog to a digital signal, turns the switching transistor on and off. This same signal is sent to the ECM as the HEI reference pulse.

When the ECM sees a REF pulse signal of about 200 RPM or more for about five seconds, it decides that the engine is running. It then sends a 5-volt signal through the bypass wire to the bypass relay. This causes the double contacts to move. The upper contact disconnects the base of the main switching transistor from the pickup coil and connects it to the EST wire. The lower contact simultaneously disconnects the EST wire from ground. The system is now in EST mode, and timing is being controlled by the ECM. The ECM considers barometric pressure, manifold pressure, coolant temperature, engine speed, and crankshaft position and then sends to the HEI module the optimum spark timing command.

Note: Some Chevrolet Chevette and Pontiac T-1000 CCC systems do not have EST.

Electronic Spark Control (ESC)

Engines that have strong potential for spark knock are equipped with ESC. The ESC

FIGURE 3-25 HEI circuit with EST. *Courtesy of GM Product Service Training.*

actually becomes a subsystem of the CCC system when it is used on a CCC vehicle. It has its own electronic module that works in conjunction with the ECM. Its function is to retard ignition timing when detonation (uncontrolled, rapid burning of the fuel charge) occurs. Two slightly different versions of ESC systems have been used.

The 1981 system, Figure 3-26, passes the EST and the bypass commands from the ECM through the ESC module on their way to the HEI module. In the event of spark knock, the detonation sensor notifies the ESC module that detonation is occurring. The ESC module then modifies the EST command. Spark timing is retarded about 4° per second until the detonation clears up. When the detonation sensor no longer "hears" detonation occurring, the ESC module begins restoring the spark advance it took away. The spark advance is restored at a slower rate than it was removed.

In the 1982 and later version of ESC, Figure 3-27, the EST and bypass commands do not pass through the ESC module. Instead, when the knock sensor produces a voltage signal that indicates that detonation is occurring, the ESC module sends a request to the CCC ECM for it to retard timing. It works as follows. Ignition voltage is applied to terminal F of the ESC module. The module feeds this voltage, reduced to about 10 volts, to terminal J. It is then carried to terminal L of the ECM. When a detonation signal is applied to terminal B of the ESC module, the module reduces the voltage at terminal J to less than 1 volt. The ECM retards timing until voltage reappears at terminal L. Either of the ESC systems can easily be tested by tapping on the intake manifold with a metal object while watching the timing with a timing light or timing meter. Be sure to consult the service manual for specific directions.

Air Management Valve

As mentioned earlier making the three-way catalytic converter work at maximum efficiency is one of the major goals of the CCC system. It achieves this primarily by controlling the air/fuel ratio. On applications using a dual-bed, three-way catalytic converter, the system achieves making the catalytic converter work at maximum efficiency by the timely introduction of air from the air pump into the oxidizing

Chapter 3 General Motors' Computer Command

FIGURE 3-26 ESC circuit, pre-1982. *Courtesy of GM Product Service Training.*

FIGURE 3-27 ESC circuit, 1982 and later. *Courtesy of GM Product Service Training.*

side of the converter. The ECM-controlled air management system is responsible for controlling air injection into the exhaust system.

CCC vehicles equipped with dual-bed, three-way converters also have a dual air management valve. Those with single-bed, three-way converters use a single air management valve, Figures 3-28 and 3-29.

Single Valve. The single valve system can direct air to one of two places. During engine warm-up air is directed to the exhaust manifold. This will:

- reduce HC and CO by burning it in the exhaust system.
- raise O_2 sensor temperature more quickly.
- raise the catalytic converter temperature more quickly; the converter needs to

FIGURE 3-28 Air management with dual-valve and dual-bed catalyst. *Courtesy of GM Product Service Training.*

FIGURE 3-29 Air management with single valve and catalyst. *Courtesy of GM Product Service Training.*

be between 205°C (410°F) and 260°C (500°F) to operate effectively.

As the coolant temperature approaches 65°C (150°F), the temperature at which the CCC system goes into closed loop, the ECM commands the divert valve to go into *divert mode*. Depending on the particular application, the air is diverted to either the air cleaner or to the atmosphere. To continue pumping air into the exhaust manifold after the system goes into closed loop is misleading to the O_2 sensor. CCC systems with single-bed converters are always in divert mode except during warm-up.

Dual Valve. On vehicles equipped with a dual-bed converter, air can be directed to one of three places. During engine warm-up the divert valve directs air to the air switching valve, Figure 3-28. The switching valve directs the air to the exhaust manifold. As the engine approaches closed-loop temperature, the ECM commands the switching valve to direct air to the oxidizing bed of the catalytic converter. Air being pumped into the oxidizing bed of the converter achieves the following:

- It prevents the additional oxygen from flowing across the O_2 sensor. This would cause it to give an incorrect signal to the ECM.
- It lowers the temperature in the exhaust manifold. Continuing to pump oxygen into the exhaust manifold after the engine has reached normal operating temperature can produce additional NO_x.
- It makes the oxidizing bed of the catalytic converter operate at maximum efficiency without interfering with the efficiency of the reducing bed.

During WOT operation and deceleration, the ECM commands the divert valve to dump the air, or use the divert mode, because there is no other safe place to put it. Either of these driving conditions results in an increased amount of HC and CO in the exhaust system. If during these conditions air were to continue being pumped into any part of the exhaust system, the converter would quickly overheat. Certain failures within the CCC system, such as in the M/C solenoid circuit or EST, also cause the ECM to put air management into divert mode.

Air Management Valve Designs. Many different valve designs have been used on various CCC system applications; thus, a discussion of each different valve is impractical. Most of them, however, are more alike than different. They all direct airflow by moving an internal valve to open or block a passage. The valves can be moved by a vacuum diaphragm, an electric solenoid, or a combination of the two. Figures 3-30, 3-31, and 3-32 show a typical dual valve in different modes of operation.

In Figure 3-30 the vehicle is decelerating. The high vacuum signal has pulled the diaphragm of the divert valve (lower valve) up. The passage to the switching valve (upper valve) is blocked, and air is diverted to the air cleaner. The air cleaner serves as a silencer. Diverting the injection pump air to it has no significant effect on air/fuel ratio.

In Figure 3-31 the ECM has put the air management system into divert mode by de-energizing the solenoid. This allows pump air pressure to be routed to the decel timing chamber. With the help of the manifold vacuum on the other side of the diaphragm, the pressure forces the diaphragm and divert valve up.

Figure 3-32 shows the air management valve directing air to the exhaust manifold during engine warm-up. Manifold vacuum is normally not strong enough to pull the divert valve diaphragm up; and the ECM is keeping the divert solenoid energized, which blocks pump air pressure from entering the decel timing chamber. The spring holds the divert valve down and allows air to flow up to the switching valve. The switching valve solenoid, energized by the ECM, allows vacuum to be applied to the upper diaphragm. This pulls it and the attached switching valve to the right to block the con-

Chapter 3 General Motors' Computer Command

FIGURE 3-30 Divert mode (deceleration). *Courtesy of GM Product Service Training.*

FIGURE 3-31 Divert mode (ECM commanded). *Courtesy of GM Product Service Training.*

FIGURE 3-32 Air directed to exhaust ports. *Courtesy of GM Product Service Training.*

verter air passage and thus opens the exhaust port passage.

In Figure 3-33 the engine has warmed up and the CCC system is ready to go into closed loop. The ECM has de-energized the switching valve solenoid, which blocks vacuum to the diaphragm. The spring moves the valve to the left, opening the converter passage and blocking the exhaust port passage.

All air management valves used with CCC systems have a pressure relief valve. At high engine speeds, the pressure relief valve exhausts excess air to reduce system pressure. This can easily be mistaken for a divert mode function.

Pulsair System. Some vehicles use a *Pulsair* system instead of an air injection system. Between each exhaust pulse in the exhaust manifold, a low pressure develops. This low pressure can be used to syphon air into the exhaust ports. The air is drawn through a tube having a check valve. The check valve blocks reverse airflow during the exhaust pulse.

Chapter 3 General Motors' Computer Command

FIGURE 3-33 Warm engine mode (closed loop). *Courtesy of GM Product Service Training.*

FIGURE 3-34 Pulsair shutoff valve. *Courtesy of GM Product Service Training.*

Pulsair Shutoff Valve. When a Pulsair system is combined with the CCC system, the ECM uses a Pulsair shutoff valve to control air flowing into the exhaust system, Figure 3-34.

During engine warm-up the solenoid is energized and vacuum is applied to the diaphragm. The valve is pulled down and air is allowed to flow into the exhaust system. With the exception of how air is made to flow, this system is very similar to the single-valve air management system.

Idle Speed Control (ISC)

Idle Speed Control Motor. In order to achieve the established goals for exhaust emissions, fuel economy and driveability, an articulate idle speed control is necessary. This is achieved by using a permanent magnet field reversible electric motor that attaches to the side of the carburetor, Figure 3-35.

The ECM normally has a fixed idle speed that it wants to maintain during idle. It moves the throttle as necessary with the ISC motor; however, it commands a higher idle speed in

FIGURE 3-35 ISC motor assembly. *Courtesy of GM Product Service Training.*

response to any one of the following three conditions:

- During closed choke idle, the fast idle cam holds the throttle blade open enough to lift the throttle linkage off of the ISC plunger. This allows the ISC switch to open so that the ECM does not monitor idle speed. As the choke spring allows the fast idle cam to fall away and the throttle returns to warm idle position, the ECM notes the still low coolant temperature and commands a slightly higher idle speed.
- If the engine starts to overheat, the ECM commands a higher idle speed to increase coolant flow.
- If system voltage falls below a predetermined value, the ECM commands a higher idle speed in order to increase generator speed and output.

During warm idle if the automatic transmission is pulled into gear (forward or reverse), the park/neutral switch signals the ECM of the impending load on the engine. It commands the ISC motor to extend the plunger. The throttle blade opens at about the same time that the transmission engages, and no appreciable change in engine speed occurs. If the air-conditioning is turned on, the ECM receives a signal from the air-conditioning switch. The ECM commands a wider throttle blade opening to compensate for the additional load of the air-conditioning compressor.

Starting in 1982 models with smaller engines were equipped with an ISC relay. The ISC relay uses a solid-state, fixed time delay device to keep the ISC motor activated in a retract mode after the key is turned off. This allows the throttle to completely close each time the engine is turned off and thus prevents dieseling.

There is no idle speed adjustment capability on CCC systems using an ISC motor. Attempting to adjust idle speed by adjusting the plunger screw results in the ECM moving the plunger to compensate for the adjustment. Idle speed does not change until the ISC plunger motor uses up all of its plunger travel trying to compensate for the adjustment, at which point the system is completely out of calibration. Proper calibration can be restored, however, by following the *minimum* and *maximum authority* adjustment procedures as outlined in the service manual. When idle speed driveability problems occur, the ISC system is usually responding to or being affected by the problem not causing it.

Note: Be sure the ignition is turned off before connecting or disconnecting the ISC motor; otherwise, damage to the ECM can occur.

Idle Stop Solenoid. Some engine applications do not use an ISC motor. Instead they use an idle stop solenoid. The idle stop solenoid is essentially the same as an antidiesel solenoid, which was used for years and dates back to the first emission control systems. When the ignition is turned on, the solenoid is energized and extends its plunger to hold the throttle lever off of the idle stop screw. The extended plunger provides the curb idle position for the throttle blade. When the ignition is turned off, the plunger retracts and allows the throttle blade to more fully close to prevent dieseling. The same device can be connected to the A/C switch. In this case the idle stop screw provides curb idle, and the solenoid increases idle speed slightly when the A/C is turned on.

Idle Load Compensator (ILC). Beginning in 1982 the Oldsmobile limited control CCC system used an ILC instead of an ISC motor, Figure 3-36. The ILC is not ECM controlled; it is controlled by engine vacuum. The ILC is very similar in construction to a vacuum advance unit on an older style distributor. It contains a diaphragm with a spring pushing against one

Chapter 3   General Motors' Computer Command

FIGURE 3-36   Idle load compensator. *Courtesy of GM Product Service Training.*

FIGURE 3-37   Vacuum-operated throttle kicker circuit (2.8 L). *Courtesy of GM Product Service Training.*

side. The other side attaches to a plunger that sticks out of the unit in the form of an adjustable screw. The head of the plunger screw acts as the throttle stop to control idle speed. Manifold vacuum is applied to the spring side of the diaphragm. When the automatic transmission is pulled into gear, the load on the engine causes engine vacuum to go down slightly. This allows the spring to push the diaphragm and plunger a little farther and thus opens the throttle blades slightly wider. The same thing occurs for any load or condition that causes engine vacuum to diminish.

Idle speed adjustments can be made to the ILC; however, the adjustment is made at the vacuum side of the diaphragm not at the plunger screw. The idle adjusting screw in the stem on the back of the unit changes the preload on the diaphragm spring. Refer to an appropriate service manual for adjustment procedures.

Throttle Kicker. Some later model CCC applications use either a vacuum or electronic *throttle kicker*. The functions of a throttle kicker are explained in Figures 3-37 and 3-38.

Torque Converter Clutch (TCC)

The TCC improves fuel mileage by eliminating hydraulic slippage and heat production in the converter once a cruise speed has been achieved.

Clutch. The major component of the TCC system is the clutch itself. It becomes a fourth element added to a conventional torque converter, Figure 3-39. The clutch plate, which is

FIGURE 3-38 Electrically operated throttle kicker circuit. *Courtesy of GM Product Service Training.*

FIGURE 3-39 Torque converter with TCC. *Courtesy of GM Product Service Training.*

splined to the turbine, has friction material bonded to its engine side and near its outer circumference. The converter cover has a machined surface just inside where the converter drive lugs are attached. This machined surface mates up to the friction material on the disc when the clutch is applied. When hydraulic pressure is applied to the turbine side of the disc, the disc is forced against the converter cover. The friction between the clutch's friction material and the cover causes the complete assembly to turn as a solid unit. The pressure source is converter feed oil coming out of the pressure regulator valve in the transmission valve body. In non-TCC applications converter feed oil is used to charge and cool the converter by circulating through the converter and then the transmission cooler in the radiator. On TCC applications, however, converter feed oil must pass through a converter clutch apply valve on its way to the converter and as it returns from the converter. The apply valve, a small valve in the transmission, controls the direction of oil flow through the torque converter, Figure 3-40.

With the apply valve in its at-rest position and transmission in neutral, or the car moving at low speed, converter feed oil is directed into the converter through the release passage by way of the hollow turbine shaft. This oil is fed into the converter between the converter cover and the clutch disc. The oil forces the disc to move away from the cover and thus releases the clutch. Converter feed oil flows over the circumference of the disc and circulates through the converter. It exits through the apply passage, a passage between the pump drive hub and the stator support shaft.

Two criteria must be met before the clutch is applied. The transmission must be ready hydraulically; and the ECM must be satisfied with engine temperature, throttle position, engine load, and vehicle speed. When the transmission is in the right gear (this varies from transmission to transmission), hydraulic pressure is supplied at one end of the clutch apply valve, Figure 3-41. This hydraulic pressure (converter apply signal) has the potential to move the apply valve into the apply position. The apply valve is moved, however, only if the ECM is ready for the clutch to be applied.

The ECM controls the position of the clutch apply valve with a solenoid that can open or close an exhaust port at the converter apply signal end of the apply valve. If the solenoid is not energized, the exhaust port is open;

FIGURE 3-40 TCC apply circuit in release position. *Courtesy of GM Product Service Training.*

and converter apply signal oil exhausts from the signal end of the apply valve as fast as it arrives. Sufficient pressure does not develop to move the apply valve. Converter feed oil continues to flow into the converter through the release passage, Figure 3-40. When the solenoid is energized, the exhaust port is blocked. The converter apply signal oil develops pressure, and the apply valve is moved and held in the apply position. Converter feed oil now flows into the converter through the apply passage, Figure 3-41. The clutch is applied and prevents oil from exhausting through the release passage, and static pressure exists.

Figure 3-42 shows a typical TCC solenoid control circuit. The power source is the ignition switch, by way of the gauge fuse. The brake switch is in series on the high-voltage side of the circuit. Anytime the brakes are applied, the solenoid is de-energized. After the current passes through the apply solenoid, it arrives at the 4-3 pulse switch, used only on automatic transmissions with fourth-speed overdrive. The 4-3 pulse switch momentarily disengages the TCC to allow a smooth 4-3 downshift. It is closed at all other times. From there the current goes to the ECM, where the circuit is grounded when the ECM is ready to apply the TCC. When

Chapter 3 General Motors' Computer Command

FIGURE 3-41 TCC apply circuit in apply position. *Courtesy of GM Product Service Training.*

the transmission goes into fourth gear, the fourth-gear switch opens to alert the ECM that the transmission is in overdrive. In overdrive the ECM holds the TCC on through a much wider range of throttle position than it does in third or second gear. To put it in engineering language, the ECM holds the TCC on through a wider throttle position *window*. The term *window* refers to a range within a range. For instance, if the TPS signal to the ECM were 0.5 volt at idle and 4.5 volts at WOT, the ECM might select 2.0 to 3.5 as the window in which it would keep the TCC applied.

TCCs and Four-speed Automatic Transmissions

On vehicles using a four-speed automatic transmission, the TCC must be applied while the transmission is in fourth gear. Any failure that results in the clutch not applying during prolonged fourth-gear operation can result in transmission damage as a result of oil overheating. This can occur in a relatively short period of steady-state fourth-gear operation.

Chapter 3 General Motors' Computer Command

FIGURE 3-42 TCC electronic control circuit. *Courtesy of GM Product Service Training.*

The TCC test lead, which comes from the low-voltage side of the circuit, can be used to tell when the ECM has electrically applied the TCC circuit or to ground the circuit and thus override the ECM. Its use is covered in more detail under **System Diagnosis and Service** in this chapter.

Exhaust Gas Recirculation (EGR) Valve

The primary purpose of the EGR valve is to control oxides of nitrogen (NO_x) by lowering combustion temperature. It also is used to control detonation. The pressing need for better fuel economy and driveability combined with the development of computer-controlled ignition timing has pushed ignition timing to the ragged edge of detonation. Without the cooling effect of EGR on combustion temperature, detonation occurs in most cases. Even so overuse of EGR causes serious driveability problems especially on cold engines. It makes sense then to have the ECM in charge of EGR operation. On CCC applications the EGR is turned on or off or on some later models is modulated (the amount of opening is controlled) by one or more ECM-controlled solenoids. The exact method and amount of EGR control varies considerably with engine and model year application. Generally, however, EGR systems controlled by CCC can be summarized by one of the following statements:

- The ECM operates a solenoid that when energized blocks vacuum. When not energized it passes ported vacuum to the EGR valve, Figure 3-43.

Chapter 3 General Motors' Computer Command

FIGURE 3-43 Typical EGR valve. *Courtesy of GM Product Service Training.*

FIGURE 3-45 EGR bleed and control solenoids. *Courtesy of GM Product Service Training.*

FIGURE 3-44 EGR bleed solenoid. *Courtesy of GM Product Service Training.*

FIGURE 3-46 Two EGR control solenoids. *Courtesy of GM Product Service Training.*

- The ECM operates a solenoid that when activated bleeds atmospheric pressure into the ported vacuum signal to the EGR valve. A bleed solenoid can be used to turn the EGR valve off or to partially close (modulate) it, Figure 3-44.
- The ECM operates two solenoids, one to block or pass ported vacuum to the EGR valve and the other as bleed solenoid to modulate it, Figure 3-45.
- The ECM operates two blocking solenoids in series. Each one is operated according to different sets of engine calibration criteria, Figure 3-46.

- A pulse width modulating solenoid (rapidly turned on and off with the on-time variable) precisely controls the amount of vacuum that is allowed to the EGR valve, Figure 3-47.

One other EGR control system that can be encountered on a CCC vehicle should be discussed at least briefly, the *aspirator-controlled EGR*, Figure 3-48. The aspirator, mounted in the air cleaner, contains a small venturi. Air

Chapter 3 General Motors' Computer Command

FIGURE 3-47 Pulse width modulated EGR control solenoid. *Courtesy of GM Product Service Training.*

FIGURE 3-48 Aspirator-assisted EGR valve. *Courtesy of GM Product Service Training.*

from the air pump is directed to the aspirator. During periods of low engine vacuum, the aspirator control valve allows air to pass through the venturi portion of the aspirator to produce a vacuum that keeps the EGR valve open. When engine vacuum is high, the aspirator control valve closes to prevent air from passing through the aspirator venturi. The EGR valve is now operated by ported vacuum.

Note: Generally electrical failures in any of the EGR control systems result in the EGR valve being open.

Chapter 3 General Motors' Computer Command

FIGURE 3-49 Vacuum-actuated EFE valve. *Courtesy of GM Product Service Training.*

Early Fuel Evaporation (EFE)

The EFE system applies heat to the intake manifold area beneath the carburetor to help evaporate the fuel and to keep it in a vapor state. CCC systems use one of two different types of EFE systems.

Exhaust Heat Type. An EFE valve (known years ago as a heat riser valve and located where the exhaust pipe connects to the exhaust manifold on one side of the engine) is closed by a vacuum motor, Figure 3-49. This forces exhaust gases to flow through a passage in the intake manifold beneath the space over which the carburetor is mounted (plenum). Vacuum to the vacuum motor is controlled by a solenoid that is controlled by the ECM, Figure 3-50. This system varies from earlier, non-CCC, EFE systems in that they used a *vacuum control valve* (VCV) to control vacuum to the EFE actuator.

Electric Grid Type. The electric EFE uses a ceramic-encased heating element under the carburetor, Figure 3-51. The heater element is powered by a relay, Figure 3-52. When the ignition is on, voltage is available at one of the normally open relay contacts. Ignition voltage is also applied to the relay coil waiting for ground by the ECM. When coolant temperature is low, the ECM grounds the coil, the relay

FIGURE 3-50 EFE vacuum control circuit. *Courtesy of GM Product Service Training.*

Chapter 3 General Motors' Computer Command

contacts close, and the heater element is powered. Electric EFE circuits vary slightly with engine application and model year.

Controlled Canister Purge (CCP)

On many CCC vehicles, the ECM operates a solenoid that controls purge vacuum to the *purge valve* on the charcoal canister. When the CCC system is in open loop, or before a predetermined time period has elapsed since startup, or below a specific RPM, the purge solenoid is energized and blocks purge vacuum. In closed loop, after a specified elapsed time, and above a specific RPM, the purge solenoid is de-energized and canister purging occurs.

Check Engine Light and Lamp Driver

All CCC systems use a check engine light. If the ECM sees a fault in one of the circuits that it monitors for malfunctions, it turns on the check engine light located on the instrument panel. This warns the driver that a malfunction exists. By grounding the test terminal (one of the terminals in the assembly line communication link, located under the dash), the ECM causes the check engine light to flash. These flashes indicate one or more codes that identify the circuit or circuits in which the fault exists. This is explained more fully under **System Diagnosis and Service** in this chapter.

In 1982 a remote lamp driver was added to the system. It is a separate module, located under the dash in most cases, that powers the check engine light and gives it greater reliability.

E-cell

Some Oldsmobile and Chevrolet eight-cylinder engine CCC applications incorporate an E-cell, often referred to as *green engine calibration unit*. The E-cell slightly modifies some engine calibrations during engine break-in. It is not necessary to replace the E-cell even if it fails prematurely.

FIGURE 3-51 Electrically heated EFE valve. *Courtesy of GM Product Service Training.*

FIGURE 3-52 Electric EFE control circuit. *Courtesy of GM Product Service Training.*

115

Chapter 3 General Motors' Computer Command

SYSTEM DIAGNOSIS AND SERVICE

Self-diagnosis

The ECM monitors the major input sensors, the M/C solenoid, EST, and their respective circuits for proper operation. If the ECM sees a fault, such as an open, a short, or a voltage value that stays too high or too low for too long, in any of the circuits it monitors, it will turn on a check engine light on the instrument panel and record a code number in its diagnostic memory. Some later models have a service engine soon light and a service engine now light, either of which can be selected depending on the severity of the problem. The code number identifies the circuit in which the fault exists. The check engine light warns the driver that a fault exists. The check engine light should come on, however, anytime the ignition is on without the engine running.

Diagnostic Memory. A portion of the ECM's random access memory is devoted to diagnostic memory, sometimes referred to as long-term memory. The diagnostic memory enables the ECM to store code numbers, referred to as *fault codes,* that identify the type of fault and the circuit in which the fault exists. A technician can obtain the stored codes by either of two methods. Grounding the test terminal in the assembly line communication link (ALCL), often referred to as the assembly line diagnostic link (ALDL), puts the system into diagnostics, Figure 3-53. The ECM reads out its stored codes by flashing the check engine light. All codes are stored as two-digit numbers. Two quick flashes followed by a pause of about two seconds represents 2; three more quick flashes represents the second digit—3. Figure 3-54 demonstrates a code 12 being given. Once the two-digit number is completed, a slightly longer pause separates it from the next two-digit number, or code. When a code is giv-

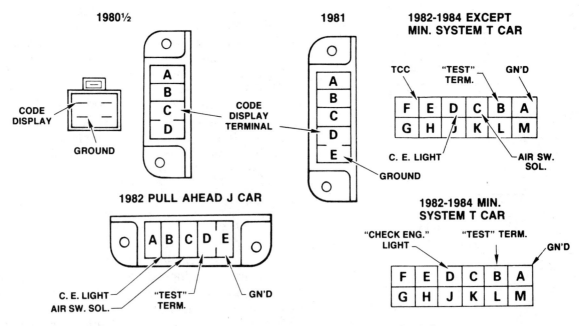

FIGURE 3-53 ALCL connectors. *Courtesy of GM Product Service Training.*

Chapter 3 General Motors' Computer Command

FIGURE 3-54 Check engine light in diagnostics. *Courtesy of GM Product Service Training.*

en, it is repeated twice before moving on to the next code. The second method of obtaining stored codes is to plug a special piece of test equipment, manufactured by several different companies and ranging in size from large engine analyzers to small handheld units, into the ALCL. With the ignition on, the test unit receives any stored codes from the ECM and displays them on its digital screen.

Note: The ALCL is most often located just under the instrument panel, somewhere between the steering column and the radio.

Note: Diagnostic procedures for 1987 and later are written to include the use of an ALCL tool, often referred to as a "scanner," where it is applicable. Previously, diagnostic procedures were written around the use of a voltmeter to obtain such measurements as sensor readings.

Intermittent Faults. If the ECM sees that a fault has turned on the check engine light and set a code in its memory, the check engine light will stay on until the ignition is turned off or until the ECM no longer sees the fault. If the perceived fault clears up, the check engine light will go off, but the code will still be stored in memory. If the fault condition does not appear again within the next fifty ignition cycles (starting the engine and turning it off), the code will automatically be erased.

Clearing the Memory. The code can also be erased by disconnecting the power supply from terminal R of the ECM for ten seconds. This can most easily be done by either pulling the fuse marked ECM from the fuse panel, disconnecting the negative battery lead, or on some later models disconnecting a fusible link near the positive battery terminal.

Behavior in Diagnostics

When the test terminal in the ALCL is grounded and the engine is not running, the check engine light should begin flashing code 12. Code 12 means no reference pulse is coming from the distributor, which is expected since the engine is not running. Any additional stored codes are displayed in numerical order after code 12.

Note: The ignition should always be on before grounding the test terminal; otherwise, the check engine light does not function properly.

There are some other noteworthy points about code 12. It is not a storable code (it is not stored in memory). It is most often used to indicate that the self-diagnosis function of the ECM is working. If you put the system into diagnostics with the engine off and do not get a code 12, the self-diagnosis function is not working properly and must be repaired before you can continue. The only time that the presence of code 12 indicates a fault is when the engine is running. In this case it means that the REF pulse is not coming in from the distributor.

While the codes are being displayed, the ECM also:

- sends a steady 30° dwell command to the M/C solenoid.
- energizes all ECM-controlled solenoids.
- pulses the ISC motor in and out.

If the test terminal is grounded while the engine is running, the ECM will:

- take out the open-loop timer, which maintains the time laps before the system can go into closed loop.
- take out some of the adaptive enrichment modes.
- send a 30° dwell command to the M/C solenoid if the system is in open loop and is not in an enrichment mode.
- cause EST to be set at a fixed spark advance.

While in diagnostics (test terminal grounded), the ECM does not store new codes.

The 1982 and later T car (Chevrolet Chevette and Pontiac T-1000) has no diagnostic memory. It can only display codes concerning faults that it can currently see. In order to receive fault codes from that minimum function system, the engine must be running when the test terminal is grounded.

Diagnostic Procedures

The diagnostic procedures are contained in several different charts or groups of charts. They are as follows:

- diagnostic circuit check
- customer complaint (or in newer manuals, driveability symptoms)
- system performance check
- diagnostic charts without trouble codes
- diagnostic charts with trouble codes
- diagnostic charts on related components

It is important to carefully follow the directions when any of the diagnostic procedures or flowcharts are being used. Failing to do so or taking shortcuts usually results in inaccurate conclusions.

Note: When using diagnostic procedures, be sure to use procedures or charts that apply to the specific engine application and year. Procedures can vary for different engines and are often changed from year to year.

Diagnostic Circuit Check, Figure 3-55. This chart, or procedure, should always be used before attempting to diagnose any ECM-controlled system. It is primarily responsible for sorting out why a code is stored. It also aids in discovering why the check engine light is not working properly if it is not. There can be several reasons for a code being stored:

- *Existing problem.* A fault can exist, in which case its corresponding code is called a *hard code*.
- *Intermittent problem.* A fault can develop and clear up by itself, in which case the resulting code is *intermittent*.
- *No real problem.* Someone previously working on the car can open a CCC system circuit that the ECM monitors, while the ignition is on, and cause a code to be set. Or a strong radio signal, such as can be encountered near an airport, can cause the code to set. This type of code is often referred to as a *phantom code*.

It is important that this type of code identification be made. The trouble code charts are written assuming that a fault exists. If one of the trouble code charts is used to pursue a phantom or intermittent code, the chart most often leads to an invalid conclusion because it is beginning with an invalid assumption, that a fault currently exists when in fact none does.

Chapter 3 General Motors' Computer Command

FIGURE 3-55 Diagnostic circuit check. Charts 5 and 6, which are referred to in the diagram, are found in the manufacturer's service manual. *Courtesy of GM Product Service Training.*

Chapter 3 General Motors' Computer Command

DRIVER COMMENT Revised: 8-31-82

Engine performance troubles (stalling, detonation, surge, fuel economy, etc.)

IF THE "CHECK ENGINE" LIGHT IS NOT ON, NORMAL CHECKS THAT WOULD BE PERFORMED ON CARS WITHOUT THE SYSTEM SHOULD BE DONE FIRST.
IF GENERATOR OR COOLANT LIGHT IS ON WITH THE CHECK ENGINE LIGHT, THEY SHOULD BE DIAGNOSED FIRST.
INSPECT FOR POOR CONNECTIONS AT COOLANT SENSOR, M/C SOLENOID, ETC., AND POOR OR LOOSE VACUUM HOSES AND CONNECTIONS. REPAIR AS NECESSARY.

- Intermittent check engine light but no trouble code stored.
    - Check for a loose connection in the circuit from:
        - Ignition coil to ground and arcing at spark plug wires or plugs.
        - Bat. to ECM term's. C and R
        - ECM terms. A and U to engine ground
        - EST wires should be kept away from spark plug wires, distributor housing, coil and generator. Wires from ECM term. 13 to dist. and the shield around EST wires should be a good ground.
        - Open Diode across A/C Compressor Clutch.

- Loss of long term. memory
    - Grounding dwell lead for 10 seconds with "test" term. ungrounded with engine running should give code 23. This code should be retained in long term. memory after the engine is stopped and restarted. If it is not, ECM is at fault.

- Stalling, Rough Idle, or Improper Idle Speed
    - See idle speed control . . . Page 4A-60

- Denotation (Spark Knock)
    - Check — ESC performance, if applicable . . . Page 4A-55
        MAP or Vacuum Sensor output . . . Pages 2A-20 and 2A-21
        EGR Check . . . Pages 4A-56 and 4A-57
        TPS enrichment operation . . . Chart No. 4
        HEI operation . . . Page 4A-54

- Poor Performance and/or Fuel Economy and Surging
    - See — Carb. on car service — Divisional Service Manual
        EFE check . . . Pages 4A-58 thru 4A-60
        TCC operational check . . . Page 4A-63
        EST diagnosis . . . Page 4A-54
        ESC diagnosis if applicable . . . Page 4A-55

- Poor Full-Throttle Performance
    - See Chart 4 if equipped with TPS

- Intermittent No-Start
    - Incorrect pick-up coil or ignition coil. See "Cranks, But Won't Run" chart.
    - Intermittent ground connections on ECM.

- All Other Comments.
    - Make system performance check on warm engine (upper radiator hose hot)

The system performance check should be performed after any repairs to the CCC system have been made

FIGURE 3-56 Driver complaint chart. Pages and charts referred to in this illustration are found in the manufacturer's service manual. *Courtesy of GM Product Service Training.*

Driver Comment Chart/Driveability Symptoms Section, Figure 3-56. This chart or section, depending on the model year, acts as a guide to aid the technician in determining the cause of problems that either do not have trouble codes or have intermittent trouble codes.

System Performance Check, Figure 3-57. This procedure should always be used after repairing any part of the CCC system or any component that the ECM controls. It verifies that the heart of the system, fuel mixture control, is working properly. If the fuel mixture is not being controlled properly, it refers to another chart for further diagnosis.

A trouble code found in memory and shown by the Diagnostic Circuit Check to be intermittent should be dismissed as a phantom code when no driver complaint or performance problem is identified and the System Performance Check shows satisfactory results.

Diagnostic Charts without Trouble Codes. This is a series of charts contained in the service manual. Any of the charts within the series can assist in finding the cause of any one of several specific fault conditions that do not have a corresponding trouble code. Figure 3-58 is shown as an example. These charts should only be used when referred to by one of the preceding charts (diagnostic circuit check, driver comment, or system performance check) or when a specific condition addressed by one of them has been identified. As mentioned previously, improper use of the charts most often leads to mistakes, sometimes costly ones.

Diagnostic Charts with Trouble Codes. This is a series of charts contained in the service manual. These charts can be used to find the cause of a fault once the fault has been identified by a code stored in the ECM's memory and after the diagnostic circuit check has been performed to verify that the fault currently exists. Figure 3-59 is shown as an example.

Diagnostic Charts on Related Components. This series of charts, contained in the service manual, helps with the diagnosis of components and related circuits controlled by the ECM but not monitored by it for proper operation. These components and circuits include the air management valve and the torque converter clutch. Figure 3-60 is shown as an example.

Dwell Diagnosis

If the system is in closed loop and the dwell is varying between 10° and 50°, then we know that the system is able to maintain the desired air/fuel ratio, Figure 3-61. If the dwell is not varying and is between 10° and 50°, it is not in closed loop. If the dwell is stuck below 10° or above 50°, a condition exists that is causing the air/fuel mixture to be either lean or rich and for which the system cannot compensate. It is possible, of course, that the system electronics are malfunctioning and are therefore unable to recognize the existing air/fuel ratio.

Let's assume, for example, that we are getting a steady dwell of 54°, which would indicate a rich condition. To determine if an electronic malfunction is occurring or if the engine is actually running rich, run the engine for two minutes at a fast idle to be sure that it is in closed loop. Next pull a large vacuum hose (just small enough to keep from killing the engine), and observe the dwell. It should begin to go down. If it does not respond, there is an electronic problem; the system is unable to recognize the change in the air/fuel mixture.

Note: Some dwellmeters do not work well on CCC systems. If a particular dwellmeter causes any change in engine performance when it is connected, do not use it.

Carburetor and M/C Solenoid Adjustment

Varying degrees of adjustment capability exist on the four different carburetors used on

Chapter 3 General Motors' Computer Command

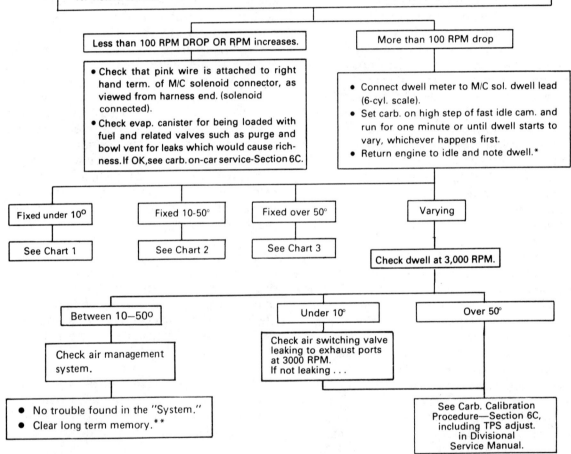

FIGURE 3-57 System performance check. Charts referred to in this illustration are found in the manufacturer's service manual. *Courtesy of GM Product Service Training.*

Chapter 3 General Motors' Computer Command

CHART #3

Revised: 4-21-82

(Rich Exhaust Indication)

DWELL FIXED OVER 50°

*Use divisional service manual.

FIGURE 3-58 Typical diagnostic chart. *Courtesy of GM Product Service Training.*

Chapter 3 General Motors' Computer Command

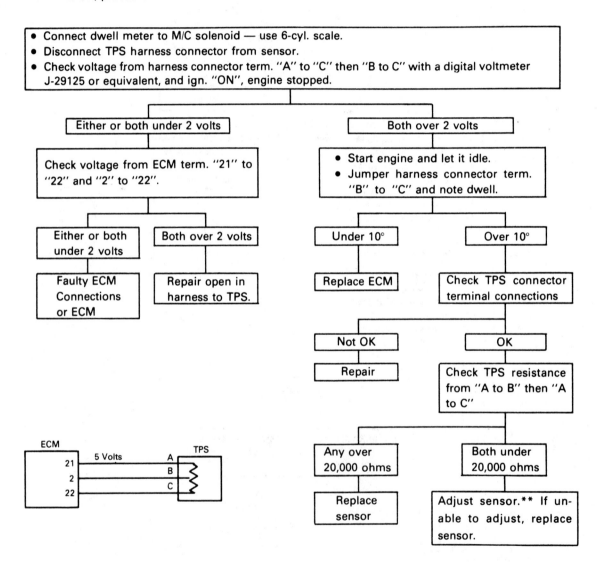

FIGURE 3-59 Typical trouble code chart. *Courtesy of GM Product Service Training.*

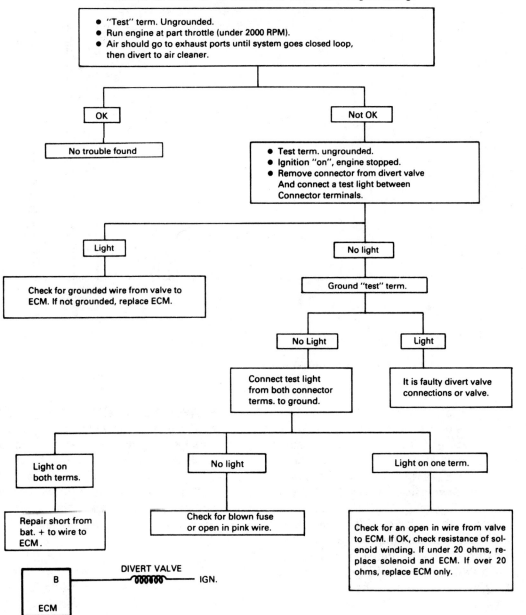

FIGURE 3-60 Typical diagnostic chart for nonmonitored functions. *Courtesy of GM Product Service Training.*

Chapter 3 General Motors' Computer Command

FIGURE 3-61 Dwell reading and system performance

CCC applications. The 6510-C Holley, used on the T car, has only its idle mixture screw as an external mixture adjustment. Mixture adjustments are made at idle only using a dwellmeter. The E2SE *Varajet* has an idle mixture screw, and down inside its bowl is a lean mixture screw, Figure 3-62. The lean mixture screw is a main metering adjustment and is made at 3,000 RPM using a dwellmeter. The idle mixture screw is adjusted at idle using a dwellmeter. The E4ME *Quadrajet* and the E2ME *Dualjet* carburetors have much more adjustment capability. Their adjustments are identical and include idle mixture screws, an idle air bleed valve, and adjustments on the M/C solenoid.

PERSONAL SAFETY: When performing mixture adjustments at high engine speeds, be certain that all instrument leads are clear of the coolant fan, that you have no loose clothing such as shirt or coat sleeves that can get caught in the coolant fan or drive belts, and that the drive wheels are firmly blocked.

Since the CCC system's introduction, several revisions have been made in adjustment specifications and procedures. A significant procedure revision made in 1983 should be observed when servicing E4ME and E2ME carburetors from 1981 through 1984. The E4ME

Chapter 3 General Motors' Computer Command

FIGURE 3-62 Rochester Varajet (E2SE). *Courtesy of GM Product Service Training.*

and E2ME carburetors on many 1985 and later models have some slight internal changes that prevent earlier procedures from applying to them. Although such carburetor adjustments should only be performed when clearly indicated, driveability problems can sometimes be eliminated by restoring the proper float level and carefully making updated adjustments.

PERSONAL SAFETY: No attempt should be made to adjust the rich mixture screw on the E2SE carburetor shown in Figure 3-62. Removing and replacing the covering plug produces a potential fuel leak and fire hazard.

ECM and PROM Service

Considerable care should be taken when installing a PROM. A finger coming into contact with the PROM's pins can discharge static electricity into it and damage it, Figure 3-63. To insert the PROM in the ECM, position the PROM in the carrier. The carrier only fits in the PROM cavity of the ECM in one position; however, the PROM fits in the carrier in either of two ways. If the PROM is installed in the ECM incorrectly, it will be damaged electrically. To avoid this, identify the small half-circle notch on one end of the PROM. Now look in the PROM cavity and identify a similar notch on one end of the PROM seat, Figure 3-64. To avoid bending the pins when installing the

Chapter 3 General Motors' Computer Command

FIGURE 3-63 PROM

FIGURE 3-65 PROM pins/PROM carrier shoulder

FIGURE 3-64 PROM, PROM carrier, and PROM socket

PROM, be sure that the PROM is positioned in the carrier with the tips of the pins above the shoulder of the carrier, as shown in Figure 3-65, before placing the PROM and carrier in the PROM cavity. This allows the carrier to guide each pin into place as the PROM is pressed down.

The PROM and ECM are serviced separately; therefore, remove the PROM before exchanging the ECM for a new one.

Weather-pack Connectors

All CCC system harness connections under the hood are made with weather-pack connectors, Figure 3-66. They are designed to provide environmental protection for electrical connections so that corrosion and contamination buildup is held to a minimum. Many of the circuits in the CCC system do not carry more than 1 or 2 milliamps at about 100 millivolts (1/10 volt). The resistance caused by oxidation or contamination of an electrical connection can have significant impact on the perfor-

Chapter 3 General Motors' Computer Command

1. OPEN SECONDARY LOCK HINGE ON CONNECTOR

2. REMOVE TERMINALS USING SPECIAL TOOL

3. CUT WIRE IMMEDIATELY BEHIND CABLE SEAL
4. SLIP NEW CABLE SEAL ONTO WIRE (IN DIRECTION SHOWN) AND STRIP 5.00mm (.2") OF INSULATION FROM WIRE. POSITION CABLE SEAL AS SHOWN.

FIGURE 3-66 Weather-pack connector. *Courtesy of GM Product Service Training.*

mance of the system. It is important, then, that the integrity of the weather-pack connectors be maintained. Don't break the seal by inserting a probe into the connector. Voltage readings can be obtained by opening the connector and temporarily installing short jumper wires made for that purpose (many called-for voltage checks require that the circuit be electrically intact). If the weather-pack connector has been damaged, it should be replaced as shown in Figure 3-66 or the connection should be soldered and taped as in Figure 3-67.

Diagnostic and Service Tips

TPS. The TPS is the one sensor that can be adjusted. Sometimes driveability complaints can be cured by adjusting the TPS to exact specifications, which vary with engine and model year application. Be sure to check the applicable service manual for correct procedures and specifications.

TCC Test Terminal. When using the TCC test terminal, the following points should be remembered:

- Voltage is present when the TCC is not applied and is near zero when the TCC is applied, Figure 3-42, because applying the TCC causes the voltage to be dropped across the apply solenoid.
- The TCC can be applied by grounding the TCC test terminal.
- Applying voltage to the TCC test terminal damages the ECM if the ECM tries to apply the TCC while the voltage is applied. (On earlier models with five-terminal ALCLs, the TCC test terminal is in the fuse panel and thus increases vulnerability to this mistake.)

Surging Complaints. Surging complaints are occasionally heard concerning CCC-equipped vehicles at steady-state speeds

---

CCC System Behavior

Some of the behaviors of the CCC system can be amusing or frustrating, depending on whether or not they are understood. For instance, driving up a long hill in the mountains at near full throttle can set a code and turn on the check engine. Changing altitude rapidly or driving backwards for about a quarter of a mile can have the same effect. There is nothing wrong; the system is just not programmed for those driving conditions.

129

Chapter 3 General Motors' Computer Command

TWISTED/SHIELDED CABLE

1. Remove outer jacket.
2. Unwrap aluminum/mylar tape. Do not remove mylar.

3. Untwist conductors. Strip insulation as necessary.

4. Splice wires using splice clips and rosin core solder. Wrap each splice to insulate.

5. Wrap with mylar and drain (uninsulated) wire.
6. Tape over whole bundle to secure as before.

TWISTED LEADS

1. Locate damaged wire.
2. Remove insulation as required.

3. Splice two wires together using splice clips and rosin core solder.

4. Cover splice with tape to insulate from other wires.
5. Retwist as before and tape with electrical tape to hold in place.

FIGURE 3-67 Wire repair. *Courtesy of GM Product Service Training.*

around 35 miles per hour or above. This is a fairly common behavior and is often caused by the EGR valve, referred to as *EGR chuggle*. The inert gases introduced in the intake manifold by the EGR valve cause an increased frequency of cylinder misfires. Although this has been occurring since the EGR valve's introduction in the early seventies, it was not noticeable to the driver until a clutch was placed in the torque converter thus making it a solid coupling, compounded by significantly lower axle numerical ratios. This behavior is sometimes interpreted, however, as a malfunction of the TCC. In order to isolate the condition, temporarily disconnect and plug the vacuum line from the EGR valve. Drive the vehicle to see if the behavior still occurs. If it does it is not EGR chuggle.

CAUTION: Do not leave the EGR valve disconnected. To do so is not only against the law but can result in engine damage due to detonation.

REVIEW QUESTIONS

1. What input values are used by the ECM to calculate fuel mixture in open loop? (See **Blended Enrichment**.)
2. The Hall-effect switch generates what type of signal?
3. What influence does the P/N switch have over engine operation?
4. What two conditions must exist before the TCC is applied?
5. Name at least two ways in which closed-loop operation can be identified.
6. An O_2 sensor voltage of 0.3 causes the ECM to issue a _____ command.
7. What inputs are considered by the ECM when determining the spark timing command?
8. With a dual-bed catalytic converter, where is the air directed under each of the following conditions?
   a. during warm-up
   b. in closed loop
   c. at wide-open throttle
9. What results if the air switching valve fails to switch air away from the exhaust manifold when the system goes into closed loop?
10. How does the ECM know when to control idle speed?
11. How does the ECM control TCC apply?
12. What can happen if the TCC fails to apply during fourth-gear operation with a four-speed automatic?
13. List two likely results if the EGR valve fails to open.
14. List two types of EFE systems.
15. During what operating mode is the canister purge solenoid de-energized, allowing the charcoal canister to be purged?
16. A dwell reading between 10° and 50° and varying suggests what?
17. How is the CCC system put into diagnostics?
18. Why is a high-impedance digital voltmeter required for many CCC system tests?
19. Why must trouble codes be identified as hard codes before the trouble code charts can be used?

ASE-type Questions. (Actual ASE test questions will probably not be this product specific.)

20. Technician A says that the CCC system only responds to the O_2 sensor when the system is in closed loop. Technician B says that the CCC system only uses the O_2 sensor input for fuel control. Who is correct?
    a. A only
    b. B only
    c. both A and B
    d. neither A nor B

21. Technician A says that when the M/C solenoid is down, the mixture is rich and the solenoid is turned off (no power applied to it). Technician B says that when the M/C solenoid is down, power is being applied to it and that the ECM is responding to a lean signal from the O_2 sensor. Who is correct?
    a. A only
    b. B only
    c. both A and B
    d. neither A nor B

22. A car equipped with a CCC system is being tested for dwell. The dwellmeter, connected to the M/C solenoid circuit and set on the six-cylinder scale, indicates a dwell reading of 50°.
    Technician A says that the car is running rich. Technician B says that the car is running lean. Who is correct?
    a. A only
    b. B only
    c. both A and B
    d. neither A nor B

23. A car equipped with TCC is being driven with a test light connected to the TCC test terminal. As the car reaches cruising speed, the light goes out. This indicates that:
    a. either the TCC circuit has failed or the fuse has blown.
    b. the TCC has applied.
    c. the TCC has failed to apply.
    d. the TCC has released.

24. Technician A says that while in diagnostics, the CCC system does not set fault codes. Technician B says that while in diagnostics, the CCC system displays fault codes only if the engine is not running. Who is correct?
    a. A only
    b. B only
    c. both A and B
    d. neither A nor B

GLOSSARY

In addition to the following terms, refer to Figure 3-68 for a list of the abbreviations used in this chapter.

ASPIRATOR A device using airflow through a restriction in a tube to create a vacuum in another tube.

BLEED A controlled leak.

DETONATION Uncontrolled, rapid burning of the fuel charge that results in spark knock.

DIVERT MODE When the ECM directs the air management system to dump air into the air cleaner or the atmosphere.

DUALJET A two-venturi carburetor that basically consists of the primary bores and circuitry of a Quadrajet; usually referred to by General Motors as E2ME, the last E meaning it is equipped for CCC system application.

E-CELL A module containing a cathode and an anode, one of which is gradually sacrificed as current passes through them, causing the circuit to become open.

FAIL-SAFE A feature of system design that provides continued operations (with less than normal performance capability) in the event of certain component or system failures or that activates or deactivates certain systems in order to prevent damage to other components due to component or system failure.

KAM Keep-alive memory.

LIGHT-EMITTING DIODE A semiconductor device that generates a small amount of light when forward biased (voltage applied with the polarity such that current flows across the PN junction).

MAXIMUM AUTHORITY A term describing the maximum idle speed that the ISC motor can provide.

MILLIAMP The most correct form is milliampere, meaning one thousandth of an amp.

MILLIVOLT One thousandth of a volt.

MINIMUM AUTHORITY A term describing minimum idle speed, achieved by retracting the ISC plunger until the throttle lever rests on the idle stop screw.

PARAMETER One of the factors that helps to define operating conditions, such as coolant temperature, engine load, air/fuel mixture, and so forth.

PULSAIR The name General Motors applies to their pulse injection system.

Chapter 3 General Motors' Computer Command

A/C	Air-Conditioning
AIR	Air Injection Reaction
ALCL	Assembly Line Communication Link
BARO	Barometric
ECM	Electronic Control Module
EFE	Early Fuel Evaporation
EGR	Exhaust Gas Recirculation
ESC	Electronic Spark Control
EST	Electronic Spark Timing
HEI	High-energy Ignition
ILC	Idle Load Compensator
ISC	Idle Speed Control
MAP	Manifold Absolute Pressure
M/C	Mixture Control
PCV	Positive Crankcase Ventilation
P/N	Park/Neutral
PROM	Programmable Read-only Memory (engine calibration unit)
PWM	Pulse Width Modulated
TCC	Torque Converter Clutch
TERM.	Terminal
TEST LEAD or TERMINAL	Lead or ALCL connector terminal grounded to obtain a trouble code
TPS	Throttle Position Sensor
VAC.	Vacuum
VIN	Vehicle Identification Number
VSS	Vehicle Speed Sensor
WOT	Wide-open Throttle

FIGURE 3-68 Explanation of abbreviations

PURGE VALVE A vacuum-controlled valve that controls the removal of stored HC vapors from the charcoal canister.

THROTTLE KICKER A device operated either by vacuum or electrically that when activated during idle bumps the throttle open slightly more.

TIMER CORE A star-shaped wheel attached near the top of the distributor shaft in an HEI distributor. It has one point per engine cylinder.

VACUUM CONTROL VALVE A switch that opens and/or closes its ports to supply vacuum in response to coolant temperature change. It is sometimes referred to as a *ported vacuum switch* or a *thermal vacuum switch*. Thermal vacuum switch, however, is more accurately used to refer to a switch located in the air cleaner that opens or closes a port in response to air temperature.

VARAJET A two-venturi carburetor consisting basically of one primary and one secondary bore and related circuitry from a Quadrajet; usually referred to by General Motors as E2SE, the last E meaning it is equipped for CCC system application.

WOT Wide-open throttle.

GENERAL MOTORS' ELECTRONIC FUEL INJECTION

Objectives

After studying this chapter, you will be able to:

- list the six most important inputs needed to control engine calibration.
- define the term *pulse width*.
- describe how the computer knows engine temperature, engine load, engine speed, and throttle position.
- name at least two EFI system failures that cause the engine to stop.
- describe the circuit that controls fuel pump operation.
- list the first two steps of EFI diagnostic procedure.
- name several precautions that can prevent damage from occurring to the computer.

Electronic fuel injection (EFI), not to be confused with a fuel injection system used by Cadillac in the late seventies and that was also called EFI, is a single-point injection system. It is often referred to as *TBI* (throttle body injection); that name, however, more accurately describes a major component of the system, Figure 4-1. This system was introduced by General Motors in 1982.

When the EFI system was introduced in 1982, it appeared in two versions: one containing a single TBI unit for four-cylinder engines and a high performance version (Crossfire) with two separate TBI units for eight-cylinder engines. In 1985 and in 1986, a dual TBI (two injectors in one TBI unit) was introduced on some V6 engine applications.

FIGURE 4-1 TBI unit. *Courtesy of GM Product Service Training.*

135

Chapter 4   General Motors' Electronic Fuel Injection

ELECTRONIC CONTROL MODULE

The ECM is usually located above or near the glove compartment. On some models, such as the Pontiac Fiero, it is in the console. On some Corvette applications, it is in the battery compartment behind the driver's seat. Its inputs and the functions it controls are shown in Figure 4-2.

ECM for 2.5 liter, EFI Operates Cruise Control

The ECM for selected engines such as the 2.5 liter engine was given more internal circuitry to enable it to operate the cruise control system in 1988.

Throttle Body Backup (TBB)

The *TBB* is a fuel backup circuit within the ECM. It is primarily responsible for providing fuel pulses to the injector solenoid in the event of an ECM failure, which prevents it from running its program. The ECM used on the 2.0 liter engine has a removable *CALPAK* that provides fuel backup. The CALPAK plugs into the ECM in the same way that the PROM does.

Block Learn

The EFI ECM has a learning ability. It accepts and corrects for gradually changing input values that result from sensor wear or driving conditions, such as changing altitude. It even learns to slightly modify output commands to complement a particular driver's driving habits. The aspect of this learning ability of most concern to technicians is covered under **System Diagnosis and Service** in this chapter.

Keep-alive Memory (KAM)

The ECM has a KAM like that discussed in Chapter 3.

Inputs

Parameters Sensed*
- A/C System Enable
- Barometric Pressure
- Brake Pedal Engagement
- Engine Coolant Temperature
- Manifold Air Temperature
- Engine Crankshaft Position
- Engine Crank Mode
- Engine Detonation
- Exhaust Oxygen
- Fuel Pump Voltage
- Manifold Absolute Pressure
- Park/Neutral Position
- Throttle Position
- Ignition Switch
- Transmission Gear Position
- Power Steering Signal
- Vehicle Speed
- Battery Power
- ALDL Diagnostic Request

Electronic Control Module (ECM)

Outputs

Parameters Controlled*
- Air Control Valve Signal
- Air Switching Valve Signal
- Canister Purge Control Signal
- EGR Control Signal
- Bypass
- Electronic Spark Timing Signal
- Idle Speed Control
- Throttle Body Injector(s)
- Torque Converter Clutch
- A/C Clutch Control
- Air Door Control
- Cooling Fan Control
- Fuel Pump Relay Power
- Check Engine/Service Engine Soon Light
- Serial Data Terminal (ALDL)
- Cruise Control

*Some features not used on all engines.

FIGURE 4-2   EFI system overview

OPERATING MODES

The EFI system features the typical closed-loop and open-loop modes. During open loop the ECM is programmed to provide an air/fuel ratio most suited to driveability as well as economy and emissions concerns.

Synchronized Mode

In synchronized mode the injector is pulsed once for each reference pulse from the distributor. To put it another way, the injector sprays once for each time a cylinder fires. On dual-TBI units the injectors are pulsed alternately. All closed-loop operation is in synchronized mode as is most open-loop operation.

Nonsynchronized Mode

In nonsynchronized mode the injector is pulsed every 12.5 milliseconds, independent of distributor reference pulses. On dual-TBI systems each injector is pulsed every 12.5 milliseconds; but because they are pulsed alternately, the engine receives a spray every 6.25 milliseconds. Nonsynchronized pulses only occur in response to one or more of the following operating conditions:

- When the injector is on or off (open or closed), time becomes too small for accurate control (about 1.5 milliseconds), as occurs at near full throttle or during deceleration. Even though the injector only opens a few thousandths of an inch, it is a mechanical device opened by a magnetic field and closed by a spring and has a finite speed as contrasted to the speed of current flow. When it is asked to open and close or to close and open at a rate that approaches its response capability limits, the ECM stops making the injector keep up with the reference pulse and allows it to go into a slower mode of operation.
- During prime pulses. On the 1982 Crossfire EFI system, when the ignition was turned on, if the coolant temperature was low enough, the ECM commanded the injector to deliver a spray or two into the manifold to prime the engine for starting. This action was similar to pumping the throttle before starting a carbureted engine. The Crossfire system was discontinued in 1983.

Cranking Mode

While the engine is cranking, the injector delivers an enriched air/fuel ratio especially suited for starting conditions (high manifold pressure and low airflow velocity). The pulse widths depend on coolant temperature. At –36° C (–32.8° F) the pulse width is calibrated to provide maximum rich air/fuel ratio of 1.5 to 1. At 94° C (201.5° F) the pulse width is calibrated to provide a maximum lean air/fuel ratio of 14.7 to 1. The higher the temperature, the shorter the pulse width or injector on-time, unless the engine is overheated. In case of overheating, the pulse width is widened to provide a slightly richer mixture.

Clear Flood

If the engine floods, depressing the throttle to wide-open throttle (WOT) or within 80% of WOT puts the system into clear flood mode. While cranking in this mode, the ECM issues a pulse command that produces an air/fuel ratio of 20 to 1. It stays in this mode until the engine starts or until the throttle is moved to less than 80%. If the engine is not flooded and the throttle is depressed 80% or more, it is highly unlikely that the engine will start.

Run

When the ECM sees a reference pulse from the distributor that indicates an RPM of 600 or more, it puts the system into open loop. In open loop the ECM does not use the inputs from the oxygen sensor to calculate air/fuel mixture commands. It does, however, watch to see if the oxygen sensor is ready to go into closed

Chapter 4 General Motors' Electronic Fuel Injection

1 Coolant temperature sensor
2 Harness connector to ECM
3 Locking tab

FIGURE 4-3 Engine coolant sensor (new style). *Courtesy of Buick Motor Division.*

loop. Three criteria must be satisfied before the system can go into closed loop:

- The oxygen sensor must produce voltage signals that cross 0.45 volt (450 millivolts). In order to do this, it must reach a temperature of at least 300° C (570° F).
- The coolant temperature must reach a specified temperature, about 65.6° C (150° F).
- A specified amount of time must elapse since the engine was started.

The values for each of these criteria are contained in the PROM. When all of the criteria are met, the ECM puts the system into closed loop. Only the input from the oxygen sensor is now used to make decisions about air/fuel mixture commands.

Semi-closed Loop

In an effort to achieve greater fuel economy, the ECM for some throttle body injected engines has been programmed to go out of closed loop during cruise conditions at highway speeds. During these periods, the air/fuel mixture may go as lean as 16.5 to 1. The ECM monitors the following parameters and will put the system into this fuel control mode only when they are within predetermined values:

- engine temperature
- spark timing
- canister purge
- constant vehicle speed

The ECM will periodically switch back to closed loop in order to check on engine operating parameters. If parameters are still within required limits, it will go back to the lean calibration, open-loop mode. Operation in the lean calibration, open-loop mode and switching back and forth between closed loop and open loop are smooth and should not be detectable to the driver.

INPUTS

Engine Coolant Temperature (ECT) Sensor

This is the same sensor as that used in the computer command control system and the other General Motors' systems. On later models it had a slightly different appearance also, Figure 4-3.

Manifold Absolute Pressure (MAP) Sensor

The MAP sensor on EFI applications, Figure 4-4, is basically the same as that on CCC applications; but the ECM has been programmed to use the sensor as a barometric pressure sensor also. When the ignition is turned to "run," on its way to "crank," the ECM looks at the MAP sensor reading. The engine is not running (the crankshaft has not moved yet); therefore, manifold pressure is atmospheric pressure. The ECM records this read-

FIGURE 4-4 EFI MAP sensor. *Courtesy of Buick Motor Division.*

Chapter 4 General Motors' Electronic Fuel Injection

14 THROTTLE POSITION SENSOR
15 TBI UNIT

FIGURE 4-5 Throttle position sensor. *Courtesy of Buick Motor Division.*

ing as a barometric pressure reading and uses it in calculating fuel mixture while in open loop and spark timing. The BARO reading is held and used until the ignition is turned off or until the throttle is pushed to WOT. At this time engine vacuum goes to atmospheric pressure, and an updated BARO reading is taken.

Oxygen (O_2) Sensor

This is the same single-wire sensor as used on other General Motors' systems.

Throttle Position Sensor (TPS)

The TPS is a variable resistor (potentiometer) mounted on the TBI unit, connected to the end of the throttle shaft, Figure 4-5. The TPS on some EFI applications is adjustable; on others TPS adjustment is not provided.

Distributor Reference Pulse (REF)

The REF, sometimes referred to as REF pulse, tells the ECM what engine speed is and what position the crankshaft is in. On most EFI engine applications, the distributor pickup coil provides the REF signal, as described in Chapter 3. It is important to note that if while the engine is running the REF signal fails to arrive at the ECM, fuel injection is stopped.

The earlier 2.5-liter EFI engines used a Hall-effect switch to supply the REF signal, Figure 4-6. The pickup coil was still used for starting, but that was its only function.

FIGURE 4-6 HEI module with EST and Hall-effect switch. *Courtesy of GM Product Service Training.*

Chapter 4 General Motors' Electronic Fuel Injection

FIGURE 4-7 Knock sensor

Detonation Sensor

A piezoelectric crystal detonation or knock sensor is used on some EFI systems, Figure 4-7.

Vehicle Speed Sensor (VSS)

Two types of VSSs have been used on EFI vehicles, as described in Chapter 3. On EFI vehicles the VSS input is also used by the ECM to help identify a deceleration condition. This is discussed further under **Injector Assembly** in the **Outputs** section of this chapter.

Ignition Switch

The ignition switch is one of two power supplies to the ECM. When the ignition is turned on, the ECM initializes (starts its program sequence) and is ready to function. The ignition switch also powers most of the actuators that the ECM controls.

Park/Neutral (P/N) Switch

The park/neutral switch, usually located near the shifter assembly, closes when the vehicle is shifted into park or neutral. This signals the ECM as to whether or not the transmission is in gear. The ECM uses this information to control engine idle speed. If the P/N switch is disconnected or significantly out of adjust-

FIGURE 4-8 Fuel pump control circuit

ment, idle quality can suffer while in park or neutral.

Transmission Switches

Most transmissions used in CCC-equipped vehicles have one or more hydraulically operated electric switches screwed into the valve body. These switches provide the ECM with signals that tell it what gear the transmission is in. This information enables the ECM to more effectively control torque converter clutch operation.

Air-conditioning (A/C) Switch

On EFI cars equipped with air-conditioning, the A/C switch is connected by a wire to the ECM. When the A/C is turned on or off, the ECM is informed and uses this information in its control of idle speed.

Ignition Crank Position

The circuit that energizes the starting motor solenoid also signals the ECM that the engine is being cranked. The ECM responds by issuing cranking mode injector commands.

Fuel Pump Voltage Signal

Some EFI systems have a wire from the positive side of the fuel pump power circuit to the ECM. This wire signals the ECM that the pump has been activated, Figure 4-8. Some systems do not use it.

OUTPUTS

Throttle Body Injection (TBI) Unit

The TBI unit is made up of three castings: the throttle body, the fuel meter body, and the fuel meter cover, Figure 4-9. The throttle body contains the throttle bore and valve. It provides the vacuum ports for EGR, canister purge, and so forth, similar to what is found in a carburetor. Mounted to it are the idle air control motor and the TPS. The fuel meter body contains the injector and the fuel pressure regulator, Figure 4-10.

Injector Assembly

The injector is a solenoid-operated assembly. When it is energized by the ECM (pulsed), a spring-loaded metering valve is lifted off its seat a few thousandths of an inch. Fuel under pressure passes through a fine screen filter that fits around the tip of the injector and is sprayed in a conical pattern against the walls of the throttle bore, just above the throttle blade.

11	THROTTLE BODY
12	FUEL BODY ASM.
13	FUEL METER COVER (WITH PRESSURE REGULATOR BUILT-IN)
14	IDLE AIR CONTROL (IAC) VALVE
15	THROTTLE POSITION SENSOR (TPS)
16	FUEL INJECTOR

FIGURE 4-9 TBI unit. *Courtesy of Buick Motor Division.*

The TBI

The fascinating thing about the EFI system is that the TBI unit performs every function that a carburetor does and does it with more control. When the TPS is moved rapidly, the injector momentarily sprays more fuel (increased pulse width), similar to an accelerator pump function. On cold starts it sprays additional fuel to duplicate the function of a choke and provides a fast idle. It has no choke, fast idle cam, accelerator pump, and so on; it is all done electronically. On deceleration it leans out the air/fuel ratio (decreased pulse width) This compensates for the fuel that begins to evaporate from the manifold walls as a result of the higher vacuum. On hard deceleration it stops fuel injection completely. Hard deceleration is identified by a closed throttle, a sharp drop in manifold pressure, and a rapid decrease in vehicle speed.

Chapter 4 General Motors' Electronic Fuel Injection

FIGURE 4-10 Idle air bypass circuit. *Courtesy of GM Product Service Training.*

Fuel Pressure Regulator

Within the TBI assembly is a fuel pressure regulator, Figure 4-10. The spring side of the diaphragm is exposed to atmospheric pressure; therefore, fuel pressure varies slightly with changes in atmospheric pressure. At lower atmospheric pressures, the fuel pressure is reduced and thus causes slightly less fuel to be injected in a given pulse width. The pressure regulator should provide a pressure ranging from 9 to 13 psi.

On Crossfire systems the rear TBI unit has a pressure regulator, and the front TBI unit has a pressure compensator, Figure 4-11. It works much the same way as the pressure regulator does, except its function is to make up for momentary fuel pressure drops between the front and rear TBI units to maintain constant operating pressure. Such drops are usually caused by the diaphragm and valve in the pressure regulator moving to a more closed position and thus allowing less fuel to the front TBI unit.

Idle Air Control (IAC)

The IAC assembly controls idle speed. It consists of a small, reversible, electric stepper motor and a pintle valve, Figure 4-12 (number 11). As the motor's armature turns, the pintle valve is extended or retracted depending on the direction the motor is turning. During idle the throttle blade is in a fixed, nearly closed position and thus allows a constant amount of

Chapter 4 General Motors' Electronic Fuel Injection

idle air into the intake manifold. The pintle valve allows additional air through the bypass passage. The ECM uses the stepper motor to position the pintle valve for desired idle speed. During warm engine operation, the ECM attempts to maintain a fixed idle speed by adjusting the IAC valve position for load variation (transmission in or out of gear, air conditioner on or off, etc.).

If during idle or at low vehicle speed (below 10 MPH) engine speed drops below a specified RPM, the ECM puts the IAC into antistall mode. The IAC motor retracts the pintle to allow additional air into the intake manifold to raise engine speed. It momentarily raises engine speed above base idle.

The stepper motor is unique; its armature has two separate windings. The direction in which the armature turns depends on which winding is powered. Power is applied in short pulses. Each pulse rotates the armature about 35° and extends or retracts the pintle valve a

FIGURE 4-11 Crossfire TBI units. *Courtesy of GM Product Service Training.*

143

Chapter 4 General Motors' Electronic Fuel Injection

FIGURE 4-12 IAC motor assembly in TBI unit. *Courtesy of GM Product Service Training.*

proportionate amount. The ECM applies as many pulses as necessary to achieve its desired idle RPM.

The ECM keeps count of the pulses applied and therefore knows where the pintle valve is at all times. There are 12 pulses (positions) per revolution and 255 total positions. Fully extended (closed bypass passage) is the reference position, zero. Fully retracted (wide-open bypass passage) is step 255.

In order to maintain accurate tracking of the IAC assembly's position, the ECM references itself frequently. When the engine comes off idle and the car starts to move, the ECM begins looking for a reading from the VSS that represents 30 MPH. The first time it sees a 30-MPH signal, it moves the pintle to the park (fully closed) position to reorient the zero reference. It then moves the pintle to a preprogrammed distance from closed. If for any reason the VSS is disabled, the ECM will not be able to reestablish the correct idle speed.

Fuel Pump Control

Fuel and fuel pressure are supplied by an electric pump inside the fuel tank. The fuel pump assembly contains a check valve that prevents fuel from bleeding from the fuel line back into the tank. If the check valve leaks, the engine must crank long enough before starting to allow the line to be purged of vapor.

FIGURE 4-13 Fuel pump and A/C relay. *Courtesy of Buick Motor Division.*

Chapter 4 General Motors' Electronic Fuel Injection

The fuel pump is turned on and off by a fuel pump relay usually mounted in the engine compartment. The fuel pump relay is controlled by the ECM. Be aware that the fuel pump relay is identical in appearance to and usually mounted beside the air-conditioning relay, which is also controlled by the ECM, Figure 4-13. On some models the cooling fan relay is also mounted in the same location and also looks like the other two.

When the ignition is turned on, the ECM activates the fuel pump relay, Figure 4-14. If the ECM does not receive a REF signal within two seconds, it deactivates the relay and reactivates it only when it does receive the REF or when the ignition is turned off and back on again. Once the engine is running, the ECM deactivates the relay when the ignition is turned off or anytime REF signals stop coming to the ECM. Note also in Figure 4-14 that the oil pressure switch, which operates the oil pressure indicator on the instrument panel, is electrically in parallel with the fuel pump relay. This is done to provide a backup to the fuel pump relay. The engine starts after a fuel pump relay failure if cranked long enough to generate about 4 PSI of oil pressure.

Fuel Pump Test Terminal. A wire is connected to the circuit that powers the fuel pump. The open end of this wire has a terminal to which a jumper lead can be connected to power the fuel pump. A voltmeter connected to the test terminal quickly shows if either the fuel pump relay or oil pressure switch is supplying power to the fuel pump circuit. The terminal is located in the engine compartment or in the

FIGURE 4-14 Fuel pump control circuit. *Courtesy of Buick Motor Division.*

145

Chapter 4 General Motors' Electronic Fuel Injection

assembly line communication link under the instrument panel, depending on the model and year.

Voltage Correction. During vehicle operation, system voltage can vary considerably as a result of the combination of electrical accessories being used, battery state of charge, and so forth. Variations in system voltage can cause the injector's opening and closing time to vary slightly. This affects the amount of fuel injected during a pulse width. Because of this the ECM monitors system voltage and will multiply the pulse width by a voltage correction factor, Figure 4-15. As system voltage goes down, the pulse width increases; as system voltage goes up, pulse width decreases. If voltage goes down to a criterion value, the ECM increases dwell to ensure good ignition performance and increases idle speed.

Electronic Spark Timing (EST)

EST maintains optimum spark timing under all conditions of engine load, speed, temperature, and barometric pressure. Two basic functions are incorporated within the system. Dwell control is provided to allow sufficient energy in the ignition coil for proper ignition voltage output. Spark is provided at just the right time to start combustion at the moment of peak pressure in the compression stroke. The EST function of the ECM eliminates the vacuum and centrifugal advance mechanisms. The HEI module accepts spark timing commands from the ECM once the engine has started.

During cranking, Figure 4-16, the solid-state switching circuit connects the main switching transistor of the HEI module to the pickup coil. The pickup coil signal turns on and off the main switching transistor, which controls the primary ignition circuit. This is called the *module mode* and provides no spark advance. On applications that do not use a Hall-effect switch, this same signal, after it is converted from an analog to a digital signal by a signal converter circuit in the module, is sent to the ECM as the REF signal. On applications that do use the Hall-effect switch, it supplies the REF signal. Note in Figure 4-17 that terminal R is not used.

When the ECM sees a REF pulse signal that represents an RPM of 600 or more, it decides that the engine is running. It then sends a 5-volt signal through the bypass wire to the solid-state switching circuit. The switching circuit in turn disconnects the pickup coil from the base of the main switching transistor. The EST wire is simultaneously disconnected from ground and connected to the main switching transistor base. The system is now in EST mode, and timing is being controlled by the ECM. The ECM considers barometric pressure, manifold pressure, coolant temperature, engine speed, and crankshaft position and then sends to the HEI module the optimum spark timing command.

Electronic Spark Control (ESC)

The Crossfire EFI system is equipped with ESC. Its function is to retard ignition timing when detonation occurs. The system is essentially the same as the later version of ESC described in the **Outputs** section of Chapter 3.

Torque Converter Clutch (TCC)

The purpose of the TCC is to improve fuel mileage. It does this by eliminating hydraulic

FIGURE 4-15 Voltage correction graph. *Courtesy of GM Product Service Training.*

Chapter 4 General Motors' Electronic Fuel Injection

FIGURE 4-16 HEI module with EST. *Courtesy of GM Product Service Training.*

slippage and heat production in the converter once a cruise speed has been achieved. Its operation is essentially the same as the one described in the **Outputs** section of Chapter 3.

Manual Transmissions. On manual transmission applications, the TCC output terminal on the ECM can be used to operate a shift light on the instrument panel. The ECM looks at

FIGURE 4-17 HEI module with EST and Hall-effect switch. *Courtesy of GM Product Service Training.*

Chapter 4 General Motors' Electronic Fuel Injection

engine speed, engine load, throttle position, and vehicle speed and alerts the driver when to shift into the next gear for optimum fuel economy.

Cooling Fan Control

Transverse engine applications usually use an electrically powered cooling fan to pull air across the radiator fins. On most EFI applications, the ECM has some control over cooling fan operation; the cooling fan can be turned on by the ECM or by a coolant temperature switch. On some applications it has full control. The fan is turned on and off by a relay that the ECM operates, Figure 4-18. On most applications the fan turns on when the coolant temperature is high, when the vehicle is traveling below 30 MPH with the air conditioner on, and in some cases when the air-conditioning high side pressure reaches a certain value.

Hood Louver Control

The Crossfire EFI system features an electrically operated air door that the ECM can open using a relay and solenoid, Figure 4-19. When open the air door allows fresh air from above the engine hood to be directed through a hole in the top of the air cleaner housing to the air cleaner elements. The ECM opens the air door above a specified coolant temperature or at WOT.

Air-conditioning (A/C) Relay

The A/C relay is what actually turns the compressor clutch on and off, Figure 4-20. When the driver turns the A/C switch on, a signal is sent to the ECM and power is available at terminal B of the relay. The ECM waits a half-second and then ground terminal C of the relay. This energizes the relay and engages the compressor clutch (if the engine is at idle, the

FIGURE 4-18 Cooling fan control circuit (typical)

Chapter 4   General Motors' Electronic Fuel Injection

FIGURE 4-19   Hood louver control circuit

half-second allows the ECM time to adjust IAC position before the compressor clutch is engaged). The ECM deactivates the relay at WOT, during heavy engine load, and when IAC is in power steering stall mode. As seen in Figure 4-20, the compressor clutch circuit can be opened by the high pressure switch or the pressure cycling switch, which is a standard A/C control circuit feature.

Related Emission Controls

On all but the Crossfire EFI system, EGR, canister purge, and air management control are not controlled by the ECM. The EGR, canister purge, and air management controls used on the Crossfire system are essentially the same as those used on V8 engines with computer command control, as discussed in Chapter 3.

Catalytic Converter. Most EFI applications use a single-bed, three-way catalytic converter. No supplemental air is introduced into the converter; therefore, either no air injection is used or a Pulsair system is used with no computer control.

EGR. Many four-cylinder EFI systems use a self-modulating EGR valve with no ECM control (most use a back-pressure EGR, Figure

149

Chapter 4 General Motors' Electronic Fuel Injection

FIGURE 4-20 A/C relay circuit (typical)

FIGURE 4-21 Back-pressure EGR valve. *Courtesy of GM Product Service Training.*

150

4-21). Under the main diaphragm found in a conventional EGR valve is a second diaphragm. A small spring pushes downward on the second diaphragm. A passage in the EGR valve stem allows exhaust manifold pressure to enter into a chamber under the second diaphragm. When exhaust pressure is low, the small spring is able to hold the second diaphragm down in spite of the exhaust pressure trying to push it up. Atmospheric pressure flows through a vent in the lower part of the EGR valve housing and through a port in the second diaphragm, into the space between the two diaphragms. If the second diaphragm is in its down position, the atmospheric pressure is free to flow through another port in the main diaphragm into the vacuum chamber. With atmospheric pressure finding its way into the vacuum chamber, enough vacuum is not developed to lift the main diaphragm and open the EGR valve. As the throttle opening is increased, exhaust pressure increases. When exhaust pressure is high enough, the second diaphragm is forced up to close off the port through the main diaphragm into the vacuum chamber. The vacuum chamber is no longer vented to atmospheric pressure, and sufficient vacuum is now developed to open the EGR valve.

Canister Purge. On four-cylinder EFI engines, canister purge control is achieved with a thermal vacuum switch that blocks purge vacuum until the engine is in closed loop. In closed loop any additional fuel from the canister can be compensated for, still maintaining a 14.7 to 1 air/fuel ratio.

Service Engine Soon Light

All EFI systems use a service engine soon light; on early 1980s vehicles, it was called a check engine light. If the ECM sees a fault in one of the circuits that it monitors for malfunctions, it turns on the service engine soon light located on the instrument panel. This warns the driver that a malfunction exists. By grounding the test terminal in the assembly line communication link (ALCL) located under the dash, the ECM flashes the service engine soon light. These flashes indicate one or more codes that identify the circuit or circuits in which the fault exists. This is explained more fully under **System Diagnosis and Service** in this chapter.

SYSTEM DIAGNOSIS AND SERVICE

Self-diagnosis

The self-diagnostic capability and procedure for the EFI system is essentially the same as that of the CCC system described in the **System Diagnosis and Service** section of Chapter 3. An abbreviated discussion that points out the variations from the CCC system is presented here.

Note: Diagnostic procedures for 1987 and later are written to include the use of an ALCL tool, often referred to as a "scanner" or "scan," where it is applicable. Previously, diagnostic procedures were written around the use of a voltmeter to obtain such measurements as sensor readings.

Approach to Diagnosis

It is important that when any of the diagnostic procedures are being used that the directions be followed carefully. Failing to do so or taking shortcuts usually results in inaccurate conclusions.

Diagnostic Circuit Check. In diagnosing a problem on an EFI system, after verifying that all non-EFI engine support systems are working properly, start with a diagnostic circuit check, Figure 4-22. The diagnostic circuit check either verifies that the system is working properly or refers to another chart for further

Chapter 4 General Motors' Electronic Fuel Injection

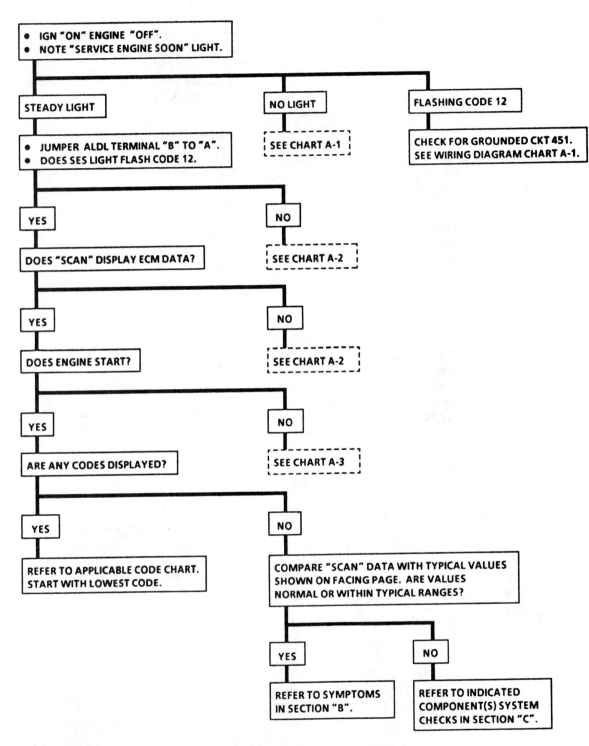

FIGURE 4-22 Diagnostic circuit check. *Courtesy of Buick Motor Division.*

Chapter 4 General Motors' Electronic Fuel Injection

FIELD SERVICE MODE: ENGINE RUNNING, DIAGNOSTIC TERMINAL GROUNDED. "CHECK ENGINE" LIGHT IS "OFF" WHEN EXHAUST IS LEAN AND "ON" WHEN IT IS RICH.
OPEN LOOP: "CHECK ENGINE" LIGHT FLASHES AT A RATE OF 2 TIMES PER SECOND.
CLOSED LOOP: "CHECK ENGINE" LIGHT FLASHES AT A RATE OF 1 TIME PER SECOND.
LEAN EXHAUST INDICATION: "CHECK ENGINE" LIGHT IS OUT ALL THE TIME OR MOST OF THE TIME.
RICH EXHAUST INDICTION: "CHECK ENGINE" LIGHT IS ON ALL THE TIME OR MOST OF THE TIME.

* OPEN LOOP AT IDLE ONLY MAY BE CAUSED BY A COLD OXYGEN SENSOR AFTER A PERIOD OF IDLING. IF THIS IS THE CAUSE, RUNNING ENGINE AT 2000-3000 RPM FOR ONE MINUTE SHOULD WARM IT UP AND GIVE CLOSED LOOP OPERATION AT IDLE.

AFTER ANY REPAIR, CLEAR CODES AND CONFIRM "CLOSED LOOP" OPERATION.

FIGURE 4-23 Field service mode. *Courtesy of GM Product Service Training.*

Chapter 4 General Motors' Electronic Fuel Injection

1982-1983
CHART 1
FUEL INJECTION
DRIVER'S COMPLAINT

- INTERMITTENT "CHECK ENGINE" LIGHT OR STORED CODES.

    NOTICE: DO NOT USE DIAGNOSTIC CHARTS FOR INTERMITTENT PROBLEMS. THE FAULT MUST BE PRESENT TO LOCATE THE PROBLEM. IF THE FAULT IS INTERMITTENT, USE OF THE CHARTS MAY RESULT IN THE REPLACEMENT OF NON-DEFECTIVE PARTS.

    MOST INTERMITTENT PROBLEMS ARE CAUSED BY FAULTY ELECTRICAL CONNECTORS OR WIRING. DIAGNOSIS MUST INCLUDE A CAREFUL VISUAL AND PHYSICAL INSPECTION OF THE INDICATED CIRCUIT WIRING AND CONNECTORS.

    - POOR MATING OF THE CONNECTOR HALVES OR TERMINALS NOT FULLY SEATED IN CONNECTOR BODY (BACKED OUT "TERMINALS").
    - IMPROPERLY FORMED OR DAMAGED TERMINALS. ALL CONNECTOR TERMINALS IN PROBLEM CIRCUIT SHOULD BE CAREFULLY REFORMED TO INCREASE CONTACT TENSION.
    - HEI DISTRBUTOR EST WIRES SHOULD BE ROUTED AWAY FROM DISTRIBUTOR, IGNITION COIL, SECONDARY WIRING AND GENERATOR.
    - CKT 419- "CHECK ENGINE LAMP" TO ECM, SHORT TO GROUND.
    - CKT 451- DIAGNOSTIC CONNECTOR TO ECM, SHORT TO GROUND.
    - CKT 450 AND 450R- CHECK ECM GROUND AT ENGINE BLOCK ATTACHMENT.
    - ELECTRICAL SYSTEM INTERFERENCE CAUSED BY A DEFECTIVE RELAY, ECM DRIVEN SOLENOID, OR A SWITCH CAUSING A SHARP ELECTRICAL SURGE. NORMALLY, THE PROBLEM WILL OCCUR WHEN THE DEFECTIVE COMPONENT IS OPERATED.
    - IMPROPER INSTALLATION OF ELECTRICAL OPTIONS, I.E. LIGHTS, 2 WAY RADIO, ETC.
    - OPEN AIR CONDITIONING CLUTCH DIODE.

- STALLING, ROUGH OR IMPROPER IDLE SPEED - SEE CHART 11.

- ENGINE CRANKS BUT WILL NOT RUN- SEE CHART 4 OR 4A.

- HARD STARTING, POOR PERFORMANCE, DRIVABILITY, OR FUEL ECONOMY- SEE CHART 7.

- DETONATION (SPARK KNOCK)
    - ESC PERFORMANCE CHART 10, IF APPLICABLE.
    - EGR CHART 8.

- POOR ENGINE PERFORMANCE WITH AIR CONDITIONING ON.
    REFER TO SECTION 1B.

FOLLOWING ANY REPAIRS OR ADJUSTMENTS, ALWAYS CLEAR CODES AND CONFIRM "CLOSED LOOP" OPERATIONS AND NO "CHECK ENGINE" LIGHT.

FIGURE 4-24 Driver's complaint chart. *Courtesy of GM Product Service Training.*

diagnosis. Be sure that the diagnostic circuit check and trouble code charts come from the same model year manual as the vehicle being worked on. On earlier systems the diagnostic circuit check is designed to sort out whether or not the fault currently exists or if it is an intermittent fault present at one time but not now. On those systems the trouble code charts are written assuming that a fault exists. If one of them is used to pursue an intermittent code, the chart most often leads to an invalid conclusion because it is beginning with an invalid assumption—that a fault currently exists when in fact none does.

For later systems the diagnostic circuit check and many of the trouble code charts are written to allow the trouble code charts to determine if the code obtained from the diagnostic memory identifies a current fault (hard code) or an intermittent code.

Field Service Mode. Grounding the test terminal with the engine running puts the system into field service mode. In this mode the ECM uses the check engine light to show whether the system is in open or closed loop and whether the system is running rich or lean, Figure 4-23.

The field service mode was incorporated into the diagnostic circuit check in 1984 and was discontinued when the diagnostic procedures were written around the use of a scanner. The scan tool can provide the information the field service mode was designed to provide.

Driver's Complaint Chart. The driver's complaint chart, Figure 4-24, acts as a guide to assist in finding the cause of problems not identified by trouble codes. It either suggests things to check for certain problems or refers to other, more specific charts. For later models this chart has been expanded into a complete section in the service manual and is referred to as *Symptoms*. There are other charts in the later service manuals that assist in diagnosing specific EFI component or system malfunctions.

1 ECM 3 PROM CARRIER
2 PROM (ENGINE CALIBRATOR) 4 CALPAC

FIGURE 4-25 PROM and CALPAK in ECM. *Courtesy of Buick Motor Division.*

ECM and PROM Service

Considerable care should be taken when installing a PROM or CALPAK (only 2.0-liter engine ECMs are equipped with a *CALPAK*). A finger coming into contact with its pins can discharge static electricity into it and damage it, Figure 4-25. The PROM carrier fits into the PROM cavity of the ECM in only one position; however, the PROM fits into the carrier in either of two ways. If the PROM is installed in the ECM end for end from its correct position, it will be damaged electrically. Consult the service manual for installation instructions.

The PROM, CALPAK, and ECM are serviced separately; therefore, remove the PROM and CALPAK before exchanging the ECM for a new one.

If the CALPAK is missing or severely damaged, the engine does not start.

Weather-pack Connectors

All EFI system harness connections under the hood are made with weather-pack connectors as discussed in the **System Diagnosis and Service** section of Chapter 3.

Diagnostic and Service Tips

TCC Test Terminal. See **TCC Test Terminal** in the **System Diagnosis and Service** section of Chapter 3.

Chapter 4 General Motors' Electronic Fuel Injection

16	FUEL INJECTOR
A	FILTER
B	LARGE "O" RING
C	STEEL BACK-UP WASHER
D	SMALL "O" RING

FIGURE 4-26 EFI injector assembly. *Courtesy of Buick Motor Division.*

Vacuum Leaks. A vacuum leak on an EFI system usually results in an increased idle speed while in closed loop.

TBI. The TBI unit has two O-rings, Figure 4-26, between the injector solenoid and the fuel meter body. The O-rings prevent pressurized fuel from leaking past the injector solenoid. A leak at either of these O-rings can result in dieseling and/or an overly rich mixture. A suspected leak can easily be checked. With the engine off and a jumper lead supplying battery voltage to the fuel pump test terminal, inspect the injector for signs of leakage.

PERSONAL SAFETY: Fuel Pressure Test. Some driveability complaints require a fuel pump pressure test. Before the fuel system is opened, pressure in the line should be relieved; the system is designed to retain operating pressure in the line after the engine is shut off. Opening the line under pressure causes a considerable spray, resulting in a fire and safety hazard. To relieve the pressure, remove the fuel pump fuse and crank the engine for several seconds. See the service manual for test procedures.

Erasing Learned Ability. Anytime battery power is removed from the ECM, its learned memory is lost. This can produce a noticeable change in performance. Learning can be restored by driving the car for usually not more than half an hour at normal operating temperature, part throttle and idle, and at moderate acceleration.

REVIEW QUESTIONS

1. What are the six inputs most involved in engine calibration control?
2. What is the second function of the MAP sensor? Under what conditions does it perform this function?
3. Name three functions that the VSS influences.
4. What influence does the A/C switch have on engine operation?
5. What introduces fuel to the engine on EFI applications?
6. How is fuel quantity controlled?
7. Describe the ECM's fuel system control response to each of the following conditions on a four-cylinder EFI application:
   a. the ignition is turned on
   b. no REF signal is sent to the ECM within two seconds

Chapter 4   General Motors' Electronic Fuel Injection

8. What variable control controls the amount of air going into the intake manifold during idle?
9. During what two operating conditions can the antistall mode be activated?
10. What signal alerts the ECM to move the IAC valve back to its parked position?
11. What is the power source for most output devices?
12. What type of device is used by the ECM to activate the fuel pump, cooling fan, and A/C clutch?
13. What are the first two steps when pursuing a driveability complaint on an EFI-equipped vehicle?
14. What is the function of the weather-pack connector?
15. What two functions does the TCC test lead serve?
16. Name six things that should be observed to avoid damage to the ECM.

ASE-type Questions. (Actual ASE test questions will probably not be this product specific.)

17. Technician A says that EFI systems that use a Hall-effect switch do not use a pickup coil in the distributor. Technician B says that the Hall-effect switch, on those EFI systems that use it, provides the signal to the HEI module to control primary circuit switching. Who is correct?
    a. A only
    b. B only
    c. both A and B
    d. neither A nor B

A/C	Air-conditioning
AIR	Air Injection Reaction
ALCL	Assembly Line Communication Link
BARO	Barometric Pressure
CCP	(Charcoal) Canister Purge
DIAGNOSTIC TEST TERMINAL	Lead or ALCL connector terminal grounded to obtain a trouble code
ECM	Electronic Control Module
EFI	Electronic Fuel Injection
EGR	Exhaust Gas Recirculation
ESC	Electronic Spark Control
EST	Electronic Spark Timing
HEI	High-energy Ignition
IAC	Idle Air Control
MAP	Manifold Absolute Pressure
PCV	Positive Crankcase Ventilation
P/N	Park/Neutral Switch
PROM	Programmable Read-only Memory (engine calibration unit)
TBI	Throttle Body Injection
TCC	Torque Converter Clutch
TPS	Throttle Position Sensor
VIN	Vehicle Identification Number
VSS	Vehicle Speed Sensor
WOT	Wide-open Throttle

FIGURE 4-27   Explanation of abbreviations

18. The engine of a car equipped with an EFI system dies while being driven. The car is brought into the garage, and it is determined that the ignition system is operating properly, the fuel tank is at least half full, and there appears to be no problem in the fuel lines. The ECM is working properly, and no fault codes are stored in it. Technician A says that the problem could be a fault in the reference pulse wire from the distributor. Technician B says that the fault could be a blown fuel pump fuse. Who is correct?
    a. A only
    b. B only
    c. both A and B
    d. neither A nor B
19. A car equipped with EFI is brought in with a complaint of poor mileage and black smoke during some driving conditions. A fuel pressure check shows that the fuel pressure is too high. The cause is most likely _____ .
    a. a faulty fuel pump
    b. a faulty fuel pressure regulator
    c. a faulty coolant temperature sensor
    d. a faulty ECM
20. Technician A says that an EFI system is put into field service mode by grounding the test terminal with the ignition on and the engine off. Technician B says that the same system is put into field service mode by grounding the test terminal with the engine running. Who is correct?
    a. A only
    b. B only
    c. both A and B
    d. neither A nor B

GLOSSARY

In addition to the following terms, refer to Figure 4-27 for a list of the abbreviations used in this chapter.

CALPAK A removable part of the ECM that plugs in like the PROM and is primarily responsible for fuel calibration.

PICKUP COIL A small coil of wire in which the signal of a pulse magnetic generator is produced, such as in most electronic distributors.

TBB (Throttle Body Backup) A backup circuit in the ECM of most EFI systems that operates the injector solenoid in the event of a partial system failure.

5

GENERAL MOTORS' PORT FUEL INJECTION

Objectives

After studying this chapter, you will be able to:

- name at least two advantages of port fuel injection over other types of fuel induction systems.
- describe two methods used by General Motors' PFI systems to measure a value equivalent to air's mass.
- describe the operation of a cold start valve.
- describe the operation of a turbocharger waste gate.
- describe at least one type of EGR diagnostic switch.
- describe how the idle air control valve controls idle speed.
- describe how the computer becomes aware that spark knock is occurring.
- describe how the EGR valve is modulated on PFI systems.

Port fuel injection (PFI) represents the fourth generation of General Motors' comprehensive computerized engine control systems, excluding the Cadillac digital fuel injection discussed in Chapter 6. It was introduced, with limited application on 3.8-liter and 1.8-liter engines, in 1984. It followed the electronic fuel control (General Motors' first closed-loop fuel control system) introduced in 1978, the CCC system introduced in 1980, and the EFI system introduced in 1982. In 1985 the PFI system was expanded to several engine applications, Figure 5-1. Some General Motors' car divisions have chosen a special name for the PFI system because of some special feature designed for a given engine or body application, such as:

- sequential fuel injection (SFI) on the turbocharged 3.8-liter engine.
- multiport fuel injection (MFI) on most other engine applications.
- tuned port injection (TPI) on the Corvette 5.7-liter engine. The TPI version offers manifold runners matched in shape, length, and cross-sectional area to help maximize *volumetric efficiency* (VE).

Because only air moves through the intake manifold (there is no concern about holding fuel in a vapor state as it passes through the runners), the heated air *(THERMAC)* and the early fuel evaporation systems are eliminated.

Chapter 5 General Motors' Port Fuel Injection

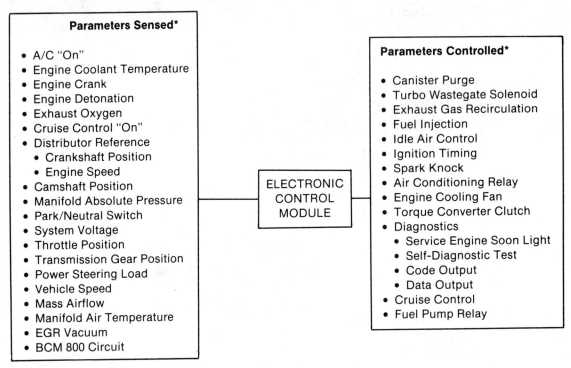

*Not all systems used on all engines.

FIGURE 5-1 PFI system overview

Most of the PFI applications, however, do run engine coolant through a passage in the throttle body to prevent throttle blade icing.

ELECTRONIC CONTROL MODULE

The ECM is usually located under the instrument panel or behind the passenger kick panel. On the P car (Pontiac Fiero) it is in the console between the seats. It contains a removable PROM and CALPAK. The CALPAK provides calibration for cold-start cranking and for fuel backup in the event of an ECM failure. The fuel backup circuit (FBC) provides an operating mode when any of the following conditions occur:

- ECM voltage is below 9 volts (most likely occurs during cranking)
- the PROM is missing or not functioning
- the ECM is unable to provide its normal computer-operated pulses (COP)

The FBC makes use of the throttle position sensor, the coolant temperature sensor, and the engine RPM signal and is powered by the ignition switch. During FBC mode the engine runs erratically and sets a code 52.

Keep Alive Memory (KAM)

The ECM has basically the same self-diagnostic capability and KAM as discussed in Chapter 3. The inputs fed into the ECM and the functions controlled for most systems are shown in Figure 5-1.

OPERATIONAL MODES

The different operating modes of the PFI system control how much fuel to introduce into the intake manifold for various operating conditions.

Starting Mode

When the ignition is turned on, the ECM turns the fuel pump relay on. If it does not see a signal within two seconds telling it that the engine is being cranked, it turns the fuel pump relay off. The fuel pump provides fuel pressure to the injectors. As the system goes into cranking mode, the ECM checks coolant temperature and throttle position to see what the air/fuel ratio should be. The air/fuel ratio ranges from 1.5 to 1 at −36°C (−33°F) to 14.7 to 1 at 94°C (201°F). The ECM varies the air/fuel ratio by controlling the time that the injectors are turned on; this is referred to as pulse width.

The 2.8-, 5.0-, and 5.7-liter engine PFI applications also use a cold-start injector. This is an additional injector in the intake manifold that improves cold-start characteristics by spraying additional fuel into the manifold during cranking, Figure 5-2. It is not ECM controlled. Spraying the additional fuel for cold-start enrichment into the manifold gives it more opportunity to evaporate than if it were sprayed into the intake ports. The injector is activated by the cranking system and controlled by a timing mechanism. It is described in more detail in the **Outputs** section of this chapter.

Clear Flood Mode

The clear flood mode works essentially the same as it does on the EFI system. As long as the engine is turning less than 600 RPM and the throttle is held to within 80% of full throttle, the ECM will hold the air/fuel ratio to 20 to 1, except on the 2.8-liter engine, where fuel is cut off completely.

FIGURE 5-2 Cold-start injector (typical). *Courtesy of GM Product Service Training.*

Run Mode

The run mode consists of the open- and closed-loop operating conditions. When the engine is started and the RPM is above 400, the system goes into open loop. In open loop the ECM ignores information from the O_2 sensor and determines air/fuel ratio commands based on input from other sensors (coolant, vacuum or mass airflow, and throttle position and engine speed if the vacuum sensor is used). The system stays in open loop until:

- the O_2 produces a varying voltage showing that it is hot enough to work properly.
- the coolant is above a specified temperature.
- a specified amount of time has elapsed since the engine was started.

These values vary with engine application and are contained in the PROM. When all three conditions are met, the ECM puts the system into closed loop. In closed loop the ECM uses only O_2 sensor input to calculate air/fuel ratio commands and keeps the air/fuel ratio at a near perfect 14.7 to 1. During heavy acceleration, wide-open throttle, or hard deceleration, the system temporarily drops out of closed loop.

Acceleration Mode

Rapid increases in throttle opening and manifold pressure or airflow signal the ECM to enrich the air/fuel mixture. This compensates for the reduced evaporation rate of the gasoline that results from the higher manifold pressure.

Deceleration Mode

Rapid decreases in throttle opening and manifold pressure or airflow cause the ECM to lean out the air/fuel mixture. If the changes are severe enough, fuel is momentarily cut off.

Battery Voltage Correction Mode

The battery voltage correction feature on the PFI system is different from that on the EFI system. On the PFI system, when battery voltage drops below a specified value, the ECM will:

- enrich the air/fuel mixture
- increase throttle opening, if the engine is idling
- increase ignition dwell

to compensate for a weakened ignition spark.

Fuel Cutoff Mode

When the ignition is turned off, the ECM immediately stops pulsing the injectors; this is done to prevent dieseling. Injection also is stopped anytime the distributor reference pulse stops coming to the ECM.

FUEL SUPPLY SYSTEM

Fuel pressure is supplied by an electric pump in the fuel tank, Figure 5-3. The pump is turned on and off by a fuel pump relay that is controlled by the ECM. A pressure relief valve in the pump limits maximum pump pressure to between 60 and 90 PSI. This pressure is realized only if flow stops and the pump is working against static pressure. The filter mounted to the bottom of the pump assembly is a 50-micron filter; downstream is a 10- to 20-micron in-line filter. Because of the high operating pressure, an O-ring is used at all threaded connections and all flex hoses have internal steel reinforcement.

Fuel pressure is controlled by the fuel pressure regulator, Figure 5-4, which is usually mounted on the fuel rail, Figures 5-5 and 5-6. Manifold vacuum is supplied to the spring side of the diaphragm. At light throttle the high vacuum helps to pull the diaphragm up and against the spring. This allows more fuel to return to the tank and thus reduces output pressure. Pressure ranges from 34 PSI at idle to 44 PSI at full throttle. A constant stream of fuel flowing from the tank to the pressure regulator and back to the tank through the return line ensures

Chapter 5 General Motors' Port Fuel Injection

FIGURE 5-3 Fuel pump and sending unit assembly. *Courtesy of GM Product Service Training.*

cool fuel available to the fuel rail and injectors. This plus the operating pressure eliminate the chances of vapor lock.

INJECTORS

An injector is installed in the intake port of each cylinder with 70 millimeters to 100 millimeters between the tip of the injector and the center of the valve on V-type engines, Figure 5-7. The nozzles spray fuel in a 25° conical pattern. O-rings are used to seal between the nozzle and the fuel rail and between the nozzle and the intake manifold. The O-rings also serve to retard heat transfer from the engine to the nozzle and to reduce nozzle vibration. The O-rings should be lubricated and/or replaced as necessary when the nozzles are removed. A damaged or hardened O-ring allows a fuel or vacuum leak; PFI systems are very sensitive to vacuum leaks.

On all PFI systems except the sequential fuel injection (SFI) version used on the Buick-built, 3.8-liter turbocharged engine in 1984 and 1985 and on all Buick-built, 3.8-liter engines in 1986, all of the injectors are pulsed simultaneously every crankshaft revolution. Fuel sprayed while the intake valve is closed is simply stored

Chapter 5 General Motors' Port Fuel Injection

FIGURE 5-4 Fuel pressure regulator. *Courtesy of GM Product Service Training.*

FIGURE 5-5 Fuel rail: 1.8 L, 3.0 L, and 3.8 L. *Courtesy of GM Product Service Training.*

Chapter 5 General Motors' Port Fuel Injection

FIGURE 5-6 Fuel rail: 2.8 L, 5.0 L, and 5.7 L. *Courtesy of GM Product Service Training.*

in the intake port until the valve opens. The SFI system pulses each injector one at a time in the firing order.

THROTTLE BODY

The throttle body on a PFI system controls the amount of air that is allowed into the engine's induction system. This is done with a throttle blade and shaft controlled by the accelerator pedal, just as in a carburetor or TBI unit, Figures 5-8 and 5-9. In this case, however, only air passes the throttle blade. An idle stop screw determines how nearly closed the throttle blade is at idle. A throttle position sensor keeps the ECM informed as to the throttle valve's position. Idle speed is controlled by an idle air control motor and valve assembly. It controls the amount of air that is allowed to bypass the throttle blade through a bypass

165

Chapter 5 General Motors' Port Fuel Injection

FIGURE 5-7 Port fuel injection (typical). *Courtesy of GM Product Service Training.*

passage and is controlled by the ECM. Most throttle body units have a cavity for engine coolant to flow through to prevent throttle blade icing.

NON-ECM EMISSION CONTROL

A trend seems to be developing whereby as the computerized engine control systems become more efficient, fewer of the traditional emission control devices are needed. This is evidenced by the following list of non-ECM-controlled emission control devices during the first or second year of PFI production.

- 1.8 liter—canister purge, PCV, and internal EGR (no EGR valve)
- 2.8 liter—PCV and deceleration valve (lets additional air in intake manifold during deceleration on manual transmissions only)
- 3.0 and 3.8 liter—PCV
- 5.0 and 5.7 liter—PCV

C^3I System

For years automotive engineers have generally felt that it did not make much difference when the fuel was sprayed into the intake port as long as precisely the right amount was sprayed. When you consider how fast all of this is happening (the intake valve opens and closes twenty-five times per second at 3,000 RPM), it would not seem to make much difference. The use of the three-coil ignition system (C^3I), introduced in 1984 by Buick, required a sequential fuel injection system because the C^3I system fires each spark plug each time the plug's respective piston approaches top dead center. As the piston finishes the exhaust stroke and starts the intake stroke, the spark plug fires while both valves (intake and exhaust) are open. This firing serves no intended purpose; it is just a by-product of the ignition system design. If, however, a pressurized, combustible mixture (the first application of the C^3I system was on a turbocharged engine) were standing on the other side of the open intake valve, a backfire could occur in the intake manifold. To avoid this, each injector pulse is timed so that the fuel is immediately purged from the manifold by its respective cylinder's intake stroke.

The additional circuitry and related expense seems to have proved to be worthwhile; however, as GM has rapidly expanded the sequential fuel injection system to other engines, Ford introduced a sequential fuel injection system in 1986 on some 5-liter applications and has expanded its use of sequential injection to other engines since then.

Chapter 5 General Motors' Port Fuel Injection

FIGURE 5-8 Throttle body assembly, 2.8 L. *Courtesy of GM Product Service Training.*

FIGURE 5-9 Throttle body assembly, 3.0 L. *Courtesy of GM Product Service Training.*

Chapter 5 General Motors' Port Fuel Injection

INPUTS

Coolant Temperature Sensor (CTS)

This is the same sensor used in the computer command control system, the electronic fuel injection system, and other General Motors' systems. The CTS input affects the control of:

- air/fuel mixture (in open loop).
- spark timing.
- spark knock control (on some engine applications).
- engine idle.
- torque converter clutch.
- canister purge (except 1.8 liter as of 1985).
- EGR (except 1.8 liter as of 1985).
- cooling fan (on some applications).

Mass Airflow (MAF) Sensor

Many General Motors' PFI systems either use or have used a MAF sensor. In the mid-1980s, two MAF sensors were used: one on Buick engines and one on Chevrolet engines. Buick engines used a MAF manufactured by AC, a division of General Motors, while Chevrolet engines used a Bosch unit. In the late 1980s, the AC unit was discontinued and a unit built by Hitachi of Japan was used on Buick engines. Chevrolet engines were no longer equipped with a MAF sensor.

AC. The AC unit, Figure 5-10, contains a screen to break up the airflow. After the air passes through the screen, it flows over an air temperature sensing resistor. A sample tube then directs some of the air to flow over a heated foil sensing element. The power to heat the sensing element comes from a fuse (a relay as on some applications), Figure 5-11. The circuitry of the MAF sensor controls current flow through the foil sensing element to maintain it at 75° C above the incoming air temperature, as measured by the temperature-sensing resistor. The power required to keep the foil sensing element 75° C above incoming air temperature is a measure of mass airflow. This value is sent to the ECM as a digital signal ranging in frequency from 30 cycles per second (30 hertz, or 30 Hz) to 150 Hz (150 Hz represents the highest mass airflow rate). The ECM compares the MAF sensor information to a preprogrammed look-up chart and finds airflow in grams per second. Using this value, engine temperature, and engine RPM, the ECM can determine precisely how much fuel is required to achieve the desired air/fuel ratio. Mass airflow readings are taken and air/fuel mixture calculations are made about 160 times per second.

Bosch. The Bosch MAF sensor works much the same way except that a wire heat element is used instead of a foil sensing element. Also, each time the ignition is turned off after the system has been in closed loop, a separate burn-off module, Figure 5-12, momentarily puts enough current through the wire heat element to make it red hot. This burns off any residue accumulation. It is very critical that its surface be clean because an accumulation of deposits will retard its ability to transfer heat.

Manifold Air Temperature (MAT) Sensor

The MAT sensor looks like and is essentially a coolant temperature sensor with its thermistor sensing element partially exposed instead of being fully enclosed like the CTS. It screws into a hole in the intake manifold, or in some cases into the air cleaner housing, with its nose extending inside. It is sometimes referred to as an air temperature sensor (ATS). It is not used on all applications. Where it is used, it influences air/fuel mixture, spark timing, and idle control.

Throttle Position Sensor (TPS)

The TPS is a potentiometer mounted on the side of the throttle body, Figure 5-9, and attached to the end of the throttle shaft. Its input affects most ECM-controlled functions,

Chapter 5 General Motors' Port Fuel Injection

FIGURE 5-10 Mass airflow sensor (A/C type). *Courtesy of GM Product Service Training.*

FIGURE 5-11 MAF relay circuit. *Courtesy of GM Product Service Training.*

Chapter 5 General Motors' Port Fuel Injection

FIGURE 5-12 Bosch MAF circuit and burn-off module. *Courtesy of GM Product Service Training.*

System	Engine	PM Reluctor	Hall Effect Switches			
		Crank Pos. Sensors	CAM/SYNC Sensors			
		On Crank	Harmonic Balancer	Harmonic Balancer	Dist. Hole	Timing Gear	18x
DIS	2.0	X					
	2.5	X					
	2.8	X					
IDI	2.3	X					
C³I-I	3.0		X	X^C			
I or II^A	3.8 SFI		X			X	
I or II	3.8 TURBO		X		X		
III^B	3.800		X			X	X

A — Type I or II may be used on 3.8 liter engines. Type I has molded one piece coil pack. Type II has independent coils.

B — Introduced in 1988, has independent coils.

C — Now called a "sync" sensor rather than a "cam" sensor.

FIGURE 5-13 General Motors distributorless ignition system applications including fuel control, spark timing, idle air control, torque converter clutch, and air-conditioning control.

Reference Pulse (REF)

On all but those engines equipped with a distributorless ignition system, the REF signal comes from the distributor pick-up coil. Without this signal the engine does not start or run because the ECM will not pulse the injectors without it. A more complete discussion of the REF signal is found in the **Inputs** section of Chapter 3.

The distributorless ignition systems use sensors to provide information to the ignition module relative to crankshaft and camshaft position and speed, Figure 5-13. The C³I (Computer-Controlled Coil Ignition) systems use Hall-effect switches while both the DIS (Direct Ignition System) and the IDI (Integrated Direct Ignition) systems use a permanent magnetic pulse generator, Figure 5-14.

C³I. There are three different versions of the C³I system: Type I, Type II, and Type III. Each of the three types uses a Hall-effect switch mounted behind the crankshaft balancer, Figure 5-15. The balancer has three vanes, which are positioned to form a circle, that extend back toward the engine. As the crankshaft turns, the vanes alternately pass through the Hall-effect switch, producing a signal that is fed to both the ignition module and the ECM. The Type I system when used on the 3.0 liter engine has a second Hall-effect switch mounted behind the crankshaft balancer with a single vane extending back, nearer the outer circumference of the balancer, that acts as the camshaft sensor. The Type I or Type II systems, when used on the 3.8 liter Turbo engine, feature a Hall-effect switch mounted in

FIGURE 5-14

Chapter 5   General Motors' Port Fuel Injection

FIGURE 5-15   Crankshaft position sensor (3.0 L, SFI). *Courtesy of GM Product Service Training.*

FIGURE 5-16   Camshaft position sensor (3.8L, SFI). *Courtesy of GM Product Service Training.*

what used to be the distributor hole to serve as a camshaft sensor, Figure 5-16. It is driven by the camshaft. These same two systems, when used on the 3.8 liter, non-turbo, SFI engine use a Hall-effect cam sensor mounted in the timing cover and triggered by a vane attached to the camshaft timing gear. The cam sensor signal is used to alert the ECM when the number one cylinder is at TDC so that it can start the firing sequence over again and to time injector pulsing. Loss of this signal will result in a code 41 being set and if the signal is lost during cranking, the engine will not start. In more recent General Motors literature, the camshaft sensor is referred to as a "sync sensor."

On the Type III system, sometimes referred to as the "Fast Start," which also has its sync sensor located in the timing cover, a third sensor is used. It is a Hall-effect switch and is located behind the harmonic balancer along with the crankshaft sensor. This sensor, referred to as the "Crank 18x," is triggered by 18 evenly spaced interruptor vanes protruding rearward from near the outer circumference of the balancer. The Crank 18x sensor provides 18 signals, Figure 5-17, per crankshaft revolution to the ignition module that, in turn, sends it on to the ECM.

On this system, the crank sensor is called the "Crank 3x" sensor, and its three interruptor vanes, nearer the center on the balancer, are not symmetrical as they are on the other C^3I systems. The windows between the vanes are 10, 20, and 30 degrees wide as shown in Figure 5-17. The signal produced by the leading and trailing edge of each window constitutes a 3x pulse. The number of 18x signals that occurs during each 3x pulse enables the ECM to determine which 3x pulse it is reading. There will be one 18x signal during the 10° window pulse, two during the 20° window pulse, and three during the 30° window pulse. This enables the system to fire a coil for the appropriate cylinders within the first 120° of crankshaft rotation. This system provides the following advantages:
- faster starts
- more accurate REF signals to the ECM, especially at low engine speed
- increased run reliability as the engine will continue to operate without the cam sensor
- potential for the ECM to read crankshaft acceleration and deceleration rates

Chapter 5 General Motors' Port Fuel Injection

FIGURE 5-17 Combination sensor for C³I-(Type 3)-Fast Start 1

DIS and IDI. Both the Integrated Direct Ignition (IDI) and the Direct Ignition System (DIS) systems use a magnetic pulse generator on the crankshaft instead of the Hall-effect switches for crankshaft reference (position and speed). The crankshaft for those engines which use either the DIS and the IDI system has a round machined steel disc that is concentric to the crankshaft's centerline for its center counterweight, Figure 5-18. This disc has seven notches cut in it and is called a reluctor. The term "reluctor" refers to any device that changes the reluctance of a material to conduct magnetic lines of force. The same reluctor is used on both four- and six-cylinder engines. The crankshaft sensor is mounted to the engine block and extends through a hole in the crankcase. Its tip, containing a permanent magnet and a wire coil, is spaced .050 inch from the reluctor.

Six of the reluctor's seven notches are spaced evenly around it at 60-degree intervals,

173

Chapter 5 General Motors' Port Fuel Injection

FIGURE 5-18 Crankshaft timing disc

Figure 5-14. The seventh notch is called the "sync" notch and is located between notches 6 and 1, 10 degrees from notch 6 and 50 degrees from notch 1. By being placed at an odd position in relation to the other notches, the sync notch provides an irregular signal that the module can identify. The module uses this signal to synchronize the coil firing sequence to the crankshaft's position in the engine cycle.

On both the DIS and the IDI systems, the notch that produces the signal that results in the appropriate coil being fired during module mode operation is often referred to as the "cylinder event" notch. The crank sensor signal that occurs 60 degrees before the cylinder event notch causes voltage in the REF wire to be low. The cylinder event notch causes voltage in the REF wire to be high. When the engine is operated in the electronic spark timing mode, the ECM uses this signal as an input for calculating:

- ignition timing while operating in EST mode.
- fuel injector pulses.

Oxygen (O_2) Sensor

This is the same single-wire sensor as used on other General Motors' vehicles.

Vehicle Speed Sensor (VSS)

Two types of VSSs are used and are the same as those discussed in Chapter 3.

EGR Diagnostic Switch

On PFI systems that control EGR operation, an EGR diagnostic switch is used to tell the ECM if the EGR is actually being applied when it is commanded to be applied. Two types are used. The four- and six-cylinder engines use the same vacuum-operated switch described in Chapter 3.

The 5.0- and 5.7-liter engines use a thermal switch for exactly the same purpose as the vacuum switch used on the smaller engines. It, however, screws into the base of the EGR valve, Figure 5-19. If the EGR valve is open, the heat of the exhaust gases, flowing past the heat-sensing tip of the thermal switch, cause the switch to close. The ECM wants to see 12 volts on circuit 935 when the EGR is supposed to be off and less than 1 volt when the EGR is supposed to be on.

Detonation Sensor

Most of General Motors' PFI systems incorporate an electronic spark control (ESC) system that uses a piezoelectric knock sensor.

Park/Neutral (P/N) Switch

The park/neutral switch, which can be located near the shifter assembly or in the transmission, is a normally open switch. When the vehicle is shifted into park or neutral, it closes. The ECM monitors the P/N switch and knows if the transmission is in gear or not. This information is used to help control engine idle and EGR control.

Air-conditioning (A/C) Switch

On PFI cars equipped with air-conditioning, the A/C control switch on the instrument

Chapter 5   General Motors' Port Fuel Injection

FIGURE 5-19   EGR valve with thermal diagnostic switch. *Courtesy of GM Product Service Training.*

FIGURE 5-20   A/C control circuit with P/S switch (typical). *Courtesy of GM Product Service Training.*

175

Chapter 5 General Motors' Port Fuel Injection

panel is connected by a wire to the ECM, Figure 5-20. When the A/C is turned on, the voltage signal travels from the A/C switch to the pressure cycling switch to the ECM. Some systems also put the high-pressure cut-out switch in series in the same circuit. This signal tells the ECM that the A/C switch is on and the pressure cycling switch is closed. The ECM uses this information to turn on the A/C relay and/or adjust the IAC valve.

Transmission Switches

Transmissions used in PFI-equipped vehicles have one or more hydraulically operated electric switches screwed into the valve body. These switches provide the ECM with signals that tell it what gear the transmission is in. This information enables it to more effectively control torque converter clutch operation.

Cranking Signal

On some engines a wire from the starter solenoid circuit feeds a cranking signal to the ECM to alert it that the engine is being cranked. This puts the system in cranking mode, and an enriched air/fuel mixture is delivered by the injectors for easier starting. On other engines the cranking signal is fed to a cold-start injector in the intake manifold and provides necessary starting fuel enrichment. The cold-start injector is not ECM controlled.

System Voltage/Fuel Pump Voltage

The ECM monitors system voltage through one of its battery voltage inputs. If voltage drops below a preprogrammed value, the system goes into battery voltage correction mode. Some PFI systems have a wire from the positive side of the fuel pump power circuit to the ECM, Figure 5-21. This signal is used to make fuel delivery compensations based on system voltage and causes a code 54 to be stored if pump voltage is lost while the engine is running.

Power Steering (P/S) Switch

A power steering switch is used in PFI systems to alert the ECM when power steering pressure is high enough to significantly affect engine idle performance, Figure 5-22. The P/S switch is normally open and is closed by high P/S pressure. When the switch is open, feedback voltage to the ECM is about 12 volts; when the switch is closed, voltage is less than 1 volt. When the switch closes, the ECM increases idle air bypass and retards ignition timing. On some systems the A/C compressor clutch is also disengaged.

Body Computer Module (BCM) 800 Circuit

Some late-model PFI applications also feature a body computer module. The BCM controls air-conditioning and other systems such as a driver information center. The driver information center displays information about fuel economy, elapsed time, distance to destination, time, date, and so forth. The BCM shares information with the ECM. For instance, the ECM may send the BCM information about vehicle speed, coolant temperature, or when the vehicle is operating at WOT. The BCM can send the ECM such information as outside air temperature, the temperature in the high-pressure side of the A/C system, or when the rear window defog system is on.

This information is transmitted over a wire between the two computers. The wire is called a *data link*. If information can be transmitted both ways over the same wire, it is called a *bidirectional data link,* or it may be called a *universal asynchronous receiver/transmitter*. General Motors refers to this data link on later model applications as the "BCM 800 Circuit." The BCM 800 Circuit also extends to modules that the BCM or ECM communicates with, Figure 5-23. See the BCM section of Chapter 6 for more information about body computers.

Cruise Control

When cruise control is engaged, the ECM is notified by the cruise control module. The ECM uses this information to modify its control of the torque converter clutch.

Chapter 5 General Motors' Port Fuel Injection

FIGURE 5-21 Fuel pump control circuit (typical). *Courtesy of GM Product Service Training.*

FIGURE 5-22 Power steering switch (typical)

OUTPUTS

Injectors

The fuel injector is a solenoid-operated nozzle controlled by the ECM, Figure 5-24. The ECM pulses it by grounding the return side (low-voltage side) of the injector circuit, Figure 5-25. Fuel under nearly constant controlled pressure is sprayed past the open needle and seat assembly into the intake port of each cyl-

Chapter 5 General Motors' Port Fuel Injection

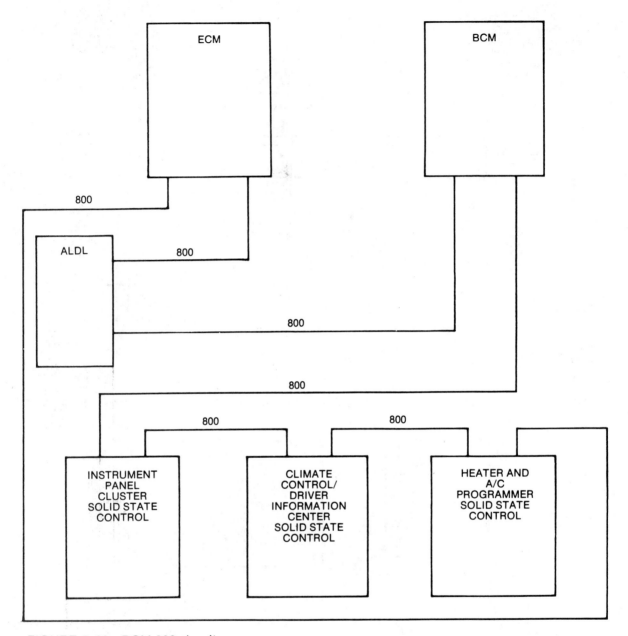

FIGURE 5-23 BCM 800 circuit

inder. On all except SFI applications, all injectors are simultaneously pulsed once, in each crankshaft revolution. The SFI system pulses each injector individually, once per engine cycle, in the firing order, Figure 5-26. During each pulse the injectors are held open for about one or two milliseconds at idle to perhaps six or seven milliseconds at WOT (1 millisecond = 1/1000 of a second).

Multec Injector. Rochester Products, a division of General Motors, has developed a new style injector for multi-point injection sys-

Chapter 5 General Motors' Port Fuel Injection

FIGURE 5-24 Injector (cutaway). *Courtesy of GM Product Service Training.*

FIGURE 5-25 Injector control circuit.

179

Chapter 5 General Motors' Port Fuel Injection

FIGURE 5-26 Injector control circuit for SFI system. *Courtesy of GM Product Service Training.*

FIGURE 5-27 Multec injector

tems, Figure 5-27. It is referred to as a "Multec," which is short for "Multiple Technology Injector." This injector features a lower opening voltage requirement (important during cold-weather cranking), fast response time, better fuel atomization, improved spray pattern control, and less susceptibility to fouling due to fuel blends or contamination. The traditional pintle, as shown in Figure 5-24, is replaced with a ball shape on the solenoid core. When the ball is lifted off its seat, fuel sprays through the seat opening and past the director plate into the intake port.

Cold-start Injector. The cold-start injector is controlled by a thermal vacuum switch screwed into the engine water jacket, Figure 5-28. When the starter solenoid is engaged, voltage is applied to circuit 806 by way of the

FIGURE 5-28 Cold-start injector control circuit (typical). *Courtesy of GM Product Service Training.*

crank fuse. Circuit 806 powers a heat element in the thermo-time switch. Branching off circuit 806 is circuit 832, which supplies voltage to the cold-start injector. Ground for the cold-start injector is provided by a bimetal switch in the thermo-time switch. If the bimetal switch is cool and closed when voltage is applied to the injector, the injector is activated. As soon as voltage is applied to the injector, however, it is also applied to the heat element in the thermo-time switch. If the coolant is at 20°C (68°F) or below, the bimetal switch remains closed for a maximum of eight seconds before the heat from the heat element causes it to open. Therefore, the cold-start injector is turned on for a maximum of eight seconds or less, depending on coolant temperature, while and only while the engine is being cranked. If the engine is started while hot, the injector does not come on.

Electric Fuel Pump

Other than providing a higher fuel pressure, the fuel pump for the PFI system operates and is controlled just like the pump for the EFI system discussed in the **Outputs** section of Chapter 4.

Electronic Spark Timing (EST)/ Distributorless Ignition Systems

EST maintains optimum spark timing under all conditions of engine load, speed, temperature, and air density or mass. The distributor type ignition systems used on PFI engines operate the same as those discussed in the **Outputs** section of Chapter 4, except that none of them use Hall-effect switches.

Given the developments that we have seen in computerized engine controls, there is no need to continue using a distributor with its tendencies toward wear and mechanical failure because:

- reference signals can be taken directly from the crankshaft and camshaft
- an ignition module can be mounted almost anywhere
- spark advance can be more effectively managed by a microprocessor
- secondary spark distribution can be managed by separate coils and a module

Chapter 5 General Motors' Port Fuel Injection

FIGURE 5-29 C³I control circuit. *Courtesy of GM Product Service Training.*

FIGURE 5-30 C³I control circuit. *Courtesy of GM Product Service Training.*

The term "distributorless ignition system" has come to be used as a generic label for an ignition system that does not use a distributor and may be abbreviated in some literature as "DIS." General Motors, however, most often uses the same abbreviation as an abbreviation for "Direct Ignition System," one of their specific distributorless ignition systems. General Motors has introduced several different distributorless ignition systems on various engine applications:

- the C^{3}I (Computer-Controlled Coil Ignition) of which there are three types
- the DIS (Direct Ignition System)
- the IDI (Integrated Direct Ignition)

Each system has its own, unique module that receives information about engine speed, and crankshaft and camshaft position from its sensors. The module monitors the information and passes it on to the ECM. A coil pack, with one ignition coil for each pair of cylinders, is mounted on and controlled by the module, Figure 5-29. The module performs the same function as an HEI module in a distributor type ignition system with the same REF, bypass, EST, and ground circuits connecting the module to the ECM, Figure 5-30. In this case the module must select and fire each coil in the correct sequence. The module and coil pack may be mounted anywhere on the engine. The end of each coil's secondary winding is connected to a spark plug by way of a spark plug wire, Figure 5-31. Each time a coil's primary circuit is opened that coil fires both of its spark plugs simultaneously.

The spark plugs that the coil fires are in companion cylinders; those cylinders arrive at TDC at the same time and are opposite each other in the firing order. For instance, if number 1 cylinder is being fired at the end of its compression stroke (assume the firing order is 1-6-5-4-3-2), number 4 will be at the top of its exhaust stroke and will be fired also. It only takes about 2kV to 3kV, (k stands for kilo, which means 1000, and V stands for volts) to fire a spark plug during the exhaust stroke (about the same as having to fire the rotor gap in a distributor-type ignition system).

The C^{3}I system was introduced in 1984 on the 3.8 liter, turbo, Buick engine. The C^{3}I II system soon followed and C^{3}I III was introduced in 1988 on the 3800 (a new 3.8-liter V-6 also introduced in 1988). The ignition module is connected to the ECM by a 14-pin connector, Figure 5-30. Ignition switch voltage at terminal "N" powers the module; ignition switch voltage at terminal "P" powers the ignition coils. The .7 ohm resistance wire in the circuit to "P" is to prevent coil overheating during high ambient temperatures. If the ignition is turned on without a cranking signal appearing within 1 to 2 seconds, the module will shut off primary circuit current to prevent coil overheating. The ECM has a spark timing range capability of 0° to 70° and will provide dwell times between 3 milliseconds at high RPM and 15 milliseconds at low RPM.

On Type I and Type II C^{3}I systems and on DIS and IDI systems, the module has to see a signal from the cam or sync sensor during cranking before it knows what position the crank is in and which coil to fire. Therefore, no spark occurs until the module sees the sync signal. Once that signal has occurred, the module will know what position the crankshaft is in and will fire a coil on the next appropriate cylinder event signal, which will be the second cylinder in the firing order. Depending on what position the crankshaft is in when cranking starts, the crankshaft may spin more than a full revolution before a spark occurs. On those systems, the first cylinder to fire will always be the second cylinder in the firing order. The C^{3}I Type III (Fast Start) system provides faster starting due to its improved sensor input.

IDI. The newest system is called "Integrated Direct Ignition" (IDI), and is used on the new 2.3-liter Oldsmobile Quad 4 engine. The IDI system is similar to the Direct Ignition System. It uses the same crankshaft-mounted, seven-notch reluctor and crankcase-mounted

Chapter 5 General Motors' Port Fuel Injection

FIGURE 5-31 Four-cylinder DIS systems schematic. *Courtesy of GM Product Service Training.*

magnetic sensor, Figure 5-14, but uses a different module and harness connectors. No part of the IDI system is visible from the exterior of the engine. The coil, module, spark plugs, and wires are mounted under the top most part on the engine, Figure 5-32.

DIS. During module mode operation (see Chapter 3 under "Electronic Spark Timing"), on four-cylinder applications, the signal produced by the sync notch tells the ignition module to skip the next signal (produced by the number 1 notch) and to fire the 2-3 coil on the signal produced by the number 2 notch, Figure 5-33. The module will then skip the signals produced by notches 3 and 4 and will fire the 1-4 coil in response to the signal produced by notch number 5. Notches number 6 and 7 pass the sensor without being responded to and the process begins again. Notice that the module is not concerned with cylinder firing order; it is concerned with coil firing order.

On the six-cylinder engine, the signal produced by notch number 7 references the ignition module to skip the signal from notch 1 and respond to the signal from notch number 2 to fire the 2-5 coil, Figure 5-34. The module then skips the signal from notch number 3 and fires the 3-6 coil on the signal from notch number 4.

Chapter 5 General Motors' Port Fuel Injection

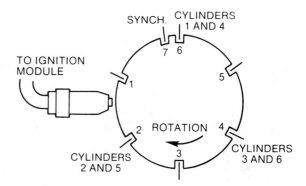

FIGURE 5-34 Six-cylinder coil firing sequence

77. IGNITION COIL AND MODULE ASM.
78. BOLTS, IGNITION COIL AND MODULE ASM. TO CAMSHAFT HOUSINGS
81. COVER, CAMSHAFT HOUSING (INTAKE SHOWN)
90. SPARK PLUG

FIGURE 5-32

The signal from notch number 5 is skipped and the 1-4 coil is fired on the signal from notch number 6.

Electronic Spark Control (ESC)

All PFI systems except the 2.8-liter applications are equipped with ESC. Its function is to retard ignition timing when detonation occurs. It has its own hybrid ESC module that works in conjunction with the ECM, Figure 5-35. Its operation is essentially the same as the later style ESC system discussed in the **Outputs** section of Chapter 3. The conditions under which ignition timing are retarded in response to spark knock are:

- engine speed is above 1,100 RPM.
- ignition voltage to the ECM is above 9 volts.
- no code 43 is present.

Idle Air Control (IAC)

The IAC assembly is a stepper motor and pintle valve just like it is on the EFI system. It is screwed into the throttle body and controls idle speed, Figure 5-36. With one exception it works like the IAC used on the EFI system (**Outputs** section of Chapter 4). The exception is that the ECM rereferences its count of IAC's pintle valve position by moving the valve to its

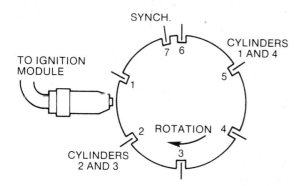

FIGURE 5-33 Four-cylinder coil firing sequence

185

Chapter 5 General Motors' Port Fuel Injection

FIGURE 5-35 ESC circuit. *Courtesy of GM Product Service Training.*

FIGURE 5-36 IAC motor control circuit. *Courtesy of GM Product Service Training.*

Chapter 5 General Motors' Port Fuel Injection

FIGURE 5-37 Divert valve control circuit (2.8 L). *Courtesy of GM Product Service Training.*

zero position (closed) each time the ignition is turned off instead of when the ECM gets a 30-MPH signal from the VSS, as the EFI system does. The ECM then issues a preprogrammed number of pulses to open the valve to a ready position for the next start-up. The following inputs can affect idle speed performance:

- battery voltage
- coolant temperature
- throttle position
- vehicle speed
- mass airflow
- engine speed
- A/C clutch signal
- power steering signal
- P/N switch signal

Air Injection Reaction (AIR) System

Only some of the PFI systems use an AIR system. They are much the same as those systems discussed in the **Outputs** section of Chapter 3 and will be discussed here only briefly.

2.8-liter Engines. The 1985 2.8-liter engine equipped with manual transmission uses an AIR system with a single-bed, three-way converter. During warm-up the ECM energizes the divert valve solenoid, Figure 5-37, which then allows vacuum to be applied to the divert valve. The pump's air is directed to the exhaust ports. As the engine approaches normal temperature (just before going into closed loop), the ECM turns off the solenoid, which in turn blocks vacuum from the divert valve. The divert valve then directs the air to the air cleaner (divert mode).

5.0- and 5.7-liter Engines. These engines use dual-bed, three-way catalytic converters. The air pump pumps air to the system control valve; the control valve, which contains two valves in one housing, can direct air to any one of three places, Figure 5-38:

- to the exhaust ports during engine warm-up.
- to the oxydizing chamber of the catalytic converter during closed-loop operation.
- to atmosphere (divert mode—during WOT, deceleration, when any failure occurs that causes the ECM to turn on the check engine light, or when RPM is high and causing the pressure relief valve inside the control valve to open, although pressure relief is not really a divert mode). Divert mode is intended to prevent converter overheating.

187

Chapter 5 General Motors' Port Fuel Injection

1. CLOSED LOOP FUEL CONTROL
2. ECM
3. REDUCING CATALYST
4. OXIDIZING CATALYST
5. O_2 SENSOR
6. CHECK VALVE
7. AIR PUMP
8. PORT SOLENOID
9. CONVERTER SOLENOID
10. ELECTRICAL SIGNALS FROM ECM
11. BY PASS AIR TO ATMOSPHERE

FIGURE 5-38 Air management system (5.0 L and 5.7 L). *Courtesy of GM Product Service Training.*

Exhaust Gas Recirculation (EGR)

The EGR valve lets small amounts of exhaust gas into the intake manifold to lower combustion temperature, thereby reducing the production or oxides of nitrogen (NO_x). The EGR valve is opened by ported vacuum (manifold vacuum on some applications), which pulls the diaphragm up against the diaphragm spring, Figure 5-39. Three types of vacuum-operated EGR valves are used: the standard ported valve as shown in Figure 5-39, a positive back-pressure valve, and a negative back-pressure valve. The other two valves are very similar in design to the standard valve except that they are sensitive to and modulated by exhaust back pressure or a combination of exhaust back pressure and manifold vacuum, respectively. They can be identified by:

1. EGR VALVE
2. EXHAUST GAS
3. INTAKE AIR
4. VACUUM PORT
5. DIAPHRAGM
6. VALVE OPEN
7. VALVE CLOSED
8. SPRING

4S 0536-6E

FIGURE 5-39 Standard EGR valve. *Courtesy of GM Product Service Training.*

1. ASSEMBLY PLANT CODE
2. PART NUMBER
3. DATE BUILT
4. LOOK HERE FOR LETTER
   P = POSITIVE BACK PRESSURE
   N = NEGATIVE BACK PRESSURE
   BLANK = PORTED VALVE

4S 0538-6E

FIGURE 5-40 EGR valve identification. *Courtesy of Buick Motor Division.*

- a P stamped on top of the positive back-pressure valve, Figure 5-40
- an N stamped on top of the negative back-pressure valve
- a blank space on top of the port valve

Chapter 5 General Motors' Port Fuel Injection

Regardless of the valve used, vacuum to it is controlled by a vacuum control assembly, Figure 5-41. The control assembly solenoid is pulsed many times per second by the ECM. When the solenoid is turned on, vacuum to the EGR port is blocked. The strength of the vacuum signal, which controls how far the EGR valve opens, is controlled by varying the pulse width. The solenoid is fitted with a vent filter, which must be replaced periodically. The ECM controls the solenoid using input concerning coolant temperature, throttle position, and mass airflow. The EGR valve is not opened unless the engine is warm and above idle.

Digital EGR Valve. A new concept in EGR valves is featured on the 3800, V-6 engine for 1988. The digital EGR valve provides the following advantages:

- vacuum is not required for its operation
- increased control of EGR flow
- faster response to ECM commands
- greater ECM diagnostic capability

The digital EGR valve is directly operated by solenoids rather than a vacuum diaphragm. Instead of trying to regulate EGR flow by controlling how far the valve opens or by duty cycling it, three different valves are used to open or close three different sized orifices. Each valve can be opened or closed independently

1. EGR VACUUM CONTROL ASSEMBLY BASE
2. EGR VACUUM DIAGNOSTIC CONTROL SWITCH
3. DIAGNOSTIC SWITCH CONNECTORS
4. EGR SOLENOID
5. FILTER

4S 1182-6EA

FIGURE 5-41 EGR vacuum control assembly. *Courtesy of GM Product Service Training.*

Increment	Orifice #1 (14%)	Orifice #2 (29%)	Orifice #3 (57%)	EGR Flow
0	closed	closed	closed	0%
1	open	closed	closed	14%
2	closed	open	closed	29%
3	open	open	closed	43%
4	closed	closed	open	57%
5	open	closed	open	71%
6	closed	open	open	86%
7	open	open	open	100%

FIGURE 5-42 Increments of EGR flow with digital EGR

Chapter 5 General Motors' Port Fuel Injection

FIGURE 5-43 *Courtesy of GM Products Service Training.*

of the others, allowing any of seven different increments of exhaust gas flow to be admitted into the induction system, Figure 5-42. Orifice number one, when open, will flow 14 percent of the maximum EGR. Orifice number two, when open, will flow 29 percent of the maximum EGR, and number three will flow 57 percent of maximum when it is open. By manipulating which valve or combination of valves are open at any given time, EGR flow can be easily and accurately controlled.

The digital EGR assembly consists of the EGR base, the EGR base plate and the solenoid and mounting plate assembly, Figure 5-43. When the base and base plate are fitted together, there is a sealed cavity between them with the base forming the floor of the cavity and the base plate forming the cavity ceiling. There are four holes in the base (floor). The center hole is always open and allows exhaust gas to enter the cavity. The other three holes are closed by the three EGR valve pintles, Figure 5-44. The pintles are attached to their pintle shafts by a

FIGURE 5-44 Digital EGR assembly. *Courtesy of GM Products Service Training.*

ball joint-type connection so that they can more readily align with the seats they rest on and maintain a more positive seal.

As shown in Figure 5-44, the pintle and its shaft are part of the armature assembly. Each pintle-shaft-armature assembly is held down by its armature return spring, which is attached to the lower portion of the shaft. When one of the solenoid windings, in the solenoid assembly, is turned on by the ECM, the resulting magnetic field lifts the corresponding armature assembly. This opens the valve and compresses the return spring. When the ECM turns the solenoid off, the spring drives the armature assembly back down and closes the valve.

There are two seals on the pintle shaft. The lower seal is pushed down against the top of the base plate. It prevents exhaust gases in the base cavity from leaking out between the stem and the base plate. On this EGR valve design, ambient air has much less tendency to leak into the induction system by way of the EGR valve stem seal than it does on the traditional EGR valve design. This is because exhaust pressure in the base cavity prevents a low pressure from developing under the seal.

The upper seal is pressed up against the bottom of the solenoid assembly. It keeps dirt and dust from entering the armature cavity of the solenoid assembly to reduce wear and provide greater reliability. A spring fitted between the seals holds them in place. The return spring also helps hold the upper seal in place.

Some engine applications use a two-solenoid digital EGR valve, which works the same way except that it only has three different EGR flow rates, Figure 5-45. When the first valve is open, about 33% of total EGR will flow. When the second valve is open, about 66% of total EGR will flow. When both valves are open, 100% of EGR is allowed to flow.

By not relying on vacuum to operate the EGR valve assembly, the ECM can be made to monitor EGR system operation for fault diagnosis without any additional external hardware. If a short or open occurs in any one of the

FIGURE 5-45 Digital EGR valve. *Courtesy of GM Products Service Training.*

solenoid circuits, a code will be set in the ECM's keep alive memory. Each solenoid has its own assigned code number.

Charcoal Canister Purge

The ECM operates a solenoid that controls vacuum to the purge valve on the charcoal canister. The solenoid is turned on and blocks vacuum to the purge valve when the engine is cold or at idle, Figure 5-46. The ECM turns the solenoid off and thus allows any stored hydrocarbons to be purged through the purge valve when:

- the engine is warm.
- the engine has run for a preprogrammed amount of time since it was started.
- a preprogrammed road speed is reached.
- a preprogrammed throttle opening is achieved.

Chapter 5 General Motors' Port Fuel Injection

FIGURE 5-46 Canister purge control solenoid circuit. *Courtesy of GM Product Service Training.*

FIGURE 5-47 Canister purge and control valves. *Courtesy of Buick Motor Division.*

In addition to the purge valve, some engine application canisters also have a non-ECM-controlled control valve, Figure 5-47. It is controlled by ported vacuum. When vacuum is applied to it, it allows canister purging to occur through another port connected to manifold vacuum.

For 1987, Chevrolet 5.0 liter and 5.7 liter engines have a revised canister purge strategy. When the engine is operating in the closed loop mode, the ECM will duty cycle the purge solenoid to control the amount of vapors that are admitted into the engine's induction system rather than just turning the purge solenoid on to stop purging and off to allow purging. The ECM will use input from the O_2 sensor to determine the volume of vapors to be purged from the canister. If a rich condition is indicated by the O_2 sensor, purge volume is reduced. This strategy is intended to provide improved driveability.

Turbocharger

One of the major factors that controls power is how much air and fuel are put into the cylinders. Putting more fuel in is a fairly simple matter; putting more air in is not so simple (the proper air/fuel ratio must be maintained). The amount of air that can be put into the cylinder is limited by atmospheric pressure unless some means is used to force air in at greater than atmospheric pressure. This is accomplished with some type of supercharger of which the turbocharger is the most popular and generally the most efficient.

The turbocharger is a centrifugal, variable-displacement air pump driven by otherwise wasted heat energy in the exhaust stream. It pumps air into the intake manifold at pressures that are limited only by the pressure that can be developed without significantly increasing the air's temperature.

Most normally aspirated engines (those that rely on atmospheric pressure to fill the cylinder) only fill the cylinder to about 85% of atmospheric pressure (they achieve 85% volumetric efficiency, or VE, at full throttle with the engine at its maximum torque speed). Turbocharging produces VE values in excess of 100%.

Turbochargers

Because a turbocharger puts more air into the cylinder, it raises the engine's compression pressure. Because it does not produce significant boost pressure at low exhaust flow rates, it allows an engine the benefits of low compression during light throttle operation and high compression during heavy throttle operation. Low compression offers lower combustion chamber temperatures, which result in less wear and lower NO_x emissions. High compression provides higher combustion chamber temperatures, which result in more complete combustion and more power for the amount of fuel consumed.

Another way to look at it is in terms of displacement. If an engine is operating at an atmospheric pressure of 14 PSI and the turbocharger is producing a boost of 14 PSI, the cylinders are being charged with a pressure of 28 PSI. The engine is consuming approximately double the amount of air that it would without the turbocharger; we have effectively doubled the engine's displacement. The engine certainly consumes more fuel but not as much as it would if it had twice the displacement. This is because the higher compression pressure and temperature produces more complete combustion.

Turbo Operation. Upon leaving the manifold, the hot exhaust gases flow through the vanes of the exhaust turbine wheel and thus

Chapter 5 General Motors' Port Fuel Injection

FIGURE 5-48 Turbocharger and wastegate control (typical). *Courtesy of GM Product Service Training.*

cause it to spin, Figure 5-48. The more exhaust coming out of the engine, the faster the turbine spins. It can achieve speeds in excess of 130,000 RPM. The exhaust turbine drives a short shaft that drives a compressor turbine. The high speed produces centrifugal force, which moves the air from between the vanes and thus causes it to flow out radially. The air is then channeled into the intake manifold. As air is thrown from the turbine vanes, a low pressure develops in its place. Atmospheric pressure pushes more air through the air cleaner, the mass airflow sensor (if used), and the throttle body.

A criterion turbine speed must be reached before a pressure boost is realized; that speed varies with turbine size and turbocharger design. If turbine speed goes to high, boost pressure (and therefore charge temperature) will go too high and will cause preignition and engine damage. To limit boost pressure, a wastegate is used to divert exhaust gases away from the exhaust turbine and thus limit its speed.

Wastegate. The wastegate is opened by a wastegate actuator. A spring in the actuator holds the wastegate closed. When manifold pressure (turbo boost) reaches approximately 8 PSI, it overcomes the actuator spring and opens the wastegate. If, however, engine operating parameters are favorable (coolant temperature, incoming air temperature, etc.), the ECM pulses a wastegate solenoid, which in turn bleeds off some of the pressure acting on the actuator. When this occurs boost pressure is allowed to rise to 10 PSI before the wastegate opens.

A code 31 is set if:

- an overboost is sensed by the MAP sensor on the 1.8 liter.
- the ECM, which monitors wastegate solenoid circuit operation, sees a malfunction in the circuit while the solenoid is being operated between a 5% and a 95% duty cycle on the 3.8 liter.

Transmission Converter Clutch (TCC)

Notice that the name has changed slightly from this unit's application on CCC and EFI systems, where TCC meant torque converter clutch; it is still the same part. The purpose of the TCC is to increase fuel economy. It does this by eliminating hydraulic slippage during cruise conditions and by eliminating heat production in the torque converter, especially during overdrive operation. For a more detailed explanation on TCC operation, see **Torque Converter Clutch** in the **Outputs** section of Chapter 3.

Electric Cooling Fan

All General Motors' vehicles with *transverse* mounted engines and some with longitudinal engines are equipped with an electric cooling fan to pull air through the radiator and A/C condensor. Control of the fan varies somewhat with engine application. In all cases, however, the fan is turned on when:

- coolant temperature exceeds a specified value. This can be done by either the ECM or a coolant temperature override switch on some applications and only by the ECM on others.

COOLANT FAN CONTROL OPERATING CONDITIONS (WITH A/C)					
A/C SW.	HEAD PRESS.	ROAD SPEED	ENG. TEMP.	FAN SPEED	FAN ON BECAUSE:
OFF/ON	UNDER 260 PSI	UNDER 45 MPH	OVER 98°C	LOW	ECM TURNED ON
OFF/ON	UNDER 260 PSI	UNDER 45 MPH	UNDER 95°C	OFF	
OFF/ON	UNDER 260 PSI	OVER 45 MPH	UNDER 106°C	OFF	
ON	OVER 260 PSI	N/A	N/A	LOW	A/C HEAD PRESS. SW. (LOW) TURNED ON
ON	OVER 300 PSI	N/A	N/A	HIGH	A/C HEAD PRESS. SW. (HIGH) TURNED ON
OFF/ON	N/A	N/A	OVER 106°C	HIGH	COOLANT TEMP. OVERRIDE SW. TURNED ON
WITHOUT A/C					
IGN. SW.		ROAD SPEED	ENG. TEMP.	FAN SPEED	FAN ON BECAUSE:
ON		N/A	OVER 98°C	LOW	ECM TURNED ON
ON		OVER 45 MPH	UNDER 108°C	OFF	
ON		OVER 45 MPH	OVER 108°C	HIGH	COOLANT TEMP. OVERRIDE SW. TURNED ON

FIGURE 5-49 Coolant fan control circuit (typical). *Courtesy of Buick Motor Division.*

Chapter 5 General Motors' Port Fuel Injection

- when A/C compressor output pressure (head pressure) exceeds a specified value. This is done by a switch in the high-pressure side of the A/C system on some engine applications. It is done by the ECM on other applications (on those applications an A/C head pressure switch feeds head pressure information to the ECM).

On most applications the ECM turns on the fan anytime the A/C is on and vehicle speed is less than a specified value. Others turn it on under a specified speed whether the A/C is on or off. Once the vehicle reaches a criterion speed, enough air is pushed through the radiator without the aid of the fan, unless overheating or high A/C head pressure conditions exist. This is a fuel economy feature. Study Figure 5-49 as a typical example.

Air-conditioning (A/C) Control

The ECM controls the relay that turns on and off the A/C clutch for two and in some cases three reasons, Figure 5-50.

- When the A/C control switch (on the instrument panel) is turned on, the ECM delays A/C clutch engagement for 0.4 second to allow time to adjust the IAC valve.
- The ECM disengages the A/C clutch during WOT operation.
- On some applications the ECM turns off the A/C clutch if power steering pressure exceeds a specified value during idle. On others the power steering switch is in series with either the A/C clutch or the control winding of the A/C relay and disengages the A/C clutch without relying on the ECM.

FIGURE 5-50 A/C control circuit (typical). *Courtesy of GM Product Service Training.*

Chapter 5 General Motors' Port Fuel Injection

SYSTEM DIAGNOSIS AND SERVICE

Self-diagnosis

The self-diagnostic capabilities and procedures of the PFI systems are essentially the same as those of the CCC system discussed in the **System Diagnosis and Service** section of Chapter 3. In 1988 General Motors made a significant increase in the self-diagnostic capability of most PFI engine applications. The ECM was designed to monitor more circuits, both input and output, and new diagnostic code numbers were assigned. It is important to note that from year to year and in some cases from engine to engine, code numbers may have different meanings. Be sure that the service literature you are using is applicable to the engine being serviced and to the model year.

TCC Test Lead. The TCC test lead is in cavity F of the ALCL. It can be used to monitor TCC circuit operation with a voltmeter or test light, or it can be used to ground the TCC test lead overriding the ECM.

Fuel Pump Test Lead. On some vehicles the fuel pump test lead is found in cavity G of the ALCL. On others it is found on the left side of the engine compartment. A voltmeter or test light can be connected to this lead to determine if the fuel pump relay or oil pressure switch has supplied power to the fuel pump power lead. Or a jumper wire can be connected from the test lead to a 12-volt source. This powers the fuel pump for fuel pressure or injector tests.

> **CAUTION:** When powering the fuel pump by the above method, be sure that there are no fuel leaks and avoid causing any sparks to occur.

Note: Diagnostic procedures for 1987 and later are written to include the use of an ALCL tool, often referred to as a "Scanner," where it is applicable. Previously, diagnostic procedures were written around the use of a voltmeter to obtain such measurements as sensor readings.

Diagnostic Procedure

Diagnostic Circuit Check. After making a thorough visual inspection, paying particular attention to possible causes of vacuum leaks, and correcting any problems found, perform the diagnostic circuit check. This procedure is outlined in the service manual and is designed to identify the type of problem causing the complaint and to refer to the next step or chart to use.

Fuel Pressure Test. Some diagnostic charts call for a fuel pressure test. Fuel pressure is critical to the performance of a fuel injection system. The fuel rail has a fuel pressure test fitting that contains a schrader valve like that used in an A/C pressure test fitting or in the valve stem of a tire.

> **PERSONAL SAFETY:** While the fuel pressure gauge hose is screwed to the test fitting, a shop towel should be wrapped around the fitting to prevent gasoline from being sprayed on the engine. Remember, the fuel is under high pressure and the residual check valve in the fuel pump can hold pressure for some time after the engine is shut off.

Injector Balance Test

In order to realize some of the advantages of PFI, the injectors must all deliver the same amount of fuel to each cylinder. The injector balance test tests injector performance and uniform fuel delivery. Check the applicable service manual for specific directions.

Integrator and Block Learn

The integrator serves General Motors' fuel-injected engines in much the same way that dwell, a measure of duty cycle, serves CCC applications. Let's assume that while in closed loop a particular fuel-injected engine is operating with a slight vacuum leak. The oxygen sensor reading indicates a lean mixture until the ECM increases injector pulse width enough to compensate for the vacuum leak and get the air/fuel mixture back to 14.7 to 1. If an ALCL tool were connected to the ALCL with "integrator" selected, a number would be shown on the ALCL tool representing the injector pulse width currently being used. In a normal situation where no compensation is required, that number would be 128. If, as in our example, pulse width has been increased to compensate for a tendency to be lean, the number would be above 128. If the pulse width has been decreased to compensate for a tendency to be rich, the number would be below 128.

Block learn functions much like the integrator, except it represents:

- pulse width adjustments that have become a trend over a longer period of time
- learned behavior on the part of the ECM that has caused it to modify its original pulse width programming.

Once the engine has been running in closed loop long enough for block learn information to be stored in memory (as a number), that information is retained in memory until it is modified by new information or until the ignition has been turned off. Some earlier Buick and Cadillac systems would hold block learn numbers in memory until battery power was disconnected from the ECM.

Block learn numbers are stored in sixteen different cells within a square block, as shown in Figure 5-51. Each cell represents a different combination of engine speed and load. Remember that block learn only functions during closed loop operation. If the pulse width that was originally programmed for each of the sixteen combinations of engine speed and load requires no adjustment in order to maintain a 14.7 to 1 air/fuel ratio, the block learn number in each cell will be 128 as in Figure 5-51A. If, for some reason, the engine tends toward a slightly lean condition, the block learn numbers may look similar to those in Figure 5-51B. If, due to some fault, the engine has a tendency toward running slightly rich, the block learn numbers might look similar to those in Figure 5-51C.

If block learn numbers are above 128 but the integrator is at 128, the system has a lean tendency that is being corrected by block learn with wider pulse widths, and the integrator is not having to make any additional corrections. In other words, block learn has modified "normal" to a wider pulse width and the integrator is not having to correct.

If block learn numbers are 128 but the integrator is above 128, the system is trying to run lean but has not been doing so long enough for block learn to have made any correction.

If block learn is well above 128 and the integrator is the same or higher, the system is running lean and the ECM is not able to compensate enough to correct it. The same number relationships would occur if the system were trying to run rich, except that the numbers would be below 128.

ECM, PROM, and CALPAK Service

If correctly used diagnostic procedures call for an ECM to be replaced, check the ECM and PROM service and part numbers to be sure that they are the right units for the vehicle. If a PROM is being replaced, check that it has not been superceded by a replacement PROM. This can be done with the aid of the dealer's parts department or with service bulletins. Some aftermarket service manuals contain service bulletins. Carefully remove the PROM and CALPAK from the ECM to be replaced. The service (replacement) ECM does not contain a PROM or CALPAK. Be careful not to touch the pins of either unit with your fingers because static electricity applied to the pins can damage them. Store them in a safe place while they are out of the ECM.

Chapter 5   General Motors' Port Fuel Injection

A
System operating at design specifications

heavy
L
O
A
D
light

128	128	128	128
128	128	128	128
128	128	128	128
128	128	128	128

low   hi
RPM

B
Result of slightly lean tendency

heavy
L
O
A
D
light

128	130	128	129
129	130	130	131
133	136	129	128
135	132	129	130

low   hi
RPM

C
Result of slightly rich tendency

heavy
L
O
A
D
light

121	124	128	128
119	120	119	121
115	117	116	121
122	119	126	120

low   hi
RPM

FIGURE 5-51   Block learn information

When the PROM is reinstalled, be sure to orient it properly to the PROM carrier, Figure 5-52. If it is installed backwards, it will be destroyed when the ignition is turned on.

Weather-pack Connectors

All PFI system harness connections under the hood are made with weather-pack connectors as discussed in the **System Diagnosis and Service** section of Chapter 3.

Service Tips

IAC Assembly. If the engine is run with the IAC harness disconnected or if the IAC assem-

1 REFERENCE END
2 PROM
3 PROM CARRIER   4S 0465-6E

FIGURE 5-52   PROM and PROM carrier. *Courtesy of GM Product Service Training.*

Chapter 5 General Motors' Port Fuel Injection

bly is removed and reinstalled, idle speed will probably be incorrect when it is reconnected. To rereference idle speed, simply start the engine with the IAC harness connected. Allow the engine to warm up and then turn the ignition off. The 3.0- and 3.8-liter engines rereference simply by cycling the ignition on and off.

If the IAC assembly is removed, be sure that the distance from the tip of the pintle valve to the motor housing does not exceed 28 millimeters (1-1/8 inch), Figure 5-53; otherwise, it will be damaged when it is installed. See service manual for directions. When it is installed, use correct torque specifications; overtorquing can distort the IAC housing and cause the motor to stick. Complaints of engine stalling as the car comes to a stop can be a result of this condition. Also, if the IAC assembly is replaced, be sure to get the right part. There are three different pintle shapes, and the correct one must be used.

Minimum Idle Speed Adjustment. Minimum idle speed is the speed achieved when the IAC valve is in its closed position; the only air getting into the intake manifold is what goes by the throttle blade. An idle stop screw is used to position the throttle blade. The head of the idle stop screw is recessed and covered with a metal plug. The plug can be removed and the idle stop screw adjusted; however, this should only be done if the throttle body is being replaced or if a service procedure calls for it. It is not a routine adjustment.

In order to make the minimum idle speed adjustment, the engine should be warm and the IAC valve fully extended (closed). Consult the appropriate service manual for specific directions and specifications.

TPS Adjustment. The TPS on PFI systems is adjustable. Adjustment is checked by connecting a voltmeter between terminals A and B at the sensor connector with the ignition on. Gain voltmeter access to the terminals by disconnecting the three-wire weather-pack connector and inserting three short jumper leads to temporarily reconnect the circuit. Performing a minimum idle speed adjustment will probably cause the TPS to need adjustment also.

Injector O-rings. The O-ring that seals between the bottom of the injector and the manifold is a potential source of vacuum leaks. Carefully examine the O-rings and/or replace them when the injectors are serviced.

3.8-liter Turbo (SFI) Injector Leads. Because the injectors are pulsed individually and in firing order on the 3.8-liter turbocharged engine, the wires that power the injectors must be attached in the firing order. These wires are color-coded as shown in Figure 5-54.

Fuel Line O-rings. Threaded connections between the fuel pump and the fuel rail use O-rings to reduce the possibility of leaks. When replaced they should only be replaced with O-rings designed to tolerate exposure to gasoline.

Fuel Flex Hose. The fuel flex hose used on PFI systems is internally reinforced with steel mesh. No attempt should be made to repair it. It should be replaced if it becomes unusable or questionable.

60 IDLE AIR/VACUUM SIGNAL HOUSING
70 IDLE AIR CONTROL VALVE (IAC)
71 IAC GASKET

05 24 84   5S 1455-6E

FIGURE 5-53 IAC motor. *Courtesy of GM Product Service Training.*

CYLINDER	WIRE COLOR
1	Black
2	Dk. Green
3	Pink
4	Red
5	White
6	Lt. Green

FIGURE 5-54 SFI injector wire color codes. *Courtesy of GM Product Service Training.*

PERSONAL SAFETY: The factory-recommended procedures should be carefully adhered to when servicing fuel lines or fuel line connections. The fuel is under high pressure, and failure to comply with factory-recommended procedures can result in a fuel leak and, consequently, fire.

REVIEW QUESTIONS

1. Name at least two systems typically found on engines with a carburetor or a TBI unit that are not used on a PFI system.
2. What is the major weakness of PFI?
3. During what two driving conditions are the above weaknesses most pronounced?
4. How are the fuel lines used on PFI systems different from those used on other systems, such as carburetion or TBI?
5. What causes the fuel pressure regulator to vary the fuel pressure from 34 PSI to 44 PSI?
6. Name at least two ECM responses to the power steering switch opening during idle or speeds below 10 MPH?
7. What circuit powers the cold-start injector?
8. List two unique features of the C^3I system?
9. If the wire from terminal C of the ESC module to terminal B7 of the ECM became open, what would the result be?
10. How is vacuum to the EGR valve controlled?
11. What opens the wastegate on PFI applications?
12. Name at least two conditions that cause the cooling fan to come on.
13. What does the check engine light do if the ECM sees a fault in a circuit it monitors and then the fault clears up?
14. What is the primary function or functions of the diagnostic circuit check?
15. Name at least one precaution that should be taken before connecting or disconnecting the fuel pressure test gauge.
16. What removable part is found in the ECM besides the PROM?
17. What is a clamping diode or resistor?
18. What can be the result of overtorquing the IAC motor assembly when installing it in the throttle body?
19. Name at least one major concern relevant to the O-ring that seals between the injector and the manifold.
20. What is the combustion air source during the minimum air adjustment?

ASE-type Questions. (Actual ASE test questions will probably not be this product specific.)

21. Technician A says that on most General Motors' PFI systems, the injectors are all pulsed at the same time. Technician B says that they are divided into two groups, with each group being pulsed alternately. Who is correct?
    a. A only
    b. B only
    c. both A and B
    d. neither A nor B
22. A 5-liter tuned port injection engine is hard to start cold. Inspection shows that the

Chapter 5 General Motors' Port Fuel Injection

cold-start injector is not working. This could be the fault of _____ .
a. the ECM
b. the thermal time switch
c. either A or B
d. neither A nor B

23. A car with a PFI system is hard to start, especially after sitting for an hour or more. A fuel pressure test shows that fuel pressure is normal but drops to zero quickly after the engine is turned off. The hard-starting problem is most likely the result of _____ .
a. a defective fuel pump
b. a plugged injector
c. a defective check valve in the fuel pump
d. none of the above

24. Technician A says that the PFI system re-references the IAC motor each time the VSS produces a signal that equals 30 miles per hour. Technician B says that the PFI system rereferences the IAC motor when the ignition is cycled. Who is correct?
a. A only
b. B only
c. both A and B
d. neither A nor B

25. Technician A says that if the wastegate fails to open, detonation will probably occur. Technician B says that if the wastegate fails to open, piston damage can occur. Who is correct?
a. A only
b. B only
c. both A and B
d. neither A nor B

GLOSSARY

In addition to the following terms, refer to Figure 5-55 for a list of the abbreviations used in this chapter.

AC A parts manufacturing and distributing division of General Motors.

C³I A name used by General Motors to describe their three-coil ignition system that uses a module instead of a distributor.

HIGH-PRESSURE CUT-OUT SWITCH A pressure-operated switch located in the high-pressure side of an air-conditioning system that opens when freon pressure goes too high. When the switch opens, it disables the A/C compressor clutch.

LONGITUDINAL Parallel to the length of the car.

MICRON A unit of measurement that describes the size of a small particle; a 1-micron particle is one millionth of an inch in diameter.

NORMALLY ASPIRATED Using atmospheric pressure to force air into the engine.

PRESSURE CYCLING SWITCH A pressure-operated switch in the low-pressure side of the air-conditioning system that opens as freon pressure drops below a specific pressure. When the switch opens, the A/C compressor clutch is disabled.

RHEOSTAT A variable resistor that usually has two leads and that serves as a voltage or current limiter in series with a load.

SPEED DENSITY A term that refers to a method of calculating the flow rate and density of air. The result reveals the air's ability to evaporate fuel.

THERMAC The term used by General Motors to identify the heated air inlet system. The word is a shortened form of thermostatic air cleaner.

THERMO-TIME SWITCH An electrical switch, usually applied to a bimetal strip, that uses heat to open or close a set of electrical contacts. The amount of heat is controlled so that the time it takes to open or close the contacts is always the same.

A/C	Air-conditioning
AIR	Air Injection Reaction
ALCL	Assembly Line Communication Link
ATS	Air Temperature Sensor
BARO	Barometric Pressure
C^3I	Computer-Controlled Coil Ignition
CO	Carbon Monoxide
DIAGNOSTIC "TEST" TERM.	Lead or ALCL connector that is grounded to get a trouble code
DVM (10 Meg.)	Digital Voltmeter with 10 million ohms resistance
ECM	Electronic Control Module (controller)
EECS	Evaporative Emissions Control System
EFE	Early Fuel Evaporation
EFI	Electronic Fuel Injection
EGR	Exhaust Gas Recirculation
ESC	Electronic Spark Control
EST	Electronic Spark Timing
HC	Hydrocarbons
HEI	High Energy Ignition
IAC	Idle Air Control
IP	Instrument Panel
IGN	Ignition
MAF	Mass Airflow
MAP	Manifold Absolute Pressure
MAT	Manifold Air Temperature
MFI	Multiport Fuel Injection. Individual injectors for each cylinder are mounted in the intake manifold. The injectors are fired in groups rather than individually.
MPH	Miles Per Hour
N.C.	Normally Closed
N.O.	Normally Open
NO_x	Oxides of Nitrogen
O_2	Oxygen (sensor)
PCV	Positive Crankcase Ventilation
PFI	Port Fuel Injection
P/N	Park/Neutral
PROM	Programmable Read-only Memory (calibrator)
RPM	Revolutions per Minute
SFI	Sequential Fuel Injection. Used on 3.8 L Turbo. Each injector is individually fired.
TACH	Tachometer
TBI	Throttle Body Injector (unit)
TCC	Transmission Converter Clutch
THERMAC	Thermostatic Air Cleaner
TPI	Tuned Port Injection
TPS	Throttle Position Sensor
TVS	Thermal Vacuum Switch
V	Volt
VIN	Vehicle Identification Number
VSS	Vehicle Speed Sensor
WOT	Wide-open Throttle

FIGURE 5-55 Explanation of abbreviations

TRANSVERSE At right angle (perpendicular) to the center line of the vehicle.

VOLUMETRIC EFFICIENCY (VE) A measure of cylinder filling efficiency. VE is expressed as the percentage of atmospheric pressure to which the cylinder is filled at the completion of compression stroke.

WASTEGATE A valve or door that when opened allows the exhaust gas to bypass the exhaust turbine of a turbocharger. It is used to limit turbocharger boost pressure.

CADILLAC'S DIGITAL FUEL INJECTION

Objectives

After studying this chapter, you will be able to:

- list the functions that the ECM controls.
- describe the sensors that feed information to the ECM.
- list the functions that the BCM controls.
- describe the relationship between the ECM and the BCM.
- describe the features of the self-diagnostic system.
- put the system into diagnostics and get any stored codes displayed.
- put the system into any of its self-diagnostic features (switch test, ECM data, BCM data, cooling fans override, or ECM outputs cycling).

Digital fuel injection (DFI) was introduced in mid-1980. By 1981 it had replaced all other fuel induction systems on eight-cylinder Cadillac engines except diesels. Prior to DFI Cadillac used a system they called electronic fuel injection (a Bendix-patented, single-function, analog computer-controlled, fuel injection system). The Cadillac EFI system was used from 1975 through 1980. In 1979 and 1980, some California 6.0-liter engine applications used an early version of the computer command control system, referred to then as computer-controlled catalytic converter. For clarification any further use of the EFI abbreviation in this chapter refers to the GM TBI type system. Since 1981 all four- and six-cylinder-equipped Cadillacs have used either CCC or EFI systems.

In 1987, Cadillac introduced a two-seater sports car called the Allante. The Allante featured the 4.1-liter engine that Cadillac introduced in 1982, but with a sequential, port fuel injection system. This system is similar to the non-turbo sequential PFI system on the Buick V-6, with the following exceptions:

- It uses an O_2 sensor in each exhaust manifold. Two O_2 sensors allow the system to control the air/fuel ratio more closely; it can adjust the pulse width on one side of the engine independently of the other side.
- It uses a distributor and HEI ignition system just as the DFI did as opposed to a C^3I ignition system, Figure 6-1.

Chapter 6 Cadillac's Digital Fuel Injection

FIGURE 6-1 4.5-liter, PFI/HEI ignition system. *Courtesy of Cadillac Motor Car Division.*

- A single, 180-degree vane, Hall-effect switch is used in the distributor as a camshaft position sensor for referencing injector pulsing.

The 4.1-liter engine was later enlarged to 4.5 liters. The DFI system was replaced by the newer PFI system on more Cadillac models until 1990 when all 4.5-liter engines were equipped with the PFI system. Also in 1990, Cadillac offered the 5.7-liter, TBI-equipped engine as used in the Chevrolet Suburban as part of an optional trailer towing package in their full-size, rear-wheel-drive car.

Cadillac's DFI system is very similar in operation to the EFI system; the EFI system probably came directly from DFI, minus a few features. The DFI system uses a dual throttle body injection (TBI) unit with two injectors. Do not confuse dual throttle body with the Crossfire EFI dual TBIs. The Crossfire system used two separate TBI units; the DFI's dual TBI is in a configuration similar to that of a conventional two-barrel carburetor, Figure 6-2.

Fuel pressure is supplied by an electric fuel pump located in the fuel tank. The pressure is controlled by a pressure regulator located in the TBI unit. Each injector is pulsed alternately to the other, with a pulse occurring each time a cylinder fires during normal driving.

The two most unique features of the DFI system are, first, the type of functions it controls and monitors in addition to engine parameters such as air-conditioning and heating, fuel mileage information, and cruise control. Secondly, the DFI system has an elaborate self-diagnostic capability that has expanded since its introduction.

The 1981 DFI system controlled the *modulated displacement,* or 8-6-4, system in which selected cylinders were disabled (by preventing the rocker arms from opening the valves) when power requirements were low. Under acceleration all eight cylinders were used. During steady-state driving, when minimal power was required, only six cylinders were used; and during very low power demand or deceleration, only four cylinders were used. That system was used on only the 6-liter engine, which was discontinued in all but limousine and commercial chassis vehicles in 1982, when the 4.1-liter V8 was introduced.

Cadillac Allante

Cadillac's new two-seater sports car, introduced in 1987, features the 4.1-liter engine with a sequential, port fuel injection system. This system is similar to the non-turbo, sequential, PFI system on Buick V-6, except that it uses an O_2 sensor in each exhaust manifold. This allows the system to control the air/fuel ratio more closely. It can adjust the pulse width on one side of the engine independently of the other side.

In 1985 Cadillac introduced a body computer (body computer module, BCM) on the new front-wheel drive C body (C designates a particular size of car body) to work in conjunction with the ECM. It took over all A/C and related functions, data display, and some diesel engine controls for those vehicles equipped with diesel engines. Although this text is aimed at computerized engine controls, a brief unit on the BCM is included in this chapter as an introduction to what is in store.

ELECTRONIC CONTROL MODULE (ECM)

The ECM for the DFI system has more capability than do those used with other General Motors' systems. It is located on the right side of the instrument panel and contains a removable PROM. All inputs fed into it and functions it controls are shown in Figure 6-3.

ECM Learning Ability

As driving conditions change, sensor input values change. Not all of them, however, change in proportion to each other. For instance, if a car that is normally driven at sea level starts on a trip that takes it to a significantly higher altitude, the ECM will see:

- that engine speed and road speed still match up as they normally do, provided the transmission converter clutch is still applied but the throttle position that normally accompanies that road and engine speed is now different.
- that engine vacuum is less than it normally is at the present throttle position.

If this trip is of long enough duration so that the new combination of values are seen consistently, the ECM will learn to accept them as normal and will issue spark timing and other commands that provide optimum driveability. If battery power is removed from the ECM, it will forget what it has learned and performance can

1 FUEL INJECTORS	4 THROTTLE POSITION SENSOR
2 FUEL METER COVER	5 IDLE SPEED CONTROL (ISC)
3 FUEL METER BODY	6 THROTTLE LEVER

FIGURE 6-2 TBI unit. *Courtesy of Cadillac Motor Car Division.*

Chapter 6 Cadillac's Digital Fuel Injection

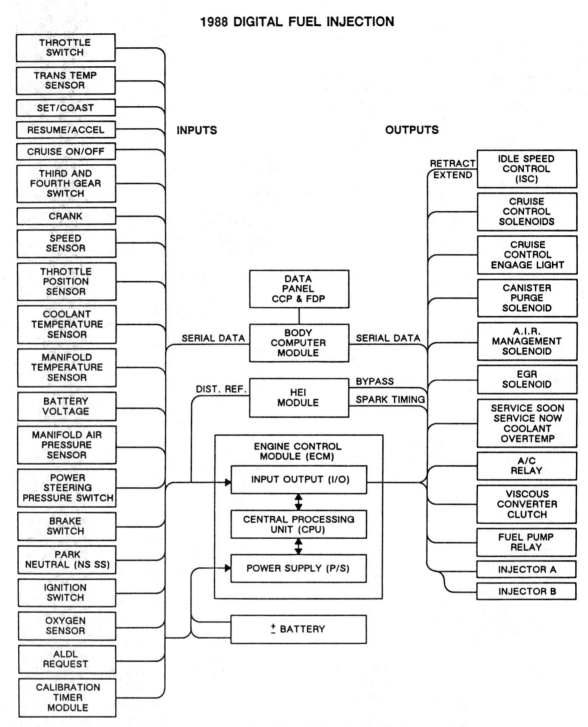

FIGURE 6-3 DFI overview. *Courtesy of Cadillac Motor Car Division.*

be noticeably affected until it relearns. To help the ECM relearn, the car should be driven at normal operating temperature, part throttle, idle, and moderate acceleration until normal performance returns.

Keep-alive Memory (KAM)

The ECM has the same KAM as discussed in Chapter 3.

OPERATING MODES

The ECM operates the DFI system in one of several operating modes, depending on prevailing conditions. Although some of the DFI modes are the same as EFI operating modes, others are slightly different.

Starting Mode

When the ignition is turned on, the ECM turns on the fuel pump relay and checks coolant temperature and throttle position. When the engine starts cranking, the injectors are pulsed so as to deliver an air/fuel ratio that ranges from 1.5 to 1 at −36°C (33°F) coolant temperature to 14.7 to 1 at 94°C (201°F) coolant temperature.

Clear Flood Mode

If for some reason the engine becomes flooded, pressing the throttle to the floor (or within 80% of full throttle) and cranking it will cause the injectors to be pulsed so that an air/fuel ratio of 25.5 to 1 is produced. This is leaner than the EFI clear flood mode. This ratio is maintained until the engine starts or until the throttle is allowed to close to less than 80%.

Run Mode

This mode is used during most of the driving time and presents one of two conditions: open loop or closed loop. When the engine is first started and exceeds 400 RPM, the system goes into open loop. The ECM calculates air/fuel requirements based on coolant temperature, manifold pressure, throttle position, and engine RPM (in that order of influence). Oxygen sensor inputs are not used.

The ECM puts the system into closed loop when it sees:

- a varying voltage from the oxygen sensor that crosses 0.45 volt. This means it is hot enough to work properly.
- that coolant temperature is above 70°C (158°F).
- that sixty seconds have elapsed since the engine was started. This allows all sensors to stabilize.

The ECM now uses only oxygen sensor information to determine injector pulse width. The injector is turned on for only a few milliseconds during maximum fuel requirements.

Acceleration Mode

If the ECM sees throttle position change rapidly toward a more open position and this change is confirmed by a like change in manifold pressure, the ECM will take the system out of closed loop and put it into acceleration mode. The higher manifold pressure produced by the greater throttle opening reduces the fuel evaporation rate. More fuel is now required than is provided for a 14.7 to 1 air/fuel ratio. If the throttle position sensor indicates wide-open throttle (WOT) operation, the system will stay in this enrichment mode until throttle opening is reduced.

Deceleration Mode

When the ECM sees rapid changes in throttle position and manifold pressure values that indicate that the throttle is closing, it reduces the amount of fuel being injected. If deceleration is severe, fuel will be cut off to compensate for the fuel that will begin to evaporate from the manifold floor and walls due to the sudden drop in pressure.

Battery Voltage Correction

If battery voltage drops significantly, the ECM will make the following compensations in

Chapter 6 Cadillac's Digital Fuel Injection

order to maintain the best possible operating efficiency:

- increase injector pulse width to maintain the proper air/fuel ratio
- increase idle RPM
- increase ignition dwell time to maintain sufficient ignition performance

INPUTS

All of the inputs used in the DFI are listed in this section even though some are listed without discussion because they are the same as those used in other General Motors systems. Those are:

- coolant temperature sensor (CTS)
- throttle position sensor (TPS)
- oxygen (O_2) sensor
- ignition switch
- crank signal

The others are discussed as needed.

Manifold Absolute Pressure (MAP) Sensor

The earliest DFI systems used Bendix-built MAP and BARO sensors located inside the car.

Distributor Reference (REF)

The DFI system has consistently used the pickup coil in the distributor as the source of the reference pulse.

Manifold Air Temperature (MAT) Sensor

The MAT sensor is the same as that discussed in Chapter 5. It measures air/fuel mixture temperature in the intake manifold.

Throttle Switch

This is the same as the ISC switch discussed in the **Inputs** section of Chapter 3 and can be studied there. The DFI system uses an idle speed control motor like the one used on CCC systems instead of the idle air control motor that EFI and PFI systems use.

Battery Voltage

The ECM monitors battery voltage to determine when it is necessary to go into voltage correction mode.

Park/Neutral (P/N) Switch

The park/neutral switch indicates to the ECM when the transmission is in park or neutral as opposed to being in gear. This information is used to control idle speed, cruise control, and the torque converter clutch or viscous converter clutch system. The front-wheel drive C body, introduced in 1985, uses the 440-T4 transaxle, which uses a viscous converter clutch. This transaxle and converter clutch were also introduced in 1985 (see **Viscous Converter Clutch** in the **Outputs** section of this chapter).

Third- and Fourth-gear Switches

These are hydraulically operated switches that are screwed into passages of the valve body. When the transmission shifts into third gear, the third-gear switch opens; when it shifts into fourth gear, the fourth-gear switch opens, Figure 6-4. By monitoring these switches, the ECM knows if the transmission is in third, fourth, or one of the lower gears. The ECM uses this information to control the viscous converter clutch.

Viscous Converter Clutch (VCC) Temperature Sensor

A thermistor is located in the transmission to monitor transmission fluid temperature, Figure 6-4. This information influences the ECM's control of the viscous converter clutch.

Generator Monitor

On some applications the ECM monitors generator voltage by means of a wire that taps into the fuel pump power circuit (ECM terminal 17, red connector P2 in Figure 6-5). Although

Chapter 6 Cadillac's Digital Fuel Injection

FIGURE 6-4 VCC control circuit. *Courtesy of Cadillac Motor Car Division.*

FIGURE 6-5 Fuel pump control circuit. *Courtesy of Cadillac Motor Car Division.*

211

Chapter 6 Cadillac's Digital Fuel Injection

FIGURE 6-6 Cruise control circuit. *Courtesy of Cadillac Motor Car Division.*

Chapter 6 Cadillac's Digital Fuel Injection

this circuit does not connect directly to the generator (alternator), it does reflect system voltage, which is a function of generator output. If the ECM sees less than 10.5 volts or more than 15.5 volts while the engine is running, it will turn on the service now light and set a code 16 in its diagnostic memory. On cars equipped with a body computer, all engine function codes are preceded by the letter E.

Power Steering (P/S) Anticipate Switch

If power steering pressure reaches a criterion pressure, it will open a hydraulically operated switch. When the ignition is on, battery voltage is passed through the switch to the ECM. If the switch opens and thus causes signal voltage at the ECM terminal to drop to zero during idle, the ECM will increase throttle blade opening to support the additional engine load. If the switch opens above 40 MPH, the ECM will turn on the service engine soon light and record a code 40. This switch is not found on DFI systems prior to 1985.

Cruise Control Enable (ON/OFF Switch)

When the cruise control switch is turned on (upper left corner of Figure 6-6), it supplies a 12-volt signal to the ECM. This alerts the ECM to operate the cruise control system. On 1986 Eldorado and Seville models, the cruise control operation was taken over by the body computer.

Cruise Set/Coast Switch

When the set/coast switch is pressed, a 12-volt signal is delivered to the ECM, Figure 6-6. If the cruise control system is in a coast mode, the ECM will engage and maintain the current vehicle speed. If it is in an engaged mode when the switch is pressed, the ECM will allow the vehicle speed to coast down.

Cruise Resume/Accelerate Switch

When the resume/accelerate switch is moved toward resume, a 12-volt signal is delivered to the ECM, Figure 6-6. The ECM resumes a controlled preset speed that was disengaged as a result of brake application. If the switch is moved toward accelerate while the system is engaged at a particular speed, the ECM will open the throttle and accelerate the vehicle as long as the switch is held. When the switch is released, the ECM maintains the speed at which it was released.

Cruise Control Brake Switch

When the ignition is on, it supplies battery voltage to the brake switch assembly, Figure 6-6. When the switch is closed (brake not applied), voltage is fed through the switch to the ECM. When the brake is applied, the switch is opened. The voltage signal at the ECM disappears and the ECM knows that the brake is applied. The ECM uses this information to disengage the cruise control if it is engaged.

ALDL Request

When the service technician grounds the test terminal of the ALDL (assembly line data link, most often referred to as assembly line communication link) under the dash, the ECM responds by putting the electronic spark timing in a fixed timing mode. This mode is used to check ignition timing.

Serial Data

Information fed to the ECM from either of the display panels or from the body computer module comes via a data link and is correctly referred to as serial data. This information can be concerning driver-selected temperature or air-conditioning mode or fuel mileage, or it can be requested information from the BCM.

OUTPUTS

Injectors

Each of the two injectors contains an electric solenoid, Figure 6-7. When the solenoid is pulsed by the ECM, the injector valve is lifted

213

Chapter 6 Cadillac's Digital Fuel Injection

FIGURE 6-7 Injector comparison. *Courtesy of GM Product Service Training.*

off of its seat a few thousandths of an inch. Fuel is sprayed out under a controlled pressure of 9 to 12 PSI. The amount of fuel sprayed is a function of how long the valve is held open (pulse width).

Each injector has its own power supply circuit and fuse. Voltage is supplied by the ignition switch. The ECM grounds each injector circuit individually to pulse the injectors.

Fuel Pressure Regulator. The pressure regulator provides an operating pressure ranging from 9 to 12 PSI to the injectors. Its operation is the same as that of EFI, which is discussed in the **Outputs** section of Chapter 4.

Fuel Pump Operation

Fuel and fuel pressure are supplied by a twin-turbine electric pump inside the fuel tank. The pump is turned on and off by a fuel pump relay mounted under the instrument panel on the relay panel. Its operation and control is like that of the EFI system pump discussed in the **Outputs** section of Chapter 4. The 1980 DFI system, however, did not have the oil pressure switch as a backup feature.

Electronic Spark Timing (EST)

The EST operation is the same as that of the EFI system discussed in the **Outputs** section of Chapter 4 with the exception that DFI does not use a Hall-effect switch.

Idle Speed Control (ISC)

The DFI system uses the same ISC motor to control idle speed that the CCC system does. Although the motor's operation and adjustment provisions are the same as those discussed in the **Outputs** section of Chapter 3, the ECM's strategy for its use varies a little from CCC. The ECM uses information from the following sensors and switches to control throttle blade position in any of three operational modes (some of these inputs are used to identify the operational mode):

- VSS
- TPS
- throttle switch
- REF
- A/C clutch
- CTS
- MAP
- P/N switch
- ignition switch

During Cranking and Initial Warm-up. During cranking the throttle is held in a preprogrammed position. When the engine has started, the throttle is held in a preprogrammed fast idle position to allow intake manifold conditions to stabilize. If the engine is started hot, the fast idle period will be short. If the engine is started cold, the fast idle period will be longer.

Curb Idle. When the engine is idling at normal operating temperature, curb idle is maintained at a preprogrammed speed. As engine loads change as a result of the A/C clutch cycling on and off, the transmission being put into gear or in neutral, or momentary high P/S pressure occurring, the ECM adjusts throttle blade opening to maintain the desired speed.

Deceleration. During deceleration manifold pressure goes very low, (high vacuum) and thus causes liquid fuel on the floor and walls of the manifold to evaporate. This results in high HC and CO production unless air is let into the manifold to mix with the additional fuel going into the combustion chambers. The ECM looks at vehicle speed and manifold pressure to determine severity of deceleration and how much throttle opening to provide. During hard deceleration it also stops fuel injection.

AIR Management

The air injection reaction (AIR) system used on DFI is essentially the same as the AIR system described in the **Outputs** section of Chapter 3 for a vehicle with a dual-bed catalytic converter. However, the DFI ECM's strategy for its use varies slightly. The ECM de-energizes the divert solenoid and thus puts the system into divert mode under any of the following conditions:

- acceleration or WOT operation. (Some DFI systems go into divert mode when the vehicle exceeds 60 MPH.)
- deceleration.
- extremely cold weather starting.
- certain failures that bring on either of the service engine lights.
- high RPM, which causes pump pressure to exceed the calibration of a pressure relief valve built into the control valve. This is not a divert valve function although it achieves the same results.

Note that any control valve solenoid or related electrical circuit failure results in a fail-safe condition (the failure leaves the valve in the safest operating mode). A mechanical failure within the valve may not work out so well.

Torque Converter Clutch (TCC) or Viscous Converter Clutch (VCC)

The TCC is a clutch within the torque converter that when applied eliminates the characteristic hydraulic slippage of a torque converter. When it is released, the converter acts like a conventional torque converter. Its apply and release are controlled primarily by the ECM. Its function is to improve fuel economy. For an in-depth description of the TCC and its operation, refer to **Torque Converter Clutch** in the **Outputs** section of Chapter 3. The parameters by which it is controlled are slightly different in the DFI system, and those are presented here.

The VCC, introduced in 1985 on the Cadillac front-wheel drive C body, works the same way that the TCC does except that when applied it does not provide 100% lock-up between the engine and the transmission input shaft as the TCC does. The VCC clutch assembly is made up of a rotor sandwiched between two parts called a cover and a body, Figure 6-8. The rotor and body surfaces that face each other have a series of fins that mesh together and allow a small space between each fin, Figure 6-9. This space is filled with silicone fluid, which drives the rotor. When the VCC is applied, it allows just enough constant slippage (about 40 RPM at 60 MPH) to eliminate some of the roughness that occurs with a 100% lock-up between the engine and transmission. It is controlled in the same way that a TCC is controlled except that one additional parameter is included in its control—transmission fluid temperature.

TCC/VCC Control Parameters. The ECM grounds the TCC/VCC apply solenoid circuit when:

- the criterion speed is reached as reported by the VSS. On some models this can be as low as 24 to 36 MPH depending on throttle position and transmission oil temperature.
- the engine is warm. If the coolant temperature is 18°C (64°F) or above when the engine is started, the ECM will apply the clutch when the coolant reaches 60°C (140°F) on VCC applications. If the coolant is colder than 18°C at start-up,

Chapter 6 Cadillac's Digital Fuel Injection

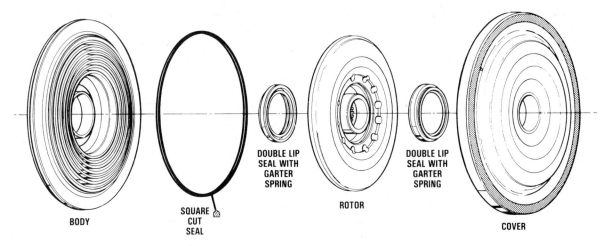

FIGURE 6-8 Viscous converter clutch (exploded view). *Courtesy of GM Product Service Training.*

the ECM will use a fixed-time delay before allowing the clutch to be applied. This is to improve low-temperature driveability.

FIGURE 6-9 Viscous converter clutch. *Courtesy of GM Product Service Training.*

- TPS is within a preprogrammed window. (The window varies depending on what gear the transmission is in.) TPS is compared to VSS, and the ECM selects different apply speeds for different combinations of TPS and VSS.

Two other switches can cause the clutch to be disengaged on VCC applications. They are the brake switch and the over-temperature protection switch, Figure 6-10. TCC applications do not use the over-temperature switch. These switches are not controlled by the ECM but are in the apply solenoid circuit and must be closed before the circuit can be completed to the ECM. They are both normally closed. The brake switch is opened when the brake is depressed. The over-temperature protection switch opens when transmission oil temperature exceeds 157°C (315°F).

The ECM on the 1987 Allante also controls the transmission 2–3 and 3–4 shifts. This is achieved by way of two ECM-controlled solenoids in the transmission's valve body. The solenoids block or open passages that allow oil to move the 2–3 and 3–4 shift valves.

Exhaust Gas Recirculation (EGR)

When open the EGR valve allows a small amount of exhaust gas (mostly carbon dioxide

Chapter 6 Cadillac's Digital Fuel Injection

FIGURE 6-10 VCC electric control circuit

and nitrogen, neither of which burn) into the intake manifold. The nonburnable exhaust gas displaces its equivalent amount of burnable material and slows down flame *propagation.* This reduces combustion temperature and thereby reduces NO_x production.

The DFI system uses a negative back-pressure EGR valve, Figure 6-11. It has some self-modulation capability. Exhaust pressure from the exhaust crossover (early fuel evaporation) passage in the intake manifold is sensed through a small hole in the lower portion of the valve itself. The hollow valve stem allows the exhaust pressure to get into a chamber beneath a second diaphragm below the main diaphragm. This pressure helps a small spring push the lower (second) diaphragm up so that it blocks a bleed hole in the main diaphragm. When the bleed hole is blocked, ambient air vented into the space between the two diaphragms is no longer allowed to bleed into the vacuum chamber above the main diaphragm. This allows sufficient vacuum to be developed in the vacuum chamber (using manifold vacuum as the source) to pull the diaphragm up to open the EGR valve. When the valve is open, the same hole that admitted exhaust pressure also admits manifold vacuum. If the net result of the engine vacuum versus exhaust pressure is a positive or near positive pressure, the lower diaphragm will be pushed up and the valve will be opened. If, however, the net result is a negative pressure, the lower diaphragm will be held down and the bleed valve will be open. This will cause the EGR valve to stay closed.

EGR Modulation. In addition to the EGR control provided by the negative back-pressure EGR valve, the DFI system on the 4.1-liter en-

Chapter 6 Cadillac's Digital Fuel Injection

1	EGR VALVE	6	VACUUM BLEED HOLE
2	EXHAUST GAS	7	SMALL SPRING
3	INTAKE AIR	8	LARGE SPRING
4	VACUUM PORT		
5	DIAPHRAGM		

FIGURE 6-11 Negative back-pressure EGR valve. *Courtesy of Cadillac Motor Car Division.*

ECM uses information from the following sensors to define prevailing driving conditions and the amount of EGR needed:

- CTS (no EGR when coolant temperature is low)
- TPS
- MAP
- ISC throttle switch (no EGR at idle)
- REF

Charcoal Canister Purge

Like most systems the DFI system uses an evaporative emission control system that consists primarily of a charcoal canister. The fuel tank is vented through the canister, and any evaporated fuel that tries to escape from the tank is stored in the activated charcoal within

1	EGR SOLENOID
2	BRACKET
3	NUT
4	HOSE ASM.
5	EGR VALVE

FIGURE 6-12 EGR solenoid. *Courtesy of Cadillac Motor Car Division.*

gine also controls vacuum to the EGR vacuum port. This control is achieved by means of a duty-cycled solenoid that while turned on bleeds atmospheric pressure into the vacuum supply hose to the EGR valve and thus causes the valve to close, Figure 6-12. When the solenoid is not turned on, it allows manifold vacuum to get to the EGR valve and thus causes it to open. By duty-cycling the solenoid, the ECM can control the amount of EGR opening and thus the EGR flow rate. The duty cycle varies for different driving conditions to provide the best possible driveability characteristics while still maintaining control of NO_x. A 10% duty cycle provides full vacuum to the EGR and a 90% duty cycle provides no vacuum to it. The

Chapter 6 Cadillac's Digital Fuel Injection

phragm down and thus closes the passage that otherwise allows air and fuel vapors to flow to the purge port. When manifold vacuum is applied to the top of the diaphragm, it is lifted and it opens the passage to the purge port. By turning on or off the solenoid, the ECM controls vacuum to the control valve. When the solenoid is energized, vacuum to the control valve is blocked. The ECM turns the solenoid off and allows purge when:

- the system is in closed loop and off idle.
- faults exist that trigger codes 13, 16, 44, or 45.

1. CHARCOAL CANISTER
2. TO FUEL TANK
3. PURGE VALVE (TO MANIFOLD)
4. CONTROL VALVE (TO MANIFOLD VACUUM)

FIGURE 6-13 Charcoal canister. *Courtesy of Cadillac Motor Car Division.*

the canister. When the engine is running in closed loop, manifold vacuum draws air through the canister and into the manifold, Figure 6-13. Any fuel vapors stored in the canister are purged by this airflow. Because the ECM has no way of knowing how much, if any, fuel is stored in the canister, it waits until the system is in closed loop so that it can compensate for any fuel introduced by the canister before it allows purging to occur.

Purge is controlled by two valves: the control valve and a solenoid-operated valve in the vacuum line that goes to the control valve, Figure 6-14. The control valve contains a diaphragm and spring. The spring pushes the dia-

1. CANISTER HOSE
2. MANIFOLD VACUUM
3. FUEL VAPOR PIPE
4. PURGE HOSE
5. VALVE ROCKER ARM COVER
6. CHARCOAL CANISTER

FIGURE 6-14 Charcoal canister hose connections. *Courtesy of Cadillac Motor Car Division.*

Chapter 6 Cadillac's Digital Fuel Injection

An electrical failure in the solenoid or its circuit causes purging to occur anytime the engine is running. This can cause:

- an overrich condition during warm-up or idle.
- an overlean condition during warm-up or idle (purging an empty canister).
- dieseling on shutdown.

Early Fuel Evaporation (EFE)

The EFE system applies heat to the area just beneath the throttle valves to prevent throttle blade icing and to enhance fuel evaporation during engine warm-up. During the warm-up period, this helps to:

- reduce the richness requirement of the air/fuel ratio.
- improve driveability.
- reduce emissions.

Rear-wheel drive DFI vehicles have used exhaust heat for this purpose without ECM control. A valve (EFE valve similar to a throttle or choke valve) is placed in the exhaust pipe on one side of the engine where it attaches to the exhaust manifold. A vacuum actuator containing a spring and a diaphragm opens and closes the valve. A thermal vacuum switch controls vacuum to the actuator. When coolant temperature is below 49° (120°F), the thermal vacuum switch routes vacuum to the EFE actuator. Vacuum applied to the actuator pulls the diaphragm against the spring and closes the EFE valve. This forces exhaust gases from that side of the engine to flow through a passage in the head to the intake manifold EFE passage. The exhaust crosses beneath the *plenum* to the other head, where it finds its way through another passage to the exhaust pipe on that side. When the coolant temperature goes above 49°C, a temperature-actuated vacuum control valve blocks vacuum to the actuator. The spring pushes the diaphragm back down and opens the EFE valve. Exhaust gas is no longer forced through the EFE passage.

| 1 | EFE HEATER ASSEMBLY | 2 | THROTTLE BODY ASSEMBLY |

FIGURE 6-15 EFE grid. *Courtesy of Cadillac Motor Car Division.*

Front-wheel drive DFI systems use a ceramic-covered electric heat element under the TBI unit for its EFE heat source, Figure 6-15. The heat element is powered by a relay (EFE relay) that is ECM controlled. The ECM activates the relay when all of the following conditions exist:

- MAT is less than 75°C (167°F).
- CTS indicates less than 106°C (223°F).
- battery voltage is more than 10 volts.

When any of the above conditions are no longer met, the EFE relay is turned off.

If EFE is off while the vehicle is being driven, the ECM will turn it back on if all of the following conditions occur:

- MAT is less than 38°C (100°F).
- CTS indicates less than 106°C.
- battery voltage is greater than 12 volts.

When any of the above conditions are no longer met, the EFE relay is turned off.

One other set of conditions causes the ECM to bring on EFE heat. If throttle position is open more than 30° (0° being closed and 90° being fully open) and MAT is less than 60°C (140°F), the EFE will be turned on for at least fifteen seconds to prevent throttle blade icing.

Chapter 6 Cadillac's Digital Fuel Injection

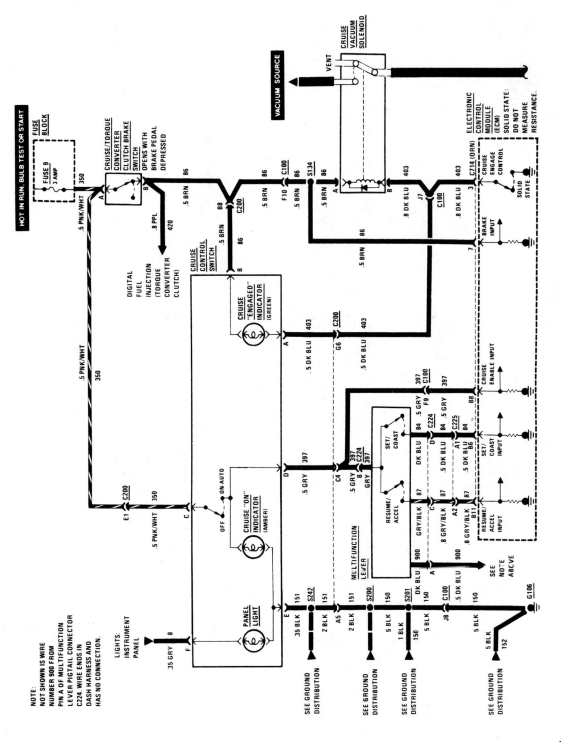

FIGURE 6-16 Cruise control circuit. *Courtesy of Cadillac Motor Car Division.* **(Continued on page 222)**

Chapter 6 Cadillac's Digital Fuel Injection

FIGURE 6-16   (Continued)

Cruise Control Vacuum Solenoid and Cruise Control Power Valve

When the ignition is turned on, power is fed to the cruise control on/off switch and to the brake switch, Figure 6-16. The cruise control on/off feeds power to:

- the resume/acceleration and set/coast switches.
- the cruise on indicator light.

The brake switch feeds power to:

- the ECM (as a signal that the brake is not applied).
- the cruise control vacuum solenoid.
- the cruise engaged indicator light.

When the set/coast switch is depressed (at speeds above 25 MPH), the selected speed is recorded in the ECM's memory, and the ECM grounds the cruise vacuum solenoid. The solenoid routes vacuum to the cruise control servo. By means of a vacuum diaphragm and spring, the servo controls throttle blade position and therefore vehicle speed. As vacuum is applied, the throttle is opened; as vacuum is removed, the spring closes the throttle. The amount of vacuum applied to the servo is controlled by the cruise control power valve, which is controlled by the ECM. The ECM supplies a pulsed voltage signal to the power valve; this is one of the few instances in which the ECM supplies the power instead of the ground. As the power valve is turned on and off, the servo is alternately exposed to vacuum or atmospheric pressure. The ECM pulses the power valve at whatever rate is necessary to make the indicated speed from the VSS match the speed it has recorded in its memory. As the ECM begins controlling vehicle speed, it also turns on the cruise control engaged light.

When the cruise control system is engaged (holding a speed), it can be disengaged by any one of three ways:

- The cruise control on/off switch can be turned off, in which case the selected speed is erased from the ECM's memory.
- The brake pedal can be depressed and thus causes the ECM to disengage the cruise control vacuum solenoid and thus vent the servo to atmospheric pressure. Moving the resume/accelerate switch to the resume position and releasing it after the brake has been released causes the system to reengage.
- Many faults that cause the ECM to turn on one of the service engine lights also cause it to disengage the cruise control system.

If while the system is engaged the resume/accelerate switch is moved and held, the system will accelerate the vehicle until the switch is released. The system will then engage on whatever speed the vehicle was traveling when the switch was released. If the set/coast switch is depressed and held while the system is engaged, it will disengage and the throttle will close. The system will reengage on whatever speed exists when the switch is released.

Air-conditioning (A/C) Relay and Cut-out

Later model DFI systems use an A/C compressor control relay to turn the compressor clutch on and off. The relay control coil is grounded by the ECM, Figure 6-17. However, the ECM often receives instructions from the Body Control Module (BCM) as to when to ground or unground the relay. The Climate Control Panel receives input from the driver. That input is used to determine the operating mode for the Climate Control System and is sent to the BCM by way of the serial data line. If the BCM determines that cooling is required, it will use its serial data line to the ECM to command that the compressor be turned on.

The BCM receives input from the A/C high side and low side temperature sensors plus the

Chapter 6 Cadillac's Digital Fuel Injection

FIGURE 6-17 Compressor clutch control circuit

Chapter 6 Cadillac's Digital Fuel Injection

FIGURE 6-18 Indicator lights circuit. *Courtesy of Cadillac Motor Car Division.*

225

Chapter 6 Cadillac's Digital Fuel Injection

A/C low pressure switch. It commands the compressor on or off according to temperature indications. It will also command the compressor to be turned off if the low pressure switch opens, indicating the refrigerant is low enough that continued operation might damage the compressor.

The ECM also has logic circuits that can control compressor operation based on input from the TPS, CTS, VSS, and power steering pressure switch.

Earlier DFI systems did not use an A/C relay, and the only A/C related concern of the DFI system is to turn the compressor off during WOT. On those systems, the A/C compressor is energized and controlled by the electronic climate control (ECC) module or the BCM depending on year and model. When the ECM sees a WOT indication from the TPS, it signals the ECC module or the BCM to turn off the compressor. This allows more power for acceleration and reduces the heat load on the engine. Refer to **Compressor Clutch Control** and **Programmer and Power Module** in the **Body Computer Module** section of this chapter for more information on compressor clutch control.

Service and Coolant Over-temperature Lights

If either of the two modules, the ECM or the BCM, sees a fault in any of the circuits they monitor, they will turn on one of the diagnostic indicator lights, Figure 6-18. The light selected depends on the nature of the problem.

BODY COMPUTER MODULE (BCM)

The BCM is a microcomputer just as is the ECM. It has the basic components (ROM, RAM, PROM, input/output interface, etc.) and about the same capability as the ECM; however, it is in charge of different functions within the vehicle. On the 1985 C body Cadillac, where it was introduced, it performs the following functions:

- controls the electronic climate control (air-conditioning system)
- controls the electric cooling fans, which draw air through the A/C condensor and radiator
- controls power to the power windows, trunk release, and *astroroof* (an electrically opened and closed window in the roof)
- provides and displays information for the driver
- controls information display panel dimming for visual clarity and comfort
- monitors for BCM system faults, stores codes to identify the faults, and in some cases provides a fail-safe operating condition to compensate for a failure

In addition to having sensors and switches for input information and actuators to send commands to, the BCM and the ECM interact with each other. In some cases information contained within the ECM is sent by way of a data link to the BCM as input information, and an output can be sent to the ECM to serve as ECM input. Figure 6-19 shows the kind of information that the two computers can exchange.

Electronic Climate Control (ECC)

In order to maintain the driver-selected temperature inside the vehicle, the BCM must select the most appropriate air inlet and outlet modes and blower speed. It also monitors the condition of ECC components.

ECC Inputs

Input information that the BCM needs to control the ECC system comes from several sources, Figure 6-20.

Climate Control Panel (CCP). The driver selects desired temperature and operating mode (automatic, hi fan, lo fan, economy, front window defog, rear window defog, etc.) on the

FIGURE 6-19 BCM/ECM interaction. *Courtesy of Cadillac Motor Car Division.*

FIGURE 6-20 BCM information sources. *Courtesy of GM Product Service Training.*

Chapter 6 Cadillac's Digital Fuel Injection

CCP. The CCP communicates this information in digital form via a data link to the BCM.

ECM. The ECM informs the BCM when the throttle is fully opened or when coolant temperature goes beyond a criterion value. In either case the BCM commands the compressor clutch to be disengaged.

Temperature Sensors. Strategically located thermistors provide information about outside air temperature, inside air temperature, and refrigerant temperature on both the low- and high-pressure sides of the A/C system. A switch located in the low-pressure side of the A/C system notifies the BCM if A/C pressure goes too low.

Air Mix Door Position. A potentiometer attached to the air mix door in the air distribution system housing provides feedback information to the BCM about the door's position. The air mix door blends heated and refrigerated air to produce the desired air temperature, Figure 6-21. By monitoring voltage at the blower motor, Figure 6-22, the BCM can determine blower speed.

ECC Outputs

Program Number. When the car is first started, the BCM compares inside air temperature to driver-selected temperature. It calculates what operational mode the ECC should be in to achieve the selected air temperature. In its calculation it considers the influence of outside air temperature. This calculation produces a program number between 0 and 100. A low number means inside air temperature is higher than selected temperature and calls for cooling. A high number means just the opposite. The program number dictates blower speed and air distribution system door positions, Figure 6-21. As inside air temperature comes closer to selected temperature, the BCM updates the program number, the blower speed decreases, and air distribution system doors are adjusted. When the temperatures are the same, the program number reaches an equili-

Chapter 6 Cadillac's Digital Fuel Injection

FIGURE 6-21 ECC mode doors. *Courtesy of GM Product Service Training.*

brium point. The BCM tries to maintain the program number at equilibrium. The program number is displayed on the CCP only when the system is in diagnostics.

Air Mix Door. The air mix door is moved by an electric motor (inside the programmer) and connecting linkage. All of the other doors are moved by vacuum motors controlled by vacuum solenoids. To determine what the air mix door position should be, the BCM checks:

- coolant temperature (requested from the ECM) to determine what heated air temperature will be.
- refrigerant temperature on the low-pressure side to determine what cold air

FIGURE 6-22 EEC blower motor control circuit. *Courtesy of GM Product Service Training.*

FIGURE 6-23 BCM interface devices. *Courtesy of GM Product Service Training.*

temperature will be. If the A/C is not operating, outside air temperature is used instead.
- the program number.

The BCM uses the air mix door feedback to see if the door is in the right position. Because the linkage between the air mix door motor and the door is adjustable, it must be carefully adjusted to be sure that its position coincides with the BCM-commanded position.

Compressor Clutch Control. The BCM controls the compressor clutch. Before applying the clutch, it checks to see that:

- outside temperature is above 0°C (32°F).
- engine coolant and/or A/C system high-pressure side temperatures are not too high.
- refrigerant pressure (indicates refrigerant charge) is not too low.
- the throttle is not wide open.

When the compressor is operating, the BCM continuously rechecks these parameters and disengages the clutch if any of them are not maintained to preprogrammed values.

To cycle the clutch, the BCM looks at outside air temperature and low-pressure side temperatures. The compressor clutch is activated when the low side temperature exceeds 10°C (50°F). The clutch is turned off again when low side refrigerant temperature drops to −1°C (30°F).

Programmer and Power Module. The BCM uses two interface devices for many ECC functions: the ECC programmer and the ECC power module, Figure 6-23. The BCM sends instructions for air door positions, heater water control, rear window defog, and compressor clutch through a data link to the programmer. The rear window defogger is a heat element in the rear window glass powered by a relay. The programmer commands the actuators for each of these functions, Figure 6-24, except the compressor clutch. The compressor clutch on/off signal is transmitted from the programmer to the power module, and the power module actually turns the clutch on and off. The compressor clutch on/off instructions, which the BCM sends to the programmer, are also sent to the ECM so that it can anticipate the change in engine load and adjust engine speed.

The BCM sends the blower motor instructions straight to the power module in the form of a variable voltage. The power module amplifies this signal and keeps it proportional to voltage received from the BCM and issues it as a command to the blower motor, Figure 6-22.

Electric Cooling Fans

Inputs. The BCM receives inputs concerning cooling fan operation from two sources: the A/C high-pressure side thermistor, located at the condensor outlet, and coolant temperature information from the ECM.

Outputs. The speed of the two coolant fans is controlled by a pulse width-modulated (PWM) signal produced by the BCM and sent to the fan control module, Figure 6-25. The module completes or opens the ground side of the fans' power circuit in response to the BCM signal. The PWM signal is cycled once for each time period (32/1,000 of a second). If the signal voltage is high during only a short portion of the period, Figure 6-26, the fan motors run at low speed because their power is supplied in short pulses and they spend most of each revolution coasting, although their momentum keeps their speed constant. The greater the portion of each period during which the pulsed voltage is high, the faster the fan motors turn.

The fans are turned on at 40% full speed when either the coolant temperature reaches 106°C (223°F) or the A/C high-pressure side temperature reaches 61°C (142°F). They are turned on at full speed when either the coolant temperature reaches 116°C (241°F) or the A/C high side temperature reaches 72°C (162°F). They are turned off when either the coolant temperature drops to 102°C (216°F) or the A/C high side temperature drops to 57°C (135°F). The lower turn-off temperatures prevent the

Chapter 6 Cadillac's Digital Fuel Injection

FIGURE 6-24 EEC programmer control of EEC mode. *Courtesy of GM Product Service Training.*

fans from indecisively turning on and off if one of the temperatures hovers at the on value.

The feedback signal sent back to the BCM from the fan control module allows the BCM to monitor fan motor performance. This signal is near 12 volts when the fans are off or stalled and near 0 volts when they are on or disconnected. The BCM compares feedback voltage to the commands it sent; if they do not match, it will set a fault code (F41).

Retained Accessory Power

After the ignition has been turned off (turned to lock) or to accessory, the BCM provides power to the power windows, astroroof, and the trunk release for ten minutes or until either a door is opened or the courtesy lights are turned on, Figure 6-27. When the ignition is turned off, the BCM grounds the retained power accessory relay for ten minutes. The relay then connects circuit 300 to battery voltage

FIGURE 6-25 Cooling fans' control. *Courtesy of GM Product Service Training.*

from an always hot source. Circuit 300 powers the trunk release, the power windows, and astroroof. If a door jam switch, the courtesy light switch, or the ignition switch is turned on, the BCM will unground the relay and the relay will reconnect circuit 300 to the ignition run terminal. Circuit 300 will now be powered only when the ignition switch is turned on.

Driver Information and Display

The BCM displays information about fuel and temperature on either the fuel data center (FDC) or the CCP. The driver selects the information to be displayed by pressing the appropriate button on one of the display panels. On vehicles equipped with digital display instrument clusters, the driver can choose to have the information displayed in either English or metric units by moving the English/metric switch on the digital cluster. Vehicles not equipped with the digital cluster display in either English or metric, depending on whether or not a wire (circuit 811) leading from terminal J2-13 of the BCM is connected to ground. If the

Chapter 6 Cadillac's Digital Fuel Injection

FIGURE 6-26 PWM command to cooling fan motors. *Courtesy of GM Product Service Training.*

FIGURE 6-28 Fuel gauge circuit. *Courtesy of GM Product Service Training.*

wire is grounded, information will be expressed in metric units; if it is not grounded, information will be expressed in English units.

Fuel Data Inputs. For fuel data information, the BCM gets information from the fuel gauge sending unit (a potentiometer), the gauge fuse, and the ECM. The potentiometer is used to measure fuel level just as fuel gauges have done for years. An arm with a float on it moves the movable contact of the potentiometer as the fuel level changes. By comparing return voltage from the potentiometer to a voltage reading taken at the gauge fuse, the BCM can calculate fuel level, Figure 6-28.

Vehicle speed and distance traveled (distance traveled can easily be calculated from VSS information because 4,004 VSS pulses equal 1 mile using the pulse generator–type VSS) plus injector pulse width information are sent from the ECM to the BCM via the data link.

FIGURE 6-27 Retained accessory power circuit. *Courtesy of GM Product Service Training.*

232

Fuel Data Outputs. Normally the BCM displays fuel level by way of the FDC as discussed above. Upon request it displays:

- *instantaneous fuel mileage.* If the INST button on the FDC panel is pressed, the BCM looks at fuel being consumed (from injector pulse width information) and vehicle speed; from this it calculates the fuel mileage being achieved at that moment.
- *average fuel mileage.* The BCM is continuously receiving and updating its information regarding fuel used and distance traveled. This information is stored in memory. If, however, the reset button on the FDC panel is pressed, previous distance traveled and fuel used information will be erased from memory and new data will be accumulated. If the AVG button is pressed, the BCM will calculate fuel mileage for the distance traveled since the last reset.
- *fuel range.* When the RANGE button is pressed, the BCM checks remaining fuel from fuel level information, calculates fuel used during the last 25 miles, then calculates and displays the distance that can be traveled on the remaining fuel.
- *fuel used.* When the FUEL USED feature is requested, the BCM calculates and displays the fuel used since the last reset.

Temperature Display. Upon request the BCM displays outside temperature as indicated to it by the outside temperature thermistor. If outside temperature is not requested, the CCP will display inside temperature.

Instrument Panel Display Dimming

The characters displayed by the FDC, the CCP, the radio, and the optional digital cluster are displayed by a group of small *vacuum fluorescent* (VF) glass tubes. *Anodes* (conductors on which positive potential voltage is placed) are placed on one side of the tube so that they form all of the vertical and horizontal bars of alpha and numerical characters. The anodes are coated with a fluorescent material. A series of thin, tungsten-coated wire strands are placed opposite the anodes on the other side of the tube. This serves as the *cathode* (a conductor through which current with negative voltage is passed). Between the anode and the cathode, a fine wire mesh called the *grid* is placed. The tube *(ampule)* is evacuated and filled with argon or neon gas.

As current is passed through the cathode, it becomes hot and causes the tungsten to give off a cloud of electrons. The BCM applies an amplified 16-volt positive potential to selected anodes. The electrons given off by the cathode are attracted to and bombard the energized anodes. As they strike an anode's fluorescent material coating, it glows and thus provides the digital display. The more electrons that strike the fluorescent material, the brighter it glows.

During daylight hours the VF displays are adjusted by the BCM to full brightness for maximum visibility. As sunlight fades the VF displays need less brightness. When the park lights are turned on (they come on with the headlights), that signal is fed to the BCM, Figure 6-29. Upon being alerted that the lights are on, the BCM looks to the signal provided by the headlight switch rheostat, which the driver can turn to select VF display brightness.

The BCM controls display brightness by controlling a pulse width-modulated voltage to the grid in the ampule. When a modulated voltage is applied to the grid, it attracts or filters some of the electrons given off by the cathode and prevents them from striking the energized anodes.

Diagnostic Testing and Fail-safe Actions

The BCM monitors many of its circuits, such as the blower motor and the cooling fan motors, for proper operation. It also has sen-

Chapter 6 Cadillac's Digital Fuel Injection

FIGURE 6-29 Vacuum fluorescent display dimming circuit. *Courtesy of GM Product Service Training.*

sors specifically used to report problems such as the low-pressure switch in the A/C accumulator. If the BCM sees a fault (voltage in a given circuit or from a sensor is not within a range expected by the BCM) in any of its related circuits, it will record a code in its diagnostic memory. It may turn on the service air-conditioning light, and it may also initiate a fail-safe action, depending on the circuit the fault is in, Figure 6-30. For example, each time that the BCM turns the A/C compressor clutch on, it checks the A/C low-pressure switch first. If the switch is open (low pressure), it will not engage the clutch. If the switch stays open for thirty seconds, code F48 will be set and the following fail-safe actions are initiated:

- The service air-conditioning light will be turned on for thirty seconds. The light will come on again each time the driver selects AUTO on the CCP and the next time the ignition is turned on.
- The compressor clutch will be disabled until the ignition is turned off.
- The ECC system will be put into ECON instead of AUTO if the AUTO mode is selected.

BCM DIAGNOSTIC CODES

CODE	CIRCUIT AFFECTED
▼ F10	OUTSIDE TEMP SENSOR CKT
▼ F11	A/C HIGH SIDE TEMP SENSOR CKT
▼ F12	A/C LOW SIDE TEMP SENSOR CKT
▼ F13	IN-CAR TEMP SENSOR CKT
▼ F14	DIESEL COOLANT SENSOR CKT
▼ F30	CCP TO BCM DATA CKT
▼ F31	FDC/DDC TO BCM DATA CKT
▼ F32	ECM-BCM DATA CKT'S
▼ F40	AIR MIX DOOR PROBLEM
▼ F41	COOLING FANS PROBLEM
☑ F46	LOW REFRIGERANT WARNING
☑ F47	LOW REFRIGERANT CONDITION
☑ F48	LOW REFRIGERANT PRESSURE
▼ F49	HIGH TEMP CLUTCH DISENGAGE
▼ F51	BCM PROM ERROR

☑ TURNS ON "SERVICE AIR COND" LIGHT
▼ DOES NOT TURN ON ANY LIGHT

COMMENTS:
F11 TURNS ON COOLING FANS WHEN A/C CLUTCH IS ENGAGED
F12 DISENGAGES A/C CLUTCH
F14 & F32 TURN ON COOLING FANS
F30 TURNS ON FT. DEFOG AT 75° F
F41 TURNS ON "COOLANT TEMP/FANS" LIGHT WHEN FANS SHOULD BE ON
F47 & F48 SWITCHES FROM "AUTO" TO "ECON"

FIGURE 6-30 BCM diagnostic codes. *Courtesy of Cadillac Motor Car Division.*

It should be noted that fail-safe actions are intended to protect system components and that the codes that initiate fail-safe actions cannot be set while the system is in diagnostics. Therefore, if a code is cleared while the system is in diagnostics and the system with a fault is operated while in diagnostics, the fail-safe actions will not occur and damage to a component might occur.

SYSTEM DIAGNOSIS AND SERVICE

The self-diagnostic function of the DFI system works basically the same way as other General Motors' computer control systems, but it is more elaborate. When introduced in mid-1980 (the same system was used through 1981), its self-diagnosis process was like that of a CCC system. Since then, however, the system's capability in that area has greatly expanded. As the ECM and/or BCM control the systems for which they are responsible, they continuously monitor operating conditions for possible faults. They compare operating conditions against expected, preprogrammed operating standards. By doing so certain circuit and component faults can be identified. When a fault is identified, a two-digit trouble code number with either an E or an F prefix is stored in memory. Codes with an E prefix are stored by the ECM, Figure 6-31; codes with an F prefix are stored by the BCM, Figure 6-30. When the system is put into diagnostics, this code number is displayed on either the CCP or the FDC, depending on the year and model.

Types of Trouble Codes

A trouble code can be stored in either of two ways: as a current code (sometimes referred to as a hard code because it currently exists) or as a history code (representing a fault that existed at one time but does not exist now). When the system is put into diagnostics, all of the codes are shown once in numerical order

ECM DIAGNOSTIC CODES

	CODE	MALFUNCTION
■■	E12	NO DISTRIBUTOR SIGNAL
☐	E13	OXYGEN SENSOR NOT READY (CANISTER PURGE)
☐	E14	SHORTED COOLANT SENSOR CIRCUIT
☐	E15	OPEN COOLANT SENSOR CIRCUIT
■■	E16	GENERATOR VOLTAGE OUT OF RANGE (ALL SOLENOIDS)
☐	E18	OPEN CRANK SIGNAL CIRCUIT
☐	E19	SHORTED FUEL PUMP CIRCUIT
■■	E20	OPEN FUEL PUMP CIRCUIT
☐	E21	SHORTED THROTTLE POSITION SENSOR CIRCUIT
☐	E22	OPEN THROTTLE POSITION SENSOR CIRCUIT
☐	E23	EST/BYPASS CIRCUIT PROBLEM (AIR)
☐	E24	SPEED SENSOR CIRCUIT PROBLEM (VCC)
☐	E26	SHORTED THROTTLE SWITCH CIRCUIT
☐	E27	OPEN THROTTLE SWITCH CIRCUIT
☐	E28	OPEN THIRD OR FOURTH GEAR CIRCUIT
☐	E30	ISC CIRCUIT PROBLEM
■■	E31	SHORTED MAP SENSOR CIRCUIT (AIR)
■■	E32	OPEN MAP SENSOR CIRCUIT (AIR)
■■	E34	MAP SENSOR SIGNAL TOO HIGH (AIR)
☐	E37	SHORTED MAT SENSOR CIRCUIT
☐	E38	OPEN MAT SENSOR CIRCUIT
☐	E39	VCC ENGAGEMENT PROBLEM
☐	E40	OPEN POWER STEERING PRESSURE CIRCUIT
■■	E44	LEAN EXHAUST SIGNAL (AIR & CL & CANISTER PURGE)
■■	E45	RICH EXHAUST SIGNAL (AIR & CL & CANISTER PURGE)
☐	E47	BCM - ECM DATA PROBLEM
■■	E51	ECM PROM ERROR
▼	E52	ECM MEMORY RESET INDICATOR
▼	E53	DISTRIBUTOR SIGNAL INTERRUPT
▼	E59	VCC TEMPERATURE SENSOR CIRCUIT
▼	E60	TRANSMISSION NOT IN DRIVE
▼	E63	CAR SPEED AND SET SPEED DIFFERENCE TOO HIGH
▼	E64	CAR ACCELERATION TOO HIGH
▼	E65	COOLANT TEMPERATURE TOO HIGH
▼	E66	ENGINE RPM TOO HIGH
▼	E67	CRUISE SWITCH SHORTED DURING ENABLE

ECM AND CRUISE CONTROL COMMENTS:

■■ TURNS ON "SERVICE NOW" LIGHT

☐ TURNS ON "SERVICE SOON" LIGHT

▼ DOES NOT TURN ON ANY TELLTALE LIGHT

( ) FUNCTIONS WITHIN BRACKETS ARE DISENGAGED WHILE SPECIFIED MALFUNCTION REMAINS CURRENT (HARD)

E16 & E24 DISABLE VCC FOR ENTIRE IGNITION CYCLE
E24 & E67 DISABLE CRUISE FOR ENTIRE IGNITION CYCLE
CRUISE IS DISENGAGED WITH CODE(S) E16, E51 OR E60 - E67

FIGURE 6-31 ECM diagnostic codes. *Courtesy of Cadillac Motor Car Division.*

Chapter 6 Cadillac's Digital Fuel Injection

from low to high; then they are shown again. This time, however, only the current codes are shown. From 1982 through 1984, DFI systems show all codes in the first two passes and show only current codes in a third pass.

Service Lights and Fail-safe

Faults that the ECM recognizes (E codes) and that require urgent attention bring on the service now light. Faults that the ECM recognizes and that need to be called to the driver's attention but are less urgent bring on the service soon light. Certain F codes pertaining to the ECC system bring on the service air-conditioning light.

If a fault that can cause damage to a circuit or component or that can result in unacceptable performance during continued operation occurs, the ECM or BCM will activate a fail-safe action. The fail-safe action is designed to either deactivate the faulty component or compensate for its malfunction.

Diagnostic Procedure

Before beginning a diagnostic procedure on a DFI-equipped vehicle, you should first be sure that you:

- have at least a general understanding of the operation of the ECM and/or the BCM and any of the subsystems in which the problem or problems seem to exist.
- are at least familiar with the self-diagnostic features of the system.
- have access to a service manual that contains the diagnostic charts.
- do a thorough visual inspection of the system in which the fault seems to exist and any related systems. Check for loose or broken vacuum lines, pinched wires, loose wire connections, and so forth.

Diagnostic Guide. Having complied with the suggestions above, continue your diagnostic procedure by answering the questions asked in the next four paragraphs.

1. *Are the on-car diagnostics working?* If the service air-conditioning, service now, or service soon lights are not working properly or if either of the data display panels fails to fully light up or respond properly, the self-diagnostic system check, Figure 6-32, will refer to the appropriate chart for correcting the problem.

2. *Are trouble codes displayed?* If any trouble codes are displayed, go to the trouble code chart of the same number. If the fault is intermittent (the code was not repeated on the second pass or third pass for earlier models), be sure to observe any notes concerning intermittent codes in the text that accompanies each trouble code chart. The procedure for displaying the trouble codes is explained later in this section. If no trouble codes are stored, proceed to the next paragraph.

3. *Do all switch tests pass?* After the trouble codes have been displayed, perform the switch test as explained under **Switch Test** in this section. If any switch tests do not pass, go to the code chart identified by the same number as the diagnostic display number (under **Switch Tests** in Figure 6-33) for the switch that did not pass. If all of the switches do pass, go to the next paragraph.

4. *Is the fuel system controlling correctly?* To answer this use the field service mode, which is a part of the DFI system check, Figure 6-34. If the fuel system is not controlling properly, the DFI system check will refer to the appropriate chart. If the fuel system is controlling properly, the DFI system check will refer to the symptom charts.

Using these simple steps at the beginning of the diagnostic procedure saves time and often prevents unnecessary parts replacement that results from misdiagnosis.

Entering Diagnostic Mode. To enter diagnostics turn the ignition on, then simultaneous-

Chapter 6 Cadillac's Digital Fuel Injection

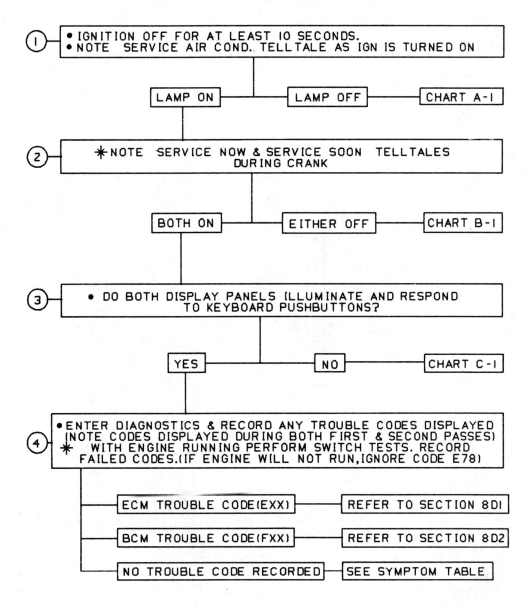

FIGURE 6-32 Self-diagnostic system check. *Courtesy of Cadillac Motor Car Division.*

Chapter 6 Cadillac's Digital Fuel Injection

FIGURE 6-33 Diagnostic chart, part 1. *Courtesy of Cadillac Motor Car Division.*

Chapter 6 Cadillac's Digital Fuel Injection

FIGURE 6-34 DFI system check. *Courtesy of Cadillac Motor Car Division.*

Chapter 6 Cadillac's Digital Fuel Injection

FIGURE 6-33 Display panel in segment check

ly depress the off and warmer buttons on the CCP. Hold them until all display panel segments of both display panels light up, Figure 6-35. If any of the digit segments do not light up (they should all make 8's), the panel will have to be replaced. An 8 being displayed by a panel with two segments inoperative can appear as a 3.

Display of Trouble Codes. After the segment check ends, the FDC displays 8.8.8 followed by ..E. The ..E indicates that the first pass of the ECM codes are coming up next and that all ECM codes will be displayed, Figure 6-35. After the first pass, ..EE is displayed and indicates that the second pass of the ECM codes is coming where only the hard codes will be displayed. If all stored ECM codes are intermittent, the ..EE will not be displayed and the second pass will not occur. If code E51 (a PROM fault) is displayed, it will continue to be displayed until the diagnostic mode is exited or until the fault is repaired and the code is cleared. While code E51 is displayed, the system will not advance to another diagnostic feature.

If no ECM codes are stored, the 8.8.8 will be followed by ..F, indicating that BCM codes are coming up. The same sequence occurs. The first pass displays all stored BCM codes followed by ..FF, and the second pass displays only hard codes. If no BCM hard codes are stored, the ..FF will not be displayed and the second pass will not occur.

After all stored codes are displayed or if there are no stored codes, .7.0 will be displayed. This indicates that the system is ready for the next self-diagnostic feature to be selected. There are several different diagnostic features that can be selected from this point:

- ECM switch tests
- ECM data display
- ECM output cycling
- BCM data display
- ECC program override
- cooling fan override
- exit diagnosis or clear codes and exit diagnosis

ECM Switch Tests. The engine must be running to do this test sequence. When the FDC displays .7.0, depress and release the brake pedal; the FDC display should advance to E.7.1. When it does, cycle the cruise control brake switch by tapping the brake pedal again within ten seconds. As the switch is cycled, the ECM monitors its operation to be sure that it is functioning properly. If the switch does not cycle within ten seconds, the ECM will record it as having failed to function. At the end of ten seconds or after the switch has cycled properly, the display advances to the next switch test code. The same sequence is repeated for each code and switch listed under **Switch Test** in Figure 6-33.

After the switch test sequence is completed, the ECM displays on the FDC the code of any switch that did not pass. The failed switch code remains on display until the switch circuit is repaired and retested. After the test sequence has been completed and all switches have passed, E.0.0 will be displayed, followed by .7.0 (the system is ready for another diagnostic feature selection).

ECM Data Display. This mode is often referred to as *engine parameters*. In this mode the ECM displays the values it is receiving from selected sensors or switches. It can be entered by pressing the lo button on the CCP while .7.0 is displayed on the FDC. When this mode is entered, E.9.0 appears on the FDC. The parameter codes, the parameters they represent, the value range, and the units in which the parameters are expressed are shown in Figure 6-36 under **Engine Data Display.** To advance to the next higher parameter code number, press the hi button on the CCP. Pushing the lo button returns to the next lower parameter code number.

These parameter values can be used as follows:

- They can be compared to those of a vehicle known to be operating properly to see if they are what they should be.
- They can be watched to see that they change as operating conditions change (for instance, the TPS parameter should change as the throttle is moved).
- Some of them are called for in the diagnostic charts, where their values are interpreted as good or at fault.

This diagnostic mode can be exited at any time and the display returned to the selection point, .7.0, by clearing the ECM or BCM codes (press off and hi for ECM codes or off and lo for BCM codes).

ECM Output Cycling. In this diagnostic mode, the ECM cycles all of its output actuators every three seconds with exception of the cruise control power valve, which cycles continuously. Cycling continues for two minutes; then the system automatically reverts to ECM Output Cycling Ready (E.9.5 displayed on the FDC). While the actuators are being cycled, you can observe that each is operating properly. To enter this output cycling mode, .7.0 must be displayed and the engine must be running.

- Turn the cruise on/off switch to the on position. This allows the cruise outputs to be cycled.
- Turn the engine off. Within two seconds turn the ignition on.
- Press hi on the CCP (E.9.5 appears on the FDC).
- Open and close the ISC throttle switch by depressing and releasing the throttle. The ECM output cycling mode is activated (E.9.6 is displayed).

This diagnostic mode can be exited at any time and the display returned to the selection point, .7.0, by clearing the ECM or BCM codes (press off and hi for ECM codes or off and lo for BCM codes).

BCM Data Display. As shown in Figure 6-36 under **BCM Data Display,** this mode displays the values sent to the BCM from most of its information sources as well as the value of some of its outputs. Figure 6-36 also shows the maximum range of each parameter and the units of measurement.

In order to go into the BCM data display mode, .7.0 must be displayed on the FDC. Depress and release the outside temp button on the CCP. The FDC displays F.8.0 briefly and then goes to parameter code P.2.0 or P.3.1 (the

Chapter 6 Cadillac's Digital Fuel Injection

E.9.0 ENGINE DATA DISPLAY

PARAMETER NUMBER	PARAMETER	PARAMETER RANGE	DISPLAY UNITS
P.0.1	Throttle Position	-10 - 90	Degrees
P.0.2	MAP	14 - 109	kPa
P.0.3	Computed BARO	61 - 103	kPa
P.0.4	Coolant Temperature	-40 - 151	C
P.0.5	MAT	-40 - 151	C
P.0.6	Injector Pulse Width	0 - 99.9	ms
P.0.7	Oxygen Sensor Voltage	0 - 1.14	Volts
P.0.8	Spark Advance	0 - 52	Degrees
P.0.9	Ignition Cycle Counter	0 - 50	Key Cycles
P.1.0	Battery Voltage	0 - 25.5	Volts
P.1.1	Engine RPM	0 - 6370	RPM ÷ 10
P.1.2	Car Speed	0 - 255	MPH
P.1.3	ECM PROM I.D.	0 - 255	Code

F.8.0 BCM DATA DISPLAY

PARAMETER NUMBER	PARAMETER	PARAMETER RANGE	DISPLAY UNITS
P.2.0	Commanded Blower Voltage	-3.3 - 18.0	Volts
P.2.1	Coolant Temperature	-40 - 215	C
P.2.2	Commanded Air Mix Door Position	0 - 100	%
P.2.3	Actual Air Mix Door Position	0 - 100	%
P.2.4	Air Delivery Mode	0 - 7	Code
	0 = Max A C 4 = Off		
	1 = A C 5 = Normal Purge		
	2 = Intermediate 6 = Cold Purge		
	3 = Heater 7 = Front Defog		
P.2.5	In-Car Temperature	-40 - 102	°C
P.2.6	Actual Outside Temperature	-40 - 93	°C
P.2.7	High Side Temperature (Condenser Out)	-40 - 215	°C
P.2.8	Low Side Temperature (Evaporator In)	-40 - 93	°C
P.2.9	Actual Fuel Level	0 - 19.0	Gallons
P.3.0	Ignition Cycle Counter	0 - 99	Key Cycles
P.3.1	BCM PROM I.D.	0 - 255	Code

ECM PROM I.D.

ECM PROM I.D. is Parameter .1.3 of Engine Data and is displayed as a numerical code as follows: X X X

- **FINAL DRIVE RATIO**
  - 2 = 3.33:1 (2.97:1 Effective Ratio)
- **EMISSIONS SYSTEM**
  - 1 = Federal
  - 2 = California
  - 3 = Export
  - 4 = Altitude
- **ECM PROM CALIBRATION**
  - Number varies with individual calibration.

BCM PROM I.D.

BCM PROM I.D. is Parameter .3.1 of BCM Data and is displayed as a numerical code as follows: X X X

- **ENGINE SYSTEM**
  - Blank = Gas
  - 1 = Diesel
- **BCM PROM CALIBRATION**
  - Numbers vary with individual calibration.

ECM STATUS LIGHT DISPLAY	LIGHT ON	IN 4th GEAR	VCC ENABLED	CLOSED THROTTLE	RICH	CLOSED LOOP
	LIGHT OFF	NOT IN 4TH GEAR	VCC DISABLED	OPEN THROTTLE	LEAN	OPEN LOOP
	INDICATOR	〰	〰	Off	Econ	Auto
	FUNCTION	4TH GEAR INPUT	VCC OUTPUT	THROTTLE SWITCH INPUT	OXYGEN SENSOR INPUT	ECM OPERATING MODE

BCM STATUS LIGHT DISPLAY	FUNCTION	A/C CLUTCH OUTPUT	COMPRESSOR LOW PRESSURE SWITCH INPUT	HEATER WATER VALVE OUTPUT	A/C-DEF MODE DOOR OUTPUT	COOLING FANS STATUS	UP/DOWN MODE DOOR OUTPUT
	INDICATOR	Outside Temp	°F	°C	Lo Fan	Auto Fan	Hi Fan
	LIGHT ON	ENERGIZED	OPEN (LOW PRESSURE)	CLOSED (NO WATER FLOW)	A C	FANS RUNNING	UP
	LIGHT OFF	DE-ENERGIZED	CLOSED	OPEN	DEF	FANS OFF	DOWN

FIGURE 6-36 Diagnostic chart, part 2. *Courtesy of Cadillac Motor Car Division.*

first and last parameters in the sequence). The parameter code is displayed for one second, then the parameter value is displayed for nine seconds. This sequence repeats with the same parameter until you select another one. To advance to another parameter, push the hi button on the CCP; to move back to a lower parameter code number, push lo on the CCP.

These parameter values can be used as follows:

- They can be compared to those of a vehicle known to be operating properly to see if they are what they should be.
- They can be watched to see that they change as operating conditions change (for instance, the commanded air mix door position should change as the selected temperature is changed).
- Some of them are called for in the diagnostic charts, where their values are interpreted as good or at fault.

This diagnostic mode can be exited at any time and the display returned to the selection point, .7.0, by clearing the ECM or BCM codes (press off and hi for ECM codes or off and lo for BCM codes).

ECC Program Override. During the BCM data display mode, the CCP displays a two-digit number that is the ECC program number. This number represents the level of heating or cooling effort by the ECC system as explained in the BCM section of this chapter. The program number automatically changes by the BCM as the inside temperature and/or the selected temperature changes.

The automatic calculation of the program number can, however, be overridden by pushing the warmer or cooler buttons (lower right of Figure 6-36). While in the BCM data display mode, holding the warmer button pushed in manually forces the program number up to a maximum of 100; 100 commands maximum heating. Pressing and holding the cooler button drives the program number down to a minimum of 0, which commands maximum cooling.

This override of the automatically calculated program number allows you to set it at any place between 0 and 100 and observe the reaction of the ECC system and the BCM data parameters. Once the program number has been overridden, it continues to be so until the BCM data display mode is exited.

Cooling Fan Override. This feature allows you to override the automatic control of the cooling fans' speed and manually command either high fans or fans off. To activate this feature the FDC must be displaying .7.0. Depress and release the econ button on the CCP. The FDC displays F.8.5. To command high fans, press and hold the hi button on the CCP until the fans achieve full speed. Releasing the hi button returns the fans to BCM automatic control. To command fans off, press and hold the lo button on the CCP until the fans stop. Releasing the lo button returns the fans to BCM automatic control. This feature can be exited at any time and the display returned to the selection point, .7.0, by clearing the ECM or BCM codes (press off and hi for ECM codes or off and lo for BCM codes).

Exit Diagnosis. Exit diagnosis by pushing auto on the CCP.

ECM/BCM Status Light Display. While in diagnostics the ECC mode indicators on the CCP are used to indicate the mode in which selected systems operated by either the ECM or BCM are operating (bottom portion of Figure 6-36). The operational mode is indicated by whether or not the designated light is on.

REVIEW QUESTIONS

1. How is the ECM's learning erased?
2. How is its learning restored?
3. The ECM monitors generator voltage to be sure that it is between _____ and _____ volts.
4. What is the function of the P/S switch?
5. Name at least one non-engine-related function that the ECM controls.

Chapter 6 Cadillac's Digital Fuel Injection

6. What is the acceptable fuel pressure for a DFI system?
7. What is the function of the black wire in the four-wire distributor connector?
8. List nine sensors or switches that are important in order for the ISC motor to operate in all driving conditions.
9. How does the ISC motor respond during deceleration?
10. List three criteria that must be met before the ECM will apply the TCC or VCC.
11. In addition to using a negative back-pressure EGR valve, how is the EGR flow rate controlled?
12. List three things that turn off the cruise control on a DFI system.
13. List six functions of the BCM.
14. What two inputs does the BCM use to control the coolant fans?
15. What does the term *fail-soft* mean?
16. How is the DFI system put into diagnostics?
17. After all of the trouble codes are displayed, .7.0 is displayed. What does this indicate?
18. What does the switch test accomplish?
19. What occurs during the ECM data display mode?
20. What is the most practical use of the ECM out cycling mode?
21. While in diagnostics what is the function of the status light display?

ASE-type Questions. (Actual ASE test questions will probably not be as product specific.)

22. A Cadillac with DFI is brought into the shop running very rough. Technician A says that the cause could be a blown injector fuse. Technician B says this cannot be the problem because if the injector fuse were blown, no fuel would be injected into the manifold and the engine would not run at all. Who is correct?
    a. A only
    b. B only
    c. both A and B
    d. neither A nor B

23. Technician A says that when a DFI system is running in module mode, ignition timing is being controlled by the pickup coil and HEI module. Technician B says that when the system is operating in module mode, the engine is at cranking speed or a system failure has occurred. Who is correct?
    a. A only
    b. B only
    c. both A and B
    d. neither A nor B

24. Technician A says that if voltage is applied to the divert valve solenoid, AIR system air will be directed to the air cleaner. Technician B says that under those conditions the air will be directed to the switching valve. Who is correct?
    a. A only
    b. B only
    c. both A and B
    d. neither A nor B

25. Technician A says that the viscous converter clutch works just like the torque converter clutch and is nothing more than a name change. Technician B says that the VCC works similarly to the TCC but that it is designed to allow a small amount of converter slippage even when applied. Who is correct?
    a. A only
    b. B only
    c. both A and B
    d. neither A nor B

26. Technician A says that the EFE system on rear-wheel drive DFI vehicles is not computer controlled and is designed to prevent throttle blade icing. Technician B says that the EFE system on front-wheel drive DFI vehicles is computer controlled and is de-

A/C	Air-conditioning
AIR	Air Injection Reaction
ALDL	Assembly Line Data Link
BARO	Barometric Pressure
CCP	Climate Control Panel
DFI	Digital Fuel Injection
ECC	Electronic Climate Control
ECM	Electronic Control Module (controller)
EECS	Evaporative Emissions Control System
EFE	Early Fuel Evaporation
EGR	Exhaust Gas Recirculation
EST	Electronic Spark Timing
FDP	Fuel Data Panel
HEI	High Energy Ignition
ISC	Idle Speed Control
MAP	Manifold Absolute Pressure
N.C.	Normally Closed
N.O.	Normally Open
O_2	Oxygen (sensor)
PCV	Positive Crankcase Ventilation
PFI	Port Fuel Injection
P/N	Park/Neutral
PROM	Programmable Read-only Memory (engine calibrator)
TBI	Throttle Body Injection (unit)
THERMAC	Thermostatic Air Cleaner
TPS	Throttle Position Sensor
TVS	Thermal Vacuum Switch
VIN	Vehicle Identification Number
VSS	Vehicle Speed Sensor
WOT	Wide-open Throttle

FIGURE 6-37 Explanation of abbreviations

signed to aid fuel evaporation during engine warm-up. Who is correct?
a. A only
b. B only
c. both A and B
d. neither A nor B

GLOSSARY

In addition to the following terms, refer to Figure 6-37 for a list of the abbreviations used in this chapter.

ACCUMULATOR A reservoir located between the evaporator (under dash unit) and the compressor of an air-conditioning system and that allows freon gas and any remaining liquid to separate.

AMPULE A strong glass tube that forms the body of an electrical component such as a vacuum tube.

ANODE A conductor with positive voltage potential applied.

ASTROROOF A term used by Cadillac to identify a power window in the roof of a vehicle.

Chapter 6   Cadillac's Digital Fuel Injection

CATHODE   A conductor with negative voltage potential applied.

FAIL-SAFE   A feature designed into a system and that provides limited continuous operation in the event of certain component failure.

MODULATED DISPLACEMENT   A system in which a series of ECM-controlled electromechanical devices can disable the rocker arms of selected cylinders and thus prevent the valves from opening. The engine operates on eight, six, or four cylinders, depending on power requirements.

PROPAGATION   The spread of the flame front across the combustion chamber of an engine as used in automotive terminology.

VACUUM FLUORESCENT   A glass tube containing anodes and a cathode. The anodes are coated with a fluorescent material and are positioned so that they form all of the segments of alpha and numerical characters. As current is passed through the cathode, it gives off electrons. If the anodes have positive voltage applied, the electrons given off by the cathode will be attracted to them and will strike the fluorescent material and cause it to glow.

7

FORD'S ELECTRONIC ENGINE CONTROL IV

Objectives

After studying this chapter, you will be able to:

- list the three operational strategies of the EEC IV system.
- identify at least five of the major sensors.
- identify at least five of the major actuators.
- identify the three types of fuel metering devices.
- identify the controlled components most directly concerned with emissions rather than engine performance or mileage.
- identify the major components of the diagnostic procedure.

The electronic engine control IV (EEC IV) system is the fifth generation of Ford's computerized engine control systems. It was first introduced in late 1982. By 1985 the EEC IV system was the only computerized engine control system Ford was using except for some microprocessor control unit (MCU) applications on 5.8-liter police and Canadian applications. By 1985 the EEC IV system was also used on most Bronco, Ranger, E-series, and F-series light trucks.

The EEC IV system's primary function is to control the air/fuel ratio. This can be achieved through the use of a carburetor, single-point injection (which Ford calls central fuel injection) or multipoint injection (which Ford calls electronic fuel injection [EFI]). The throttle body unit for CFI is shown in Figure 7-1, and the EFI unit is shown in Figure 7-2. In order to control these and in some cases other functions, the computer obtains information about engine and vehicle operating parameters from several sensors, Figure 7-3.

ELECTRONIC CONTROL ASSEMBLY (ECA)

The EEC IV system computer (ECA) has more capability than any previous Ford engine control computer, Figure 7-4. It has a keep-alive memory (an expansion of its random access memory) that enables it to store codes related to faults that it has observed but that are no longer present (this was not true for the first application on the 1.6-liter engine). The num-

Chapter 7 Ford's Electronic Engine Control IV

FIGURE 7-1 Central fuel injection unit. *Courtesy of Ford Motor Company.*

FIGURE 7-2 Multipoint injection manifold (1.6 L). *Courtesy of Ford Motor Company.*

Chapter 7 Ford's Electronic Engine Control IV

Inputs

Engine coolant temperature
Manifold absolute pressure and/or barometric pressure
Throttle position
Engine speed and crankshaft position
Exhaust gas oxygen
EGR valve position

The following sensors are unique to specific engines:

Air charge temperature
Vane air temperature
Vane airflow
Idle tracking switch
Transmission position
Inferred mileage sensor
Knock sensor
Ignition diagnostic monitor
Clutch engaged switch
Brake on/off switch
Power steering pressure switch
Air-conditioning clutch
Ignition switch

Electronic Control Assembly (ECA)

Power Relay

Outputs

Air/fuel mixture control device, carburetor solenoid or injector
Ignition timing control
Idle speed control
Thermactor airflow control
Canister purge control
EGR flow control
Torque converter clutch control
Turbocharger boost control
A/C and cooling fan controller module
Wide-open throttle A/C cut-off
Inlet air temperature control
Variable voltage choke
Temperature-compensated accelerator pump
Shift indicator light
Fuel pump relay

All components are not used on any one system.

FIGURE 7-3 EEC IV system overview

ber of codes it can recognize is greatly increased compared to the controllers of earlier Ford systems.

FIGURE 7-4 EEC IV electronic control assembly. *Courtesy of Ford Motor Company.*

Engine Calibration Assembly

The ECA's calibration assembly contains the necessary programming to fine-tune the ECA's engine calibration commands to each vehicle's weight, axle ratio, and transmission application. Unlike many other systems' computer, the calibration assembly is an internal part of the ECA and is not serviced separately.

Operating Modes

The ECA operates the engine in one of several operational modes or strategies.

Base Engine Strategy. This is the mode that the ECA uses to control warm engine calibration through the wide range of operating conditions that normally occur. Prior to 1986, the calibration for warm engine operation was

249

Chapter 7 Ford's Electronic Engine Control IV

a coolant temperature of 88°C (190°F) or higher. In 1986 that calibration was lowered to about 54°C (130°F) or higher. This mode is broken down into four submodes:

- cranking
- closed throttle operation
- part throttle operation (closed loop)
- wide-open throttle (WOT)

Input information enables the ECA to recognize these operational modes. Having recognized the existing mode, the ECA issues calibration commands to the appropriate actuators, which produce the best results in terms of emissions, fuel mileage, and driveability.

If the engine threatens to stall while in the base engine strategy mode, the ECA will employ an underspeed response feature.

MPG Lean Cruise. When predetermined criteria are met during cruise conditions on some engine applications beginning in 1988, the ECA will take the system out of closed loop to a leaner air/fuel ratio. This is done in an attempt to achieve better fuel economy.

Modulator Strategy. Operating conditions that require significant compensation in order to maintain good driveability cause the ECA to modify the base engine strategy. Conditions that cause such modifications are:

- cold engine
- overheated engine
- high altitude

Limited Operational Strategy (LOS). When a component failure prevents the system from operating in a normal strategy, the ECA enters an alternate strategy designed to protect other system components, such as the catalytic converter, and still provide enough driveability to keep the vehicle operating until it can be repaired. If the ECA's central processing unit (CPU) fails, the ECA will operate in a fixed mode: no spark advance, no EGR, and air from the *Thermactor* (Ford's air injection system) will be dumped to the atmosphere.

Adaptive Strategy

This feature, first introduced in 1985, enables the ECA to continuously adjust some of its original calibrations, which were programmed by engineers based on ideal conditions. When conditions are less than ideal due to manufacturing tolerances, wear or deterioration of sensors or any other component that affects the ECA's control of its most critical functions, the adaptive strategy puts an adjustment or adaptive factor into the calculation process. Adaptive strategy could be called a learning capability. The ECA learns from past experience so that it can better control the present. Learning starts when the engine is warm enough to go into closed loop and is in a stabilized mode.

Adaptive Fuel Control. This adaptive strategy concerns air/fuel ratio. If the injectors become fouled with deposits or if wear in the pressure regulator causes a slight loss of fuel pressure, the result is less fuel being injected into the cylinder than should be during the calibrated pulse width. When the system is in closed loop, the EGO sensor reports a lean condition to the ECA. The ECA responds by commanding a wider injector pulse width to maintain a 14.7-to-1 air/fuel ratio.

The adaptive fuel control recognizes that a wider pulse width than what the original calibration called for is needed. If this condition continues for several minutes, the adaptive fuel control modifies the original calibration to provide the wider pulse width so that the calibration is more on target and less correction has to be made in response to the EGO sensor.

How the Calibration Modification Is Achieved. The ROM contains an engine load (MAP) versus engine RPM table that is used as the starting point for calculating pulse width, Figure 7-5. The table is divided into separate cells. Each cell is assigned to a base value of "1." Cross referencing the current MAP value with the current RPM value, the microprocessor identifies the cell that best represents cur-

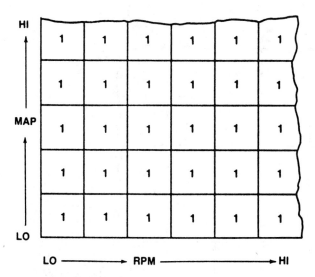

FIGURE 7-5 Adaptive fuel control table base valves

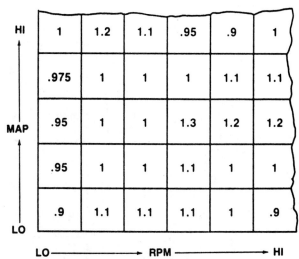

FIGURE 7-6 Adaptive fuel control table with adapted cell valves

rent engine load conditions. The value in that cell is used as a multiplier in the microprocessor's calculations to find the correct pulse width command for the injectors.

Since this table is stored in the ROM, it cannot be erased or changed. There is, however, a copy of this table in the keep alive memory (KAM). The microprocessor uses the copy to work from, and the base values in the cells of the KAM copy can be changed.

If the pulse width calculation, using the base value in a given cell, consistently does not produce the desired 14.7-to-1 air/fuel ratio, the adaptive fuel control will add to or subtract from that base value, Figure 7-6. If the air/fuel ratio tends to be too lean, the adaptive fuel control adds to the base value. If the air/fuel ratio tends to be too rich, the adaptive fuel control subtracts from the base value.

At idle, adaptive idle fuel control works the same way as it does at other engine speeds. The microprocessor uses the same table for pulse width calculations at idle, but it uses cells that are only used during idle conditions. Depending on engine application, there are either four or six cells on the MAP versus RPM table for idle fuel pulse width calculations. The table could look something like the one shown in Figure 7-7. The idle cells, however, are not selected by MAP and RPM; they are selected by the transmission being in drive or neutral, by the A/C being on or off, and in some cases by the A/C control switch being on or off.

Because the adaptive fuel control makes the correction factor in the MAP versus RPM table and the same table is used for open loop pulse width calculations, the fouled injectors are also compensated for in open loop.

Adaptive Idle Speed Strategy. Adaptive idle speed strategy is used only in CFI and EFI systems. The idle speed strategy calculates the commands that are sent to the idle speed control actuator. The device controls the amount of air that is admitted into the induction system. If the commands sent to the actuator fail to produce the calibrated idle speed, the ECA substitutes a new value in place of one of the originally calibrated values used in the calculation. The value that was replaced was obtained from an idle table in the KAM similar to the MAP

Chapter 7 Ford's Electronic Engine Control IV

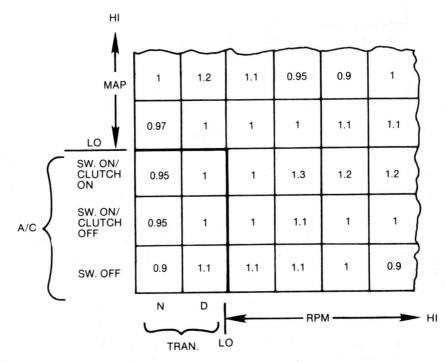

FIGURE 7-7 Idle fuel control cells on MAP versus RPM table

versus RPM table mentioned above, Figure 7-8. When idle speed is corrected, the adaptive idle speed strategy monitors the revised value needed to make the correction and records it in place of the originally calibrated value on the idle table.

Adaptive idle speed strategy learning occurs only under the following conditions:

- idle speed is stabilized during base idle conditions (warm engine, automatic transmission in drive, manual transmission in neutral, and A/C off).
- system in closed loop.
- a specific time has elapsed since idle mode began. The time varies with different engine calibrations.

Because the information learned by the adaptive strategy is stored in the KAM, it can be used or revised as needed. The learning pro-

FIGURE 7-8 Idle speed table

cess is always working during stabilized, closed loop conditions.

Other adaptive strategies incorporated into the EEC IV systems since 1985 include **adap-**

tive spark control and adaptive EGO sensor aging. Their operation is similar to those just described.

Adaptive strategy also tests regularly for the possibility that corrupt data (false values) could have gotten into the KAM. If any data corruption is found, those values are erased and replaced with the original corresponding values from ROM.

Power Relay

A power relay is used to supply operational power to the ECA, Figure 7-9. The control coil of the relay is powered by the ignition switch.

FIGURE 7-9 Power relay

When the ignition is turned on, the normally open contact points of the relay are closed and thus supply power to the ECA. The control coil of the relay is in series with a diode. If the battery's polarity is reversed, the diode will block current flow through the control coil and will prevent the contact points from closing. This protects the power circuits of the ECA from damage resulting from reversed polarity.

Beginning in 1986 on some engine applications and becoming more widespread in 1987, the power relay has been integrated into a unit called the **Integrated Controller.** This unit typically contains the fuel pump relay, the power relay, the cooling fan relay (Electro-Drive Fan), the High Speed Electro-Drive Fan relay and the solid state A/C cut-out relay.

The power to the KAM does not come from the power relay. It comes directly from a battery source and is always hot. If this power supply is disconnected, any service codes stored and all learned adaptive strategy will be erased. Keeping this circuit powered requires very little current (a few milliamps) and does not discharge the battery even if the vehicle is not operated for several days. Because the KAM is not protected by the power relay, its circuitry is designed to be less susceptible to damage resulting from reversed polarity. Care should always be taken, however, to prevent such mistakes from occurring.

INPUTS

Engine Coolant Temperature (ECT) Sensor

The ECT is a thermistor-type sensor, Figure 7-10. Its voltage signal to the ECA normally ranges from 4.5 volts when very cold to 0.3 volt when the engine is hot. Its resistance values are:

- −40° F = 269,000 ohms
- 32° F = 95,000 ohms
- 77° F = 29,000 ohms
- 248° F = 1,200 ohms

FIGURE 7-10 Engine coolant temperature sensor. *Courtesy of Ford Motor Company.*

The information it provides influences the ECA's calibration commands concerning:

- air/fuel ratio
- idle speed
- EGR
- thermactor air
- canister purge
- choke voltage
- temperature-compensated accelerator pump
- upshift light

Pressure Sensors

Ford's pressure sensors operate differently than those discussed in Chapter 2. They use a pressure-sensitive, variable capacitor, Figure 7-11. The capacitor is formed by two conductive plates separated by a thin air space and a thin layer of insulating material. One plate, the positive, is a thin copper plating on a round, rigid ceramic bed with a thin film of insulating material sprayed over the copper. The other plate, the negative, is a thin copper plating on the underside of a round, flexible ceramic disc. The flexible ceramic disc is mounted on top of the rigid ceramic bed with a thin adhesive strip around the entire circumference. The adhesive strip makes an airtight seal and spaces the two copper plates a few thousandths of an inch apart.

This assembly makes up the ceramic body of the sensor and is mounted in a plastic hous-

FIGURE 7-11 Typical variable capacitor pressure sensor

ing. The area below the ceramic body is sealed from the area above it. A small hole through the ceramic bed vents the area between the copper plates to the sealed, reference pressure below the ceramic bed. The space above the ceramic body is exposed to the pressure that is being sensed.

Electronic circuitry and components are attached to the bottom of the rigid ceramic bed. Each capacitor plate has a wire that connects it to the electronic circuitry. Additional wires connect the circuitry and the negative plate to the ECA.

If the pressure being sensed is lower than the reference pressure between the plates, the diaphragm flexes outward, moving the plates further apart. This lowers the capacity or capacitance of the capacitor. As the pressure being sensed increases, the diaphragm is forced inward, moving the plates closer together. This increases the capacitance. The electronic circuit, which is powered by the reference voltage (VREF) from the ECA, measures the capacitance of the variable capacitor and sends a corresponding electrical signal back to the ECA.

Manifold Absolute Pressure (MAP) Sensor.
In addition to being a variable capacitor, the MAP sensor, Figure 7-12, used on EEC IV systems is a frequency generating device. An addi-

Chapter 7 Ford's Electronic Engine Control IV

FIGURE 7-12 Manifold absolute pressure sensor. *Courtesy of Ford Motor Company.*

tional chip in the sensor's electronic circuit produces a frequency signal that corresponds to the capacitance value, capacitance being a function of manifold pressure. This frequency signal is sent to the ECA.

If you apply a voltmeter directly to the sensor's output circuit, you will see a steady voltage signal (about 2.5 volts) that will change very little regardless of manifold pressure. Actually the signal does change (more or fewer pulses per second) in response to changes in manifold pressure, but the frequency is so rapid that a voltmeter cannot respond to all of the changes. The meter just gives an average voltage reading. During lowest manifold pressure conditions (high vacuum) the sensor's signal frequency is 92 cycles per second (92 Hz). At WOT, manifold pressure is about equal to atmospheric pressure, and the signal frequency is near 162 Hz. The ECA reads the frequency as a manifold pressure value. The MAP sensor's input affects:

- air/fuel ratio
- EGR flow
- ignition timing

Those EEC IV engine applications that use a manifold pressure sensor do not use a barometric pressure sensor. Rather, the ECA uses the MAP sensor as a barometric pressure sensor during two specific conditions. During the brief period of time between the ignition being turned on and the starter being activated, the ECA takes a reading from the MAP sensor. Since the pressure in the intake manifold is atmospheric pressure during this period, the ECA stores the MAP reading in its RAM as a barometric pressure reading. During WOT operation, manifold pressure is near atmospheric pressure, and its difference from manifold pressure is very predictable for any given engine. Therefore, during WOT, the ECA will adjust the MAP reading for the difference between manifold pressure and atmospheric pressure and use it to update the barometric pressure reading stored in the RAM.

Barometric Pressure (BP) Sensor. The BP sensor is nearly identical to the MAP sensor both in operation and appearance, Figure 7-13. Its pressure-sensing port is exposed to atmo-

FIGURE 7-13 Barometric pressure sensor. *Courtesy of Ford Motor Company.*

spheric pressure instead of being connected to manifold vacuum. Its signal frequency ranges from 122 Hz at lowest atmospheric pressure to 162 Hz at highest atmospheric pressure.

The BP sensor is used most often on engines that use a vane airflow or a mass airflow sensor instead of a MAP sensor to measure air intake. A vane airflow meter responds to the pressure differential across the vane and measures the resulting airflow into the induction system. With the additional barometric pressure value that the BP sensor provides, the ECA can more accurately determine the air charge density. With a turbocharged engine, this is even more critical. Even though the supercharged 3.8-liter engine has a mass airflow meter and can accurately determine air mass without a BP sensor, a BP sensor is used to determine the barometric pressure's effect on the air mass. The BP's input affects:

- air/fuel ratio
- EGR flow

Throttle Position Sensor (TPS)

The TPS is a potentiometer that monitors throttle plate movement and provides an electrical signal indicating throttle position to the ECA.

Linear TPS. Most carbureted EEC IV engines use a linear TPS, Figure 7-14. A cam on the throttle shaft contacts a plunger on the TPS and pushes it in as the throttle shaft is rotated, Figure 7-15. The linear-type TPS can be adjusted.

Rotary TPS. The rotary-type TPS is similar to the linear type except that its resistor forms an arc instead of being straight and the wiper pivots from the center of the arc. It is mounted on the side of the carburetor or throttle body so that the throttle shaft engages and drives the wiper, Figure 7-16. Beginning in 1985, all EEC IV applications use the rotary type. Some are adjustable, some are not. For either type, the ECA sends a 5-volt reference input to the TPS.

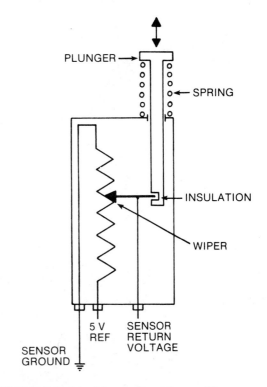

FIGURE 7-14 Linear throttle position sensor diagram

FIGURE 7-15 Linear throttle position sensor. *Courtesy of Ford Motor Company.*

Chapter 7 Ford's Electronic Engine Control IV

FIGURE 7-16 Rotary throttle position sensor. *Courtesy of Ford Motor Company.*

FIGURE 7-17 Hall-effect switch in distributor. *Courtesy of Ford Motor Company.*

It has a voltage range of 0.5 volt to 4.5 volts. Most will show a signal voltage of near 1 volt at idle and 4 volts at WOT. The ECA uses the TPS primarily to identify driving mode.

Profile Ignition Pickup (PIP) Sensor

The PIP signal is produced by a Hall-effect switch (see **Hall-effect Switch** in Chapter 2). It is called a PIP sensor and provides information to the ECA concerning engine speed and crankshaft position. The PIP sensor is in the distributor on engines equipped with a distributor, Figure 7-17.

Ford introduced sequential electronic fuel injection (SEFI) in 1986 on the 5.0-liter engine in passenger cars. In a SEFI system, the ECA pulses each injector individually in the firing order. To do that, the ECA must know where the engine is in its cycle so that it can pulse the number three injector, for example, during number three cylinder's intake stroke. To tell the ECA where the engine is in its cycle, Ford engineers made one of the vanes that triggers the PIP sensor narrower than the others. As the narrow vane passes between the magnet and the crystal, a shorter on time is produced in the PIP signal, Figure 7-18. This is called the *signature PIP*. The ECA recognizes the signature PIP as the second cylinder in the firing order. Being able to recognize a given cylinder in the firing order, combined with having the firing order in its memory and having a circuit that counts PIP signals, enables the ECA to pulse each injector in the correct sequence.

In 1989, Ford introduced a distributorless ignition system (DIS), Figure 7-19, on applications such as the 3.8-liter supercharged engine, the 3.0-liter special high-output (SHO) engine, and the 2.3-liter engine used in the Ranger truck. On these applications, the PIP sensor is located on the front of the engine just behind the crankshaft pulley and hub assembly, Figure 7-20. A vaned cup on the back of the pulley triggers the Hall-effect switch, Figure 7-21. Because the cup is mounted on the

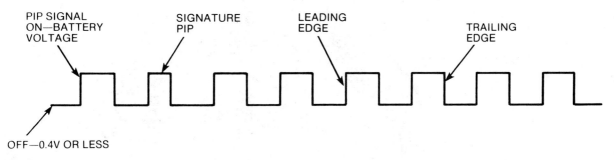

FIGURE 7-18 PIP signal for 5.0-liter sequential injection engine

FIGURE 7-19 Distributorless ignition, supercharged 3.8-liter

Chapter 7 Ford's Electronic Engine Control IV

FIGURE 7-20 DIS components; 3.0-liter SHO engine

Chapter 7 Ford's Electronic Engine Control IV

FIGURE 7-22 PIP signal

FIGURE 7-21 PIP sensor; supercharged 3.8-liter and SHO engines

crankshaft and the crankshaft makes two revolutions per engine cycle, only one vane is needed for each pair of companion cylinders. As the leading edge of the vane moves between the Hall-effect switch's magnet and crystal, the PIP signal goes to the battery voltage, Figure 7-22. The vanes are positioned on most engines so that the PIP signal will go high each time a piston reaches 10° BTDC. As the trailing edge of the vane moves out, the PIP signal goes to 0.4 volt or less.

Cylinder Identification (CID) Sensor

Distributorless ignition systems use a separate coil to simultaneously fire the spark plugs in each pair of companion cylinders. One of the plugs is fired while its cylinder is at the top of the compression stroke, and the other is fired while its cylinder is at the top of the exhaust stroke. The ECA determines the spark advance and tells the DIS module when to fire the coil, but the DIS module must decide which coil to fire. To know which coil to fire, it must know which cylinder is ready for ignition.

A Hall-effect switch is used as a CID sensor. On the 3.0-liter SHO engine, the CID sensor is mounted on the right end of the rear head and is driven by one of the overhead camshafts, Figure 7-20. On the 3.8-liter engine the CID sensor is mounted where the distributor used to be and is driven by the camshaft. On both of these engines, the camshaft drives a single vane cup, Figure 7-23. As the leading edge of the vane moves between the Hall-effect switch's magnet and crystal, the CID signal goes to battery voltage. This tells the DIS module and the ECA that the number one cylinder is 26° ATDC on the power stroke. As the trailing edge of the vane moves out from between the magnet and crystal, the CID signal goes to 0.4 volt or less. This tells the DIS module and the ECA that the number one cylinder is 26° ATDC on the intake stroke. Either transition of the CID signal, going high or low, identifies the number one cylinder. The DIS module uses this information while the engine is being cranked to determine which coil to fire first.

Because the 3.0-liter SHO engine and the 3.8-liter engines after 1988 are sequentially injected, the ECA uses the CID signal to control injector pulsing sequence just as it does the

261

Chapter 7 Ford's Electronic Engine Control IV

FIGURE 7-23   CID sensor; supercharged 3.8-liter and 3.0-liter SHO engines

FIGURE 7-24   PIP sensor; 2.3-liter, dual plug engine

signature PIP on SEFI engines that use a distributor.

On the DSI 2.3-liter engine, the CID sensor is built in combination with the PIP sensor and is mounted behind the crankshaft pulley and hub assembly, Figure 7-24. The 2.3-liter engine CID sensor works like the 3.0- and 3.8-liter CID sensors with one exception. Because the single vane that triggers CID sensors is on the back of the crankshaft pulley, it produces two high CID signals and two low CID signals per engine cycle, Figure 7-25. Therefore, all of the transitions in the CID signal do not identify the number one cylinder. When the signal goes high, either the number one or four cylinder is

10° ATDC on the power stroke. The DIS module knows to fire coil 2 in response to the next spark output command from the ECA. When the signal goes low, either cylinder number

262

FIGURE 7-25 CID and PIP sensor signals

three or two is at 10° ATDC, and the DIS module knows to fire coil 1 in response to the next spark output command from the ECA. The CID sensor is sometimes called a *cam sensor*.

PIP Signal

If the PIP signal were not generated by the Hall-effect switch, the engine would not run because there would be no ignition. If the PIP signal got to the TFI module but not to the ECA of a carbureted engine (see Figure 7-30), the engine would run because the TFI module could produce base timing from the PIP signal. If, however, the same condition occurred on an injected engine, the engine would not run because the ECA will not operate the injectors without the PIP signal.

Ignition Diagnostic Monitor (IDM)

The IDM, not used in all EEC IV systems, is simply a wire that feeds the tach signal from the negative side of the ignition coil to the ECA. On TFI IV ignition systems, the wire with a 22-kilohm resistor in it runs from the coil directly to the ECA, Figure 7-26. On engines equipped with DIS, the tach signal is picked up in the DIS module, fed through a 20-kilohm resistor, and on to the ECA, Figure 7-19. The resistor is used to protect the ECA from the coil's high-voltage surges.

The IDM is like an echo of the ECA's spark output (SPOUT) command. The ECA compares the IDM signal to the PIP signal to see that:

- The SPOUT command was carried out.
- The computed timing circuit is functioning properly.

If the ECA sees a fault, it will store a service code on its continuous monitor memory, which is explained in the **System Diagnosis and Service** section of this chapter.

Exhaust Gas Oxygen (EGO) Sensor

The EGO looks similar to the oxygen-sensing unit used on other systems and works exactly the same.

Air Charge Temperature (ACT) Sensor

The ACT sensor measures the temperature of the air that mixes with the fuel. It uses a thermistor as a temperature-sensing element just as the ECT does. On the ACT sensor, however, the sensing end of the sensor housing has openings that allow the air to come into direct contact with the thermistor coils, Figure 7-27. The ACT sensor is screwed into the intake manifold (into the air cleaner on 2.8-liter engines). The ECA uses ACT sensor information

Chapter 7 Ford's Electronic Engine Control IV

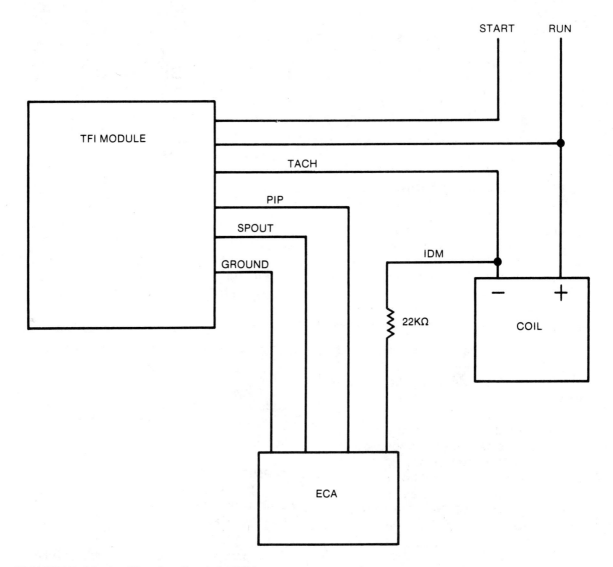

FIGURE 7-26 Ignition feedback to ECA

in the calculation of air/fuel mixture and spark timing. Voltage and resistance values for the ACT are the same as those for the ECT. The ACT sensor is used on most but not all engine applications.

EGR Valve Position (EVP) Sensor

The EVP sensor is a linear potentiometer mounted on top of the EGR valve. It is connected to the EGR valve's diaphragm and pintle assembly and tells the ECA whether the EGR is open, closed, or somewhere in between, Figure 7-28. With the EGR valve fully closed, return voltage is about 0.8 volt; fully open the return voltage is about 4.5 volts. The ECA's processor can use preprogrammed values and other sensor data to convert EGR valve position to EGR flow rate for any given set of driving conditions. The EVP sensor's input is used to:

Chapter 7 Ford's Electronic Engine Control IV

FIGURE 7-27 Air charge temperature sensor. *Courtesy of Ford Motor Company.*

FIGURE 7-28 EGR valve position sensor. *Courtesy of Ford Motor Company.*

- calculate air/fuel ratio
- calculate spark timing
- adjust EGR flow
- set a service code for an EGR valve that fails to open or close

Pressure Feedback EGR (PFE) Sensor. The PFE sensor, Figure 7-29, is a variable capacitor as described under **Pressure Sensors** in this chapter. It senses exhaust pressure in a cham-

FIGURE 7-29 Pressure feedback EGR (PFE) sensor

ber just under the EGR valve pintle, Figure 7-30. When the EGR valve is closed, the pressure in the sensing chamber is equal to exhaust pressure. When the EGR valve opens, pressure in this chamber drops because it is now exposed to the low manifold pressure, while the restriction at the other end of the chamber limits the rate at which pressure from the exhaust system can enter the chamber. The more the EGR valve opens, the more the pressure in the chamber will drop.

The PFE sensor provides an analog voltage signal to the ECA that the ECA can compare to values on a look-up table and accurately determine EGR flow.

The PFE differs from the EVP in that it monitors EGR flow while the EVP only measures EGR opening. The EVP leaves the ECA to assume that there is no carbon buildup or other restriction to restrict EGR flow.

The ECA will use PFE information to:

- fine tune its control of EGR valve opening.
- more accurately control air/fuel ratio.
- modify ignition timing.

Vane Meter

The vane meter is used on selected multi-point injection applications such as the 1.6-liter turbocharged and nonturbocharged, the 2.3-liter turbocharged engines, and the 2.2-

Chapter 7 Ford's Electronic Engine Control IV

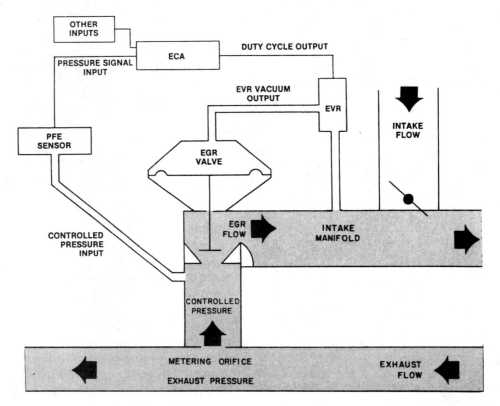

FIGURE 7-30 Pressure feedback EGR sensor and EGR flow

liter engine in the probe. It fits between the air cleaner and the throttle body and measures the velocity of air flowing into the engine's induction system and the air's temperature. Both of these measurements are made by sensors within the vane meter.

Vane Airflow (VAF) Sensor. The air inlet opening of the vane meter is closed by a spring-loaded door or vane, Figure 7-31. As the throttle valve is opened, air moving into the induction system forces the vane open; the more airflow, the wider the vane is opened. The vane is hinged on a pivot pin. As the vane is opened, it moves the wiper of a potentiometer, which acts as a sensor to the ECA. The greater the airflow, the higher the VAF sensor's signal voltage (less volt drop between reference voltage and sensor signal voltage).

FIGURE 7-31 Vane airflow sensor. *Courtesy of Ford Motor Company.*

As the vane is pushed open, its compensator flap is pushed into a specially designed cavity below the airflow passage. There is a slight lull in airflow between intake strokes, especially on four-cylinder engines; and the vane has a tendency to want to close slightly during those brief periods. The sealed space behind the compensator flap acts as a damper to reduce vane flutter.

Vane Air Temperature (VAT) Sensor. This sensor is essentially the same as the ACT sensor used on some other engine applications except that it is located in the vane meter's air inlet opening, Figure 7-32.

Speed Density Formula. The ECA compares inputs from the VAF, VAT, and BP sensors (or the MAP sensor, depending on engine application) along with those from TPS and EVP to look-up charts. This way it can determine the engine's air intake volume and flow rate. This information, along with estimates of the engine's volumetric efficiency, is processed to determine exactly how much fuel should be injected. This is Ford's application of the speed density formula. (See **Speed Density Formula** in Chapter 2).

Mass Airflow (MAF) Sensor. Beginning in 1988, selected engine applications were equipped with a MAF sensor instead of a MAP sensor (see **Mass Airflow Sensor** in Chapter 2). The Ford unit is built by Hitachi of Japan (a different MAF sensor than the Hitachi unit General Motors uses). The MAF sensor is located in the air tube between the air cleaner and the throttle body. A small air sample tube in the top of the MAF sensor air passage directs air across a hot wire and a cold wire, Figure 7-33. The cold wire contains a thermister that is used to measure incoming air temperature. The hot wire is wrapped around a piece of ceramic and is coated with a thin layer of glass. It is kept 200° C above ambient air intake temperature by a module mounted on the MAF sensor body.

Current flow through the sensor varies from 0.5 to 1.5 amps depending on the air's mass airflow rate. Voltage output varies from 0.5 volt at 0.225 kilograms (0.5 lb) of air per minute to 4.75 volts at 14.16 kilograms (31.23 lb) of air per minute. The MAF sensor's input to the ECA affects:

- air/fuel ratio
- spark timing

Idle Tracking Switch (ITS)

In the body of the idle speed control motor assembly (used on some carbureted and CFI applications) is a normally closed switch, Figure 7-34. When the throttle is closed, the throttle lever presses against a plunger that extends from the nose of the assembly. The pressure forces the plunger to move slightly back into the assembly and thus causes the switch to open. When the throttle is opened, the switch closes again. The ECA monitors the switch and can determine when the throttle is open or closed. It uses this information to control the ISC motor.

Transmission Switches

The ECA must know whether or not the vehicle is in gear. This information will affect:

FIGURE 7-32 Vane air temperature sensor. *Courtesy of Ford Motor Company.*

Chapter 7 Ford's Electronic Engine Control IV

FIGURE 7-33 MAF sensor. *Courtesy of Ford Motor Co.*

FIGURE 7-34 Idle tracking switch. *Courtesy of Ford Motor Company.*

- idle fuel and air control strategy for automatic transmission vehicles. The torque converter will create a load on the engine even if the vehicle is not being moved.
- response to rapid closing of the throttle. If it is in gear, the ECA will issue commands appropriate for deceleration conditions.

For most automatic transmissions the **neutral drive switch,** the same switch used to open the starting circuit when the transmission is in gear, serves as an input to the ECA, Figure 7-35. When the transmission is in neutral or park the switch is closed. Reference voltage is fed through a resistor in the ECA, out pin 30 to ground through the N/D switch. When the switch is closed, voltage on the switch side of the resistor is low. By monitoring this voltage, the ECA knows if the transmission is in gear or not. A diode in the switch assembly prevents battery voltage from being applied to the ECA circuit when the starter is being energized.

The **AXOD (automatic transaxle overdrive)** uses three different switches to identify what gear the transmission is in. They are:

- neutral pressure switch (NPS)
- transmission hydraulic switch (THS) 3-2
- transmission hydraulic switch (THS) 4-3

These are normally open pressure switches that are screwed into the transmission valve body so that they are exposed to the hydraulic pressure used to apply specific bands or clutches in the transmission. When hydraulic pres-

FIGURE 7-35 Transmission load switches

sure is applied, the switch closes and grounds the circuit. The ECA supplies voltage to each switch through a resistor, Figure 7-36A. By monitoring the voltage in each switch circuit, the ECA can identify each transmission gear position with the exception that first cannot be distinguished from second, Figure 7-36B.

A transmission temperature switch (TTS) is also used on some AXOD applications to alert the ECA if transmission fluid temperature goes too high, Figure 7-36A. The TTS is a normally closed switch that opens at 135°C (275°F). It is also screwed into the valve body.

If the ECA sees the TTS open during an operating condition in which it normally keeps the converter clutch off, such as a long uphill climb with a large throttle opening, it will apply the clutch to reduce the heat being produced in the torque converter.

Manual transmission applications can be taken out of gear at the clutch or in the transmission. Therefore, two switches are used, Figure 7-35. A **neutral gear switch** is attached to the shift linkage. It is closed in neutral. A **clutch engaged switch** is attached to the clutch linkage and is closed when the clutch pedal is

269

Chapter 7 Ford's Electronic Engine Control IV

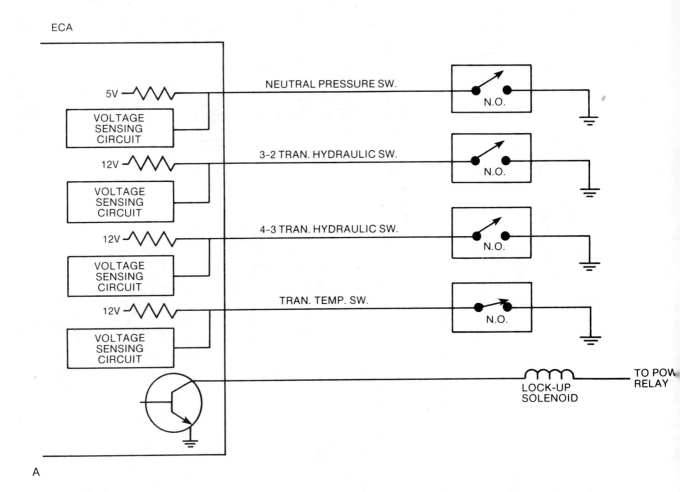

A

Gear Position	Neutral Pres. Sw.	3-2 Tran. Hyd. Sw.	4-3 Tran. Hyd. Sw.
Park/Neutral	Open (5 V)	Open (12 V)	Open (12 V)
First	Closed (0-1 V)	Open (12 V)	Closed (0-1 V)
Second	Closed (0-1 V)	Open (12 V)	Closed (0-1 V)
Third	Closed (0-1 V)	Closed (0-1 V)	Closed (0-1 V)
Fourth	Open (5 V)	Closed (0-1 V)	Open (12 V)
Reverse	Closed (0-1 V)	Open (12 V)	Open (12 V)

B

FIGURE 7-36 AXOD transmission switches

depressed (clutch disengaged). Because the neutral gear and clutch engaged switches are in parallel, they must both be open in order for the ECA to determine that the vehicle is in gear. If either switch closes, voltage at pin 30 goes low, and the ECA knows that the gear train has been opened.

Inferred Mileage Sensor (IMS)

Beginning in 1985 some light truck EEC IV applications were equipped with an electronic module (IMS) that contained an E-cell, Figure 7-37. The IMS is powered by the power relay. After a specified amount of ignition on time, the E-cell opens the IMS's feedback circuit to the ECA. The ECA responds by changing programmed calibrations to allow thermactor air to be directed to the exhaust manifold for longer periods of time and changes in EGR flow rate to compensate for engine wear.

The IMS should not be confused with the emission maintenance warranty/extended useful life (EMW/EUL) module of which there are two types, each used on truck applications. The early type contained an E-cell that was depleted after the ignition had been on for a total time that equated to about 60,000 miles. When the E-cell was depleted, it caused an indicator light on the instrument panel to be

FIGURE 7-37 Inferred mileage sensor. *Courtesy of Ford Motor Company.*

FIGURE 7-38 Knock sensor. *Courtesy of Ford Motor Company.*

turned on. This alerted the driver that the EGR valve and EGO sensor were due to be replaced. On later models the E-cell in the module was replaced by a small microprocessor that could be reset. Replacing the EGR valve or EGO sensor at regular intervals is not required on other EEC IV system applications.

Knock Sensor (KS)

The EEC IV applications that use a KS use a piezoelectric type, Figure 7-38.

Vehicle Speed Sensor (VSS)

The vehicle speed sensor, a magnetic pulse generator driven by the transmission speedometer gear, produces 8 cycles per revolution, 16 AC signals per revolution, or 128,000 signals per mile to the ECA. The ECA will modify this signal, convert it to a vehicle speed value, and store it in memory. In addition to vehicle speed control, this information may be used, depending on vehicle application, in the control of transmission converter lock-up, coolant fan control, and to identify deceleration conditions.

Brake On/Off (BOO) Switch

Vehicles equipped with the A4LD transmission, such as Rangers and Broncos, use the

Chapter 7 Ford's Electronic Engine Control IV

BOO switch (in the stop lamp switch assembly) to signal the ECA when the brakes are applied. The A4LD transmission features an ECA-controlled torque converter clutch. The ECA releases the clutch when the switch closes. If the brakes are applied during idle and the transmission is in gear, the ECA may raise idle speed. During a prolonged idle with the brakes applied, the ECA may disengage the A/C compressor clutch.

Power Steering Pressure Switch (PSPS)

The power steering switch is in the high-pressure side of the power steering system. At a pressure of 400 PSI to 600 PSI, the switch is closed. If the ECA sees this switch close during idle conditions, it will raise idle speed to compensate for the additional load on the engine.

Air-conditioning Clutch (ACC) Signal

On many engine applications, when the A/C is turned on, the same voltage signal that is sent to the A/C compressor clutch is also sent to the ECA. This signal alerts the ECA to raise idle speed slightly.

Ignition Switch

When the ignition is turned on, the power relay is activated. The power relay then supplies power to the ECA and to most of the circuits that the ECA controls. The relay's power to the ECA acts as an input and signals the ECA to start its programmed functions.

OUTPUTS

Air/Fuel Mixture Control

The EEC IV system employs three different types of fuel metering devices: carburetors referred to as feedback control carburetors, throttle body injectors referred to as central fuel injection, and port fuel injection referred to as multipoint injection.

Feedback Control (FBC) Carburetors. Three different carburetors have been used on EEC IV engine applications. They each use a duty-cycled solenoid to control the air/fuel mixture, although each uses a slightly different FBC solenoid. Each FBC solenoid is powered by the power relay, Figure 7-39; and the ECA controls the ground circuit.

FIGURE 7-39 FBC solenoid circuit

Chapter 7 Ford's Electronic Engine Control IV

FIGURE 7-40 FBC solenoid (2150-2V carburetor). *Courtesy of Ford Motor Company.*

The Motorcraft 2150A-2V, Figure 7-40, uses a solenoid that as it cycles alternately introduces and blocks air from the air cleaner into the vacuum passages, which provide bleed air to the idle and main metering systems. The air introduced by the solenoid is in addition to the air admitted by the fixed air bleeds typically used in most carburetors. If the solenoid were turned off completely (zero duty cycle), additional air would be blocked and the air/fuel mixture would be full rich. If the solenoid were turned on continuously (100% duty cycle), the air/fuel mixture would be full lean. The duty cycle is controlled by the ECA. In open loop the duty cycle is at a constant value, depending on coolant temperature, engine load, engine speed, and throttle position. In closed loop it varies constantly, depending on information being supplied by the EGO sensor.

The Carter YFA-1V carburetor is equipped with an FBC solenoid that looks different than the one used on the 2150A, but it works in the same way.

The Holley 6149-1V carburetor is controlled by a remote mounted FBC solenoid, Figure 7-41. Its control of the idle circuit air/fuel ratio is much like that of the other two solenoids. Its duty cycle controls the amount of air introduced to the idle speed circuit, Figure 7-42. The main metering mixture control, however, is somewhat different.

The solenoid controls vacuum to a diaphragm in the air horn section of the carburetor, Figure 7-43. The diaphragm is part of a fuel control valve assembly. It can easily be mistaken for a power valve assembly. A spring holds the diaphragm down. Extending down from the diaphragm into the fuel bowl is a rod with a foot at the bottom. The foot engages and pushes down on the upward extending tip of a valve (main system feedback metering valve) in the floor of the fuel bowl. With this valve pushed

Chapter 7 Ford's Electronic Engine Control IV

FIGURE 7-41 Remote duty-cycle solenoid. *Courtesy of Ford Motor Company.*

FIGURE 7-42 Idle feedback metering. *Courtesy of Ford Motor Company.*

FIGURE 7-43 Main feedback metering. *Courtesy of Ford Motor Company.*

down (open), additional fuel flows into the main metering system. When the ECA wants to lean out the air/fuel mixture, it cycles the solenoid so that more vacuum is applied to the fuel control assembly diaphragm. This causes it to lift and allows the valve in the fuel bowl floor to close at least partially.

Central Fuel Injection (CFI). Two types of CFIs are used on EEC IV applications: a high-pressure system and a low-pressure system. Each uses a throttle body unit mounted on the manifold in place of a carburetor. The solenoid-operated injectors are opened by an electrical command (pulse) from the ECA. The injector's control circuit is similar to that of the FBC solenoid in Figure 7-44; power is supplied by the power relay and ground is controlled by the ECA. Fuel is sprayed from a single injector (low-pressure system, Figure 7-42) or from two injectors (high-pressure system, Figure 7-1)

FIGURE 7-44 Low-pressure central fuel injection unit. *Courtesy of Ford Motor Company.*

positioned directly over the throttle valves. During cranking, the high-pressure CFI system pulses both injectors simultaneously in response to each PIP signal. After the engine has started, they are pulsed alternately at a fixed frequency. Because the fuel pressure is constant and the injector always opens the same amount (about 0.01 inch), air/fuel mixture is controlled by the injector pulse width.

In 1986 a slight variation of the high-pressure CFI system was introduced on the 3.8-liter engine. Each of the two injectors has its own EGO sensor and logic circuit. A problem such as a leaking exhaust manifold on one side of the engine only affects the performance of half of the cylinders.

Multipoint Injection (MPI). On most MPI system applications, the injectors are pulsed in two groups. The two groups are pulsed alternately, with each group being pulsed once per engine cycle during normal operation. They are pulsed simultaneously during cranking. In 1986 the 5-liter engine featured sequential fuel injection. Each injector is pulsed individually in the firing order in coordination with its respective cylinder's intake stroke.

Chapter 7 Ford's Electronic Engine Control IV

Fuel Supply System

Fuel Pump. The low-pressure CFI system uses a fuel pump mounted in the fuel tank. It is controlled by the fuel pump relay and delivers fuel to the fuel-charging assembly (throttle body unit) by way of the fuel pressure regulator.

The CFI high-pressure and the MPI systems use one of two fuel delivery systems. Most use a single high-pressure tank-mounted electric pump that feeds about 60 PSI to the pressure regulator. Vehicles with a greater distance between the fuel tank and the engine can use two electric fuel pumps to assure adequate fuel pressure to the injectors. A low-pressure pump is mounted in the tank. It feeds low-pressure fuel to an in-line high-pressure pump, which boosts the pressure to about 60 PSI and sends it to the pressure regulator.

Either of the high-pressure pump systems generates a pressure of about 100 PSI if the fuel line is blocked. At this point the pump is turned off by a pressure relief switch.

The carbureted EEC IV system applications use a conventional mechanical fuel pump.

Fuel Pressure Regulator. Fuel pressure is controlled by a pressure regulator similar to those described in previous chapters, Figure 7-45. Manifold vacuum is routed to the MPI system pressure regulator so that it varies fuel pressure in response to throttle position. The pressure ranges from 30 PSI during high-vacuum (low manifold pressure) conditions to about 40 PSI at WOT (high manifold pressure). On turbocharged applications the higher manifold pressure during boost conditions causes the fuel pressure to go as high as 50 PSI.

The CFI systems do not expose manifold vacuum to the pressure regulator, and the pressure stays fairly constant. The high-pressure CFI system maintains from 39 PSI to 41 PSI. The low-pressure system (used on the 2.3-liter HSC engine) maintains about 14.5 PSI at the injector tip.

Fuel Pump Relay. The fuel pump relay controls the fuel pump and is controlled by the

Chapter 7   Ford's Electronic Engine Control IV

FIGURE 7-45   Fuel pressure regulator. *Courtesy of Ford Motor Company.*

FIGURE 7-46   Fuel pump control circuit

276

ECA, Figure 7-46. When the ignition is turned on, the ECA grounds the fuel pump relay coil. If it does not get a PIP signal indicating that the engine is being cranked, within one to two seconds it will turn off the relay. Once the engine is started, the ECA keeps the fuel pump relay activated until the ignition is turned off or until it stops receiving the PIP signal.

In series with the fuel pump relay is an inertia switch, Figure 7-47. Its purpose is to disable the fuel pump if the vehicle is involved in an accident. The inertia switch consists of a steel ball sitting in a metal bowl. A magnet holds the ball in the bottom of the bowl. In the event of a jolt severe enough to dislodge the ball, the ball strikes a lever above it. The lever triggers an overcentering device that causes the contacts of the inertia switch to open and break the circuit that powers the fuel pump. This also causes a reset button to pop up. Pushing the reset button back down closes the contacts again. This should be checked anytime a fuel-injected Ford vehicle that has been running fine suddenly fails to start. The inertia switch of a parked car can be triggered by having another car back into it without any other damage occurring.

Thick Film Integrated (TFI-IV) Ignition System

The TFI-IV ignition system takes its name from the TFI module attached to the distributor housing. Thick film refers to the type of chip on which the module's circuit is located. Its function is to turn the primary ignition circuit off and on and to control primary circuit dwell. The spark output command from the ECA is the module's signal to open the primary circuit. If the spark output command fails to arrive, the TFI module will open the primary circuit in response to the PIP signal; this results in base timing only.

Spark Output (SPOUT). The ECA receives the PIP signal from the Hall-effect switch in the distributor by way of the TFI module, Figure 7-48. From this signal it determines engine RPM and crankshaft position. The ECA then modifies this signal (changes its timing) to achieve the best ignition timing results considering the engine's speed, temperature, load, atmospheric pressure, EGR flow, and air temperature. The modified signal is sent to the TFI module as the spark output (SPOUT) signal or command. The SPOUT signal is a digital sig-

FIGURE 7-47   Inertia switch. *Courtesy of Ford Motor Company.*

Chapter 7 Ford's Electronic Engine Control IV

FIGURE 7-48 TFI ignition circuit. *Courtesy of Ford Motor Company.*

nal, Figure 7-18. The leading edge of the on portion of the signal causes the TFI IV module to open the primary ignition circuit and thus causes the coil to generate the high secondary voltage needed to fire the spark plug. If the SPOUT signal fails to arrive at the TFI IV module, the module will operate the ignition from the PIP signal; that equates to base timing.

Computer Controlled Dwell (CCD). In most TFI IV ignition systems, dwell (the duration of the coil on time) is controlled by the TFI IV module. In other words, the TFI IV module decides when to turn the coil on, and the ECA tells it when to turn the coil off. The TFI IV module contains circuitry that monitors engine RPM and increases dwell as engine speed increases. Beginning in 1989, in a limited number of TFI IV applications, the ECA controlled the dwell. The trailing edge of the SPOUT signal tells the TFI IV module to close the primary circuit, and the leading edge tells it to open the primary circuit, Figure 7-18.

E Core Coil. The TFI-IV ignition system uses an E core coil that is different from the conventional coil used on prior Ford Motor Company applications. The coil core resembles two Es turned face to face. The coil windings, which are molded in epoxy, are wound

Chapter 7 Ford's Electronic Engine Control IV

FIGURE 7-49 Octane adjustment rod. *Courtesy of Ford Motor Company.*

around the Es' center horizontal bars. Because the center bars are shorter than the ones at the top and bottom, they do not quite touch each other and therefore form a small air gap between them. The smaller air gap (compared to the distance between the ends of a conventional coil core) helps the coil develop a much stronger magnetic field. With no primary ballast resistor and the small core air gap, the E core coil draws high primary circuit amperage (5.5 to 6.5 amps) and can put out as much as 50,000 volts.

Octane Adjustment Rod. The TFI-IV distributor has an octane adjustment feature. It has a rod inserted from the outside and that changes its spark advance capability, Figure 7-49. The standard rod has zero adjustment. In some cases where spark knock becomes a problem, Ford Motor Company has issued a technical bulletin that authorizes the replacement of the standard rod with a 3° or a 6° retard rod.

Distributorless Ignition System (DIS)

3.0-Liter SHO and Supercharged 3.8-Liter. In these systems, the ECA controls both dwell and timing. When the starter starts turning the engine, PIP and CID signals are fed to both the DIS module and the ECA. Once the DIS module has received a CID signal, it will turn on coil 2 by providing a ground for the coil's primary circuit, Figure 7-19, in response to the next SPOUT trailing edge that it sees, Figure 7-50. Coil 2, Figure 7-51, builds a magnetic field while the DIS module waits for SPOUT's next leading edge. When the next SPOUT leading edge occurs, the DIS module opens coil 2's primary circuit to fire spark plugs 4 and 3.

The CID signal transition that the DIS module waited for before it turned on coil 2 could have been a leading edge or a trailing edge because either transition indicates that cylinder number 1 is 26° ATDC on either the power stroke or the exhaust stroke. Following the CID transition, the DIS module selects coil 2 because coil 2 fires spark plugs 4 and 3 simultaneously and cylinders 3 and 4 are the next cylinders at TDC: one on compression and one on exhaust.

Having opened the primary circuit for coil 2, the DIS module waits for the next trailing edge of the SPOUT signal. When that occurs, it grounds coil 3. When the leading edge of the SPOUT signal occurs, it opens coil 3. Having fired coil 3, it will next turn coil 1 on and then off in response to ECA instructions (SPOUT). It is obvious then that the ECA controls when the cylinders fire by when it raises the SPOUT signal voltage, and it controls dwell by controlling how long the SPOUT signal voltage is low.

Once the number 1 cylinder has been identified, a counter circuit built into the DIS module keeps track of the coil firing sequence. The DIS module continues to monitor the CID signal as a way of verifying its counter circuit. The ECA continues to use the CID signal while the engine is running to control injector pulsing.

Chapter 7 Ford's Electronic Engine Control IV

FIGURE 7-50 DIS coil firing 3.0-liter SHO and supercharged 3.8-liter engines

FIGURE 7-51 DIS coil pack; supercharged 3.8-liter and 3.0-liter SHO engines

2.3-Liter, Dual Plug DIS. The DIS on the 2.3-liter engine works similarly to the 3.0-liter SHO and the supercharged 3.8-liter engines, but it also has some distinct features, Figure 7-52. Because the 2.3-liter engine equipped with DIS has two spark plugs per cylinder, the system has two separate coil packs: one on the right side of the engine and one on the left side, Figure 7-53. The right coil pack contains coils 1 and 2. Coil 1 fires the right side spark plugs in cylinders 1 and 4, Figure 7-54. Coil 2 fires the right side spark plugs in cylinders 2 and 3. Coil 3 fires the left side spark plugs in cylinders 1 and 4, while coil 4 fires the left side spark plugs in cylinders 2 and 3.

After the engine has started, both plugs in each cylinder are fired at the same time; coils 1 and 3 are fired together with coil 1 firing right plugs 1 and 4 and coil 3 firing left plugs 1 and 4. During cranking, only the right coil pack fires. To prevent the left coil pack from working, the ECA sends a 12-volt signal to pin 6 of the DIS module, Figure 7-53. This is called the *dual plug inhibit signal*. With 12 volts at pin 6, the DIS module does not ground coils 3 and 4. After the ECA sees that the engine has started (by monitoring RPM), it removes the 12-volt DPI signal, and the DIS module fires both coil packs. There is no spark advance while the engine is cranking.

As seen in Figures 7-50 and 7-52, when the SPOUT signal is low the primary coil circuit is on; there is a large volt drop across the primary coil winding, so the voltage at the coil negative terminal, where the IDM originates, is near zero. Therefore, the IDM signal is low. When the SPOUT signal is high, the coil primary circuit is open, and there is no volt drop across the winding; voltage at the coil negative terminal is battery voltage, and the IDM signal is high. While the 2.3-liter, dual plug engine is being cranked and the system is in dual plug interrupt mode, the IDM signal is inverted, Figure 7-52. By monitoring the IDM signal, the ECA can tell whether or not the DPI command is being responded to.

Chapter 7 Ford's Electronic Engine Control IV

FIGURE 7-52 DIS coil firing, 2.3-liter dual spark plug engine

Chapter 7 Ford's Electronic Engine Control IV

A

B

FIGURE 7-53 DIS coil packs, 2.3-liter dual plug engine

Because the 2.3-liter, dual plug engine is not sequential injection, the ECA does not need to monitor the crankshaft's position in the engine cycle. Therefore, the CID signal is not sent to the ECA.

If, with either system when the engine starts to crank, the CID signal fails to arrive at the DIS module, the module will randomly select the first coil to fire in response to the SPOUT signals. Once the first coil is selected, the other coils will follow in their normal sequence. If the correct coil is not selected, the engine will not start. Each time the cranking motor is stopped and started again, the DIS

Chapter 7 Ford's Electronic Engine Control IV

FIGURE 7-54 2.3-liter, dual plug DIS

module will make another random selection. When the random selection turns out to be the correct coil, the engine will start.

If the SPOUT signal fails, the DIS module fires the coils in response to the PIP signal. Like the TFI IV module, the DIS module has a current limiter circuit. Primary current is limited to 5.5 amps ± 0.5 amp.

Knock Sensor Response. On systems using a knock sensor, the strength of the knock sensor's signal is proportional to the severity of the spark knock. When a knock signal is received by the ECA, it retards timing 1/2° per engine revolution until the knock disappears.

Idle Speed Control

Three different systems are used for idle speed control among various EEC IV applications.

Throttle Kicker. Throttle kicker is a vacuum-actuated device containing a diaphragm, a spring, and a plunger. It is mounted on the carburetor or CFI unit. On applications where the throttle kicker is used, normal idle speed is controlled by a conventional idle stop screw. However, when vacuum is applied to the diaphragm, the plunger is extended and opens the throttle blade slightly to increase idle speed. It is sometimes referred to as a vacuum-operated throttle modulator (VOTM).

Vacuum to the throttle kicker is controlled by a solenoid controlled by the ECA. Part of the throttle kicker solenoid (TK solenoid) assembly is a normally closed vacuum valve, Figure 7-55. The term *normally closed* means that a device such as a switch or a valve is designed to remain closed unless it is opened by some force, in this case a solenoid. *Normally open* means just the opposite. When the solenoid is energized, the vacuum valve opens and allows vacuum to the throttle kicker. The throttle kicker is activated during the following operating conditions:

- when engine temperature is below a specified temperature
- when engine temperature is above a specified temperature
- when the A/C compressor clutch is engaged
- when the vehicle is above a specific altitude

THROTTLE KICKER SOLENOID

FIGURE 7-55 Throttle kicker solenoid. *Courtesy of Ford Motor Company.*

DC Motor Idle Speed Control (ISC). The ISC motor is a small, reversible, electric motor. It is part of an assembly that includes the motor, a gear drive, and a plunger, Figure 7-56. When the motor turns in one direction, the gear drive extends the plunger; when the motor turns in the opposite direction, the gear drive retracts the plunger. The ISC motor is mounted so that the plunger can contact the throttle lever. The ECA controls the ISC motor and can change the polarity applied to the motor's armature in order to control the direction it turns, Figure 7-57. When the idle tracking switch is

Chapter 7 Ford's Electronic Engine Control IV

FIGURE 7-56 ISC motor assembly. *Courtesy of Ford Motor Company.*

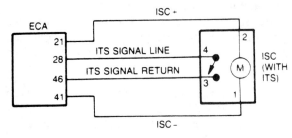

FIGURE 7-57 ISC motor circuit. *Courtesy of Ford Motor Company.*

open (throttle closed), the ECA commands the ISC motor to control idle speed. The ISC provides the correct throttle opening for cold or warm engine idle.

When the ITS is closed (throttle open), the ECA commands the ISC motor to fully extend the plunger to act as a dashpot. In other words, when the throttle tries to close again (deceleration), the plunger stops the throttle just before it finishes closing. The pressure on the plunger opens the ITS and the motor in turn retracts the plunger to normal idle position. This delay in throttle closing allows additional air into the manifold to compensate for the fuel that starts to evaporate from the manifold walls and floor as a result of the high vacuum.

When the ignition is turned off, the ECA commands the ISC motor to fully retract the plunger. This allows the throttle to fully close to prevent dieseling. When the engine stops, the ECA commands the plunger to a fully extended position in preparation for the next start-up.

Idle Air Bypass Valve Solenoid. This idle speed control device is used on multipoint injection applications. It consists of an ECA-controlled solenoid that operates a pintle-type valve, Figure 7-58. This bypass valve assembly is attached so that when it is open air is allowed

FIGURE 7-58 Throttle air bypass valve solenoid. *Courtesy of Ford Motor Company.*

to pass from in front of the throttle blade through the pintle valve to the manifold side of the throttle blade. The ECA is programmed to control this valve by duty cycling the solenoid. As the solenoid is pulsed, the valve opens. The ECA cycles the solenoid to achieve the idle speed it wants.

During a cold start, the ECA holds the solenoid at a 100% duty cycle (valve held wide open). During cranking, hot or cold, the bypass valve provides sufficient air so that it is not necessary for the operator to touch the throttle. It also acts as a dashpot during deceleration.

Thermactor Air Management

The Thermactor system delivers air to the exhaust manifold or to the catalytic converter to aid in the reduction of HC, CO, and NO_x. During operating conditions that require a rich air/fuel mixture, Thermactor air is dumped into the atmosphere in order to avoid overheating the catalytic converter. One such condition for EEC IV and other Ford electronic engine control systems is during idle. For this operation, Ford manufacturers tend to program their systems for a rich idle mixture in order to maintain a good idle quality.

FIGURE 7-59 Combination TAB/TAD valve

Chapter 7 Ford's Electronic Engine Control IV

There are two types of Thermactor systems used on EEC IV applications. One is ECA-controlled and will be discussed here. The other is a pulse injection system that uses the negative pressure existing between exhaust pulses within the exhaust manifold (most effective on four-cylinder engines) to draw air from the air cleaner past a check valve into the exhaust manifold. This system is sometimes referred to as Thermactor II and is not ECA controlled.

Thermactor Air Bypass (TAB) and Thermactor Air Divert (TAD) Valves. The pump supplies air to the bypass valve, which can either direct it to the divert valve or dump it into the atmosphere, Figure 7-59. The divert valve can either direct the air upstream to the exhaust manifold or downstream to the catalytic converter. The valves are controlled by vacuum diaphragms that are controlled by ECA-controlled solenoids.

TAB and TAD Solenoids. The TAB and TAD solenoids control whether or not vacuum is applied to the TAB and TAD valves. If the engine is started with its coolant temperature below 10°C (50°F), the ECA will leave the TAB solenoid circuit ungrounded. Vacuum to the bypass diaphragm is blocked, and the diaphragm spring holds the valve in the down position and thus bypasses pump air to the atmosphere. When engine temperature reaches 10°C, the ECA grounds the TAB solenoid circuit, Figure 7-60. Vacuum is routed to the diaphragm, which in turn lifts the bypass valve; and air is directed to the divert valve. At the same time, the ECA grounds the TAD solenoid;

FIGURE 7-60   TAB/TAD solenoid control circuit

and it directs vacuum to the divert valve diaphragm. The divert valve is pulled back to close off the passage to the catalytic converter passage and direct the air to the exhaust manifold. When the coolant reaches approximately 88°C (190°F), the system goes into closed loop, the ECA deactivates the TAD solenoid, and air is directed to the catalytic converter.

Some operating conditions put the system back into bypass mode to protect the catalytic converter from overheating. They are:

- idle
- failure of the EGO sensor signal to switch between rich and lean frequently enough (indicating that air/fuel mixture is not being controlled or that the EGO sensor is cooling)
- acceleration or WOT operation

Canister Purge

Three types of canister systems are used on EEC IV-controlled engines. Some fuel-injected engines (both CFI and multipoint) use a constant purge system with no ECA control. Some engines, especially those equipped with carburetors, use an in-line, ECA-controlled, solenoid-operated vacuum valve. The third system uses an ECA-controlled solenoid that controls vacuum to a ported vacuum switch (temperature-operated vacuum control valve), which in turn controls a canister purge valve and an exhaust heat control valve.

In-line Canister Purge (CANP) Solenoid. This is a simple, normally closed, solenoid-controlled vacuum valve that is placed in the purge line connecting the intake manifold to the charcoal canister. When the engine has reached sufficient operating temperature, the ECA duty cycles the solenoid's circuit; the duty cycle depends on the vehicle's operating conditions. This allows the canister to be purged at a controlled rate.

Canister Purge/Heat Control System. This system (limited application) gets dual use out of one ECA function. The ECA controls a normally closed, solenoid-operated vacuum valve. When the solenoid is activated, vacuum is routed to the center port of a thermal vacuum switch, Figure 7-61. When the engine is cold, the thermal vacuum switch directs the vacuum to the top port, which is connected to the heat control valve actuator. The actuator closes the heat control valve and thus forces exhaust gases to pass through the heat crossover and warm the intake manifold plenum.

When the engine warms, the wax in the lower portion of the thermal vacuum switch expands and forces the valve up. It closes the top port and opens the lower port to the vacuum signal. The heat control valve opens; then the purge control valve opens and allows the charcoal canister to be purged.

EGR Control

The ECA controls four slightly different EGR systems.

EGR Control (EGRC) and EGR Vent (EGRV) Solenoids. This system is a carryover from EEC III and is designed so that the amount of EGR flow can be controlled (valve open, closed, or anywhere between). A position sensor (EVP) on top of the EGR valve tells the ECA what position the EGR valve is in. Two solenoids are used to control vacuum to the EGR valve, Figure 7-62. The normally closed control solenoid allows manifold vacuum to the EGR valve when it is energized. The normally open vent solenoid allows atmospheric pressure into the vacuum line when it is not energized. To open the EGR valve, the ECA grounds the circuits for both solenoids; the control solenoid routes vacuum to the EGR valve, and the vent solenoid blocks the vent passage. The EVP sensor reports EGR valve position to the ECA, which in turn manipulates the two solenoids to achieve whatever EGR valve position it wants for existing operating conditions. Both solenoids are constantly being cycled to achieve and maintain desired EGR valve position.

EGR Shutoff Solenoid. This system uses one vacuum control solenoid that operates like

Chapter 7 Ford's Electronic Engine Control IV

FIGURE 7-61 Canister purge/heat control system

FIGURE 7-62 EGR control/EGR vent solenoids. *Courtesy of Ford Motor Company.*

the vacuum control solenoid just described except that a ported vacuum source is used. When the solenoid is energized, vacuum is applied to the EGR valve; when it is not energized, vacuum is blocked. With this system the EGR valve may or may not have an EVP sensor. Those with the sensor work much like the EGRV and EGRC solenoids combined into one, while those without the sensor are not modulated; they are either open or closed.

EGR Shutoff Solenoid with Back-pressure Transducer. This system uses one normally closed vacuum control solenoid like the system just described. The strength of the ported vacuum signal to the EGR valve is controlled, however, by an exhaust back-pressure variable transducer. The transducer is tied into the EGR valve's vacuum supply line and controls an air bleed into the line, Figure 7-63. Pressure signals are piped to the transducer from the exhaust system. The back-pressure signal comes from a tap in the tube that routes exhaust gas to the EGR valve. The control pressure signal comes from a point just downstream from the orifice that meters exhaust gas into the EGR exhaust supply pipe. These two pressures act on a valve in the transducer to position it to control the air bleed. When exhaust pressure is low and vacuum is high, the air bleed opens and causes the vacuum signal to the EGR valve to be weak. As exhaust pressure increases and manifold vacuum decreases, the air bleed becomes more closed and causes the vacuum signal to the EGR valve to be stronger. In this way the transducer controls how far the EGR valve opens and the resulting EGR flow. This system does not use an EVP sensor.

EGR Vacuum Regulator (EVR)

This EGR control device, introduced in 1985, is simpler than the EGRC/EGRV control system that was mentioned earlier and is just as

Chapter 7 Ford's Electronic Engine Control IV

FIGURE 7-63 Back-pressure variable transducer. *Courtesy of Ford Motor Company.*

effective. Like the EGRC/EGRV solenoid tandem, it allows the ECA to open the EGR valve to any degree of openness (modulate it). Since 1985, the EVR has become the EGR control device used on more and more Ford engine applications.

The EVR is a single unit with two vacuum ports, Figure 7-64. The lower port connects to either manifold or ported vacuum depending on the application. The upper port connects to the EGR valve. The vented cover allows air to flow through the filter and down the hollow, electromagnetic core. This allows atmospheric pressure to always be present above the metal disc at the bottom of the core. The disc is held against the bottom of the core by a weak spring. When the engine is running and the EVR is not energized, engine vacuum creates a low pressure area below the disc. With this pressure deferential, the spring cannot hold the disc against the bottom of the core tightly enough to keep atmospheric pressure from

Chapter 7 Ford's Electronic Engine Control IV

FIGURE 7-64 EGR vacuum regulator (EUR) solenoid

flooding the vacuum chamber. With atmospheric pressure in the vacuum chamber, the EGR valve closes.

When the ECA wants the EGR valve to open, it grounds the EVR. The magnetic field produced by the coil pulls the metal disc tightly against the bottom of the core, sealing off the chamber below from atmospheric pressure. Engine vacuum is now applied to the EGR valve. By rapidly cycling the EVR on and off, the ECA can control the amount of vacuum that is applied to the EGR valve and thus how far the EGR valve is opened.

The EVR is often referred to as a solenoid. Because its electromagnetic core is not movable, it is really not a solenoid but rather a simple electromagnet.

Torque Converter Clutch Control

Some Ford automatic transmissions such as the A4LD, the AXOD, the 4EAT, and the E40D feature an ECA-controlled torque con-

293

Chapter 7 Ford's Electronic Engine Control IV

FIGURE 7-65 Torque converter clutch solenoid. *Courtesy of Ford Motor Company.*

verter clutch. In each case the ECA controls the converter clutch operation with a solenoid that is located in the transmission, Figures 7-36 and 7-65. The solenoid moves a valve in the transmission valve body, which causes transmission fluid to be routed into the torque converter so as to apply the converter clutch.

The hydraulic circuitry that applies the converter clutch varies slightly between transmissions, but they are similar. Figure 7-66 shows the hydraulic clutch apply circuit for the E40D transmission in the clutch released mode. Line pressure to the converter clutch regulator valve feeds oil to the converter clutch control valve. With the converter clutch control valve in the right position as shown, line pressure oil is fed into the converter through the release passage between the transmission input shaft and the stator support. This introduces the oil into the converter on the engine side of the converter clutch plate. The pressure forces the clutch plate away from the engine side of the converter housing. The oil then flows over the clutch plate at its circumference and exits the converter through the apply passage between the stator support and then flows to the transmission oil cooler.

When the ECA energizes the converter clutch solenoid, Figure 7-67, line pressure is applied to the right end of the converter clutch control valve. It moves to the left. This reroutes the oil coming from the converter clutch regulator valve. Line pressure oil is now fed into the converter through the apply passage. With the oil introduced into the converter on the transmission side of the clutch plate, the pressure must flow over the circumference of the clutch plate to get to the exhaust passage and exit the converter. This causes the clutch plate to move toward the engine side of the converter housing. As the friction material at the outer circumference of the clutch plate contacts the machined surface of the converter housing, it seals off the flow of oil to the exhaust passage. With no pressure on the engine side of the clutch plate and line pressure on the transmission side, the converter clutch plate is forced against the converter housing with enough force to cause it to lock up and drive the transmission input shaft at engine speed.

There are slight variations in the torque converter clutch control strategy depending on the transmission and application, but in general the following conditions must exist in order for the ECA to apply the torque converter clutch:

- engine warmed up to normal operating temperature
- engine not under heavy load
- part throttle cruise
- engine within a precalibrated RPM range
- vehicle above a precalibrated speed

The solenoid will be deactivated when:

- the BOO switch indicates the brakes are applied

Chapter 7 Ford's Electronic Engine Control IV

FIGURE 7-66 Converter clutch released

- the throttle position sensor indicates WOT
- there is severe deacceleration

Turbocharger Boost Control Solenoid

Turbocharged 2.3-liter engines with EEC IV control feature an ECA-controlled, boost control solenoid that allows boost pressure to rise from the normal 10 PSI to as much as 15 PSI. Boost pressure is controlled by a wastegate that opens at the criterion pressure and allows some exhaust gas to bypass the exhaust turbine wheel. The wastegate is controlled by a spring and diaphragm assembly (actuator). In-

Chapter 7 Ford's Electronic Engine Control IV

FIGURE 7-67 Converter clutch applied

take manifold pressure is routed to the actuator diaphragm. At the criterion pressure, the diaphram compresses the spring and moves a rod that opens the wastegate.

If the ECA is satisfied with existing operating conditions, it will activate the normally closed solenoid and cycle it at 40 Hz. When the solenoid is activated, it bleeds off pressure from the wastegate actuator and allows a higher intake manifold pressure before the wastegate is opened.

A/C and Cooling Fan Control

Some engine applications use a separate electronic control module to control the A/C compressor clutch and the engine cooling fan(s). Others, typically rear-wheel drive vehicles with traditionally mounted engines and belt-driven cooling fans, use a relay to disable the A/C clutch during WOT operation.

A/C and Cooling Fan Electronic Module. The module, Figure 7-68, is usually located in the passenger compartment and gets inputs from:

- the ECA (WOT signal)
- a coolant temperature switch (closes at 105°C [221°F])
- the brake switch (brakes applied)
- the A/C switch (A/C on)

The module has a ground terminal and gets power from the battery and the ignition switch.

When the A/C switch is turned on (provided the ignition is on), the module provides power to the compressor clutch. During WOT operation the ECA signals the module, which in turn disables the A/C clutch. If the WOT condition continues, the clutch will only be disabled for thirty seconds before normal A/C

FIGURE 7-68 A/C and fan control module. *Courtesy of Ford Motor Company.*

Chapter 7 Ford's Electronic Engine Control IV

operation is resumed. If coolant temperature is below 105°C, the module will also disable the engine cooling fan in response to the WOT signal. If the coolant switch has closed, the module will override the WOT signal from the ECA by refusing to disable the cooling fan.

If the vehicle is equipped with automatic transmission and power brakes, depressing the brake pedal will cause the module to disable both the A/C clutch and the cooling fan for five seconds.

Wide-open Throttle A/C Cut-off Relay. The normally closed WOT cut-out relay is in series with the A/C switch and the clutch coil. During WOT operation the ECA grounds the WOT relay coil. The relay opens and thus disables the A/C clutch.

Inlet Air Solenoid (IAS)

A solenoid is used on some engine applications to control vacuum to the snorkel-mounted actuator that controls the air cleaner snorkel door. This function has traditionally been controlled by a bimetal vacuum valve mounted in the air cleaner. When the engine is cold, the ECA energizes the solenoid and vacuum is routed to the actuator. The actuator moves the door to close the snorkel opening to ambient air. Heated air is drawn, instead, from a shroud around the exhaust manifold, through a duct, to an opening in the bottom of the snorkel. The heated air helps evaporate the fuel as it is introduced to intake manifold.

As the engine and the air around it warm up, the heated air mode is not needed. The ECA deactivates the solenoid, vacuum to the actuator is blocked, and the door moves to open the snorkel's ambient air inlet and close the heated air inlet.

Variable Voltage Choke

On some engine applications that use a feedback carburetor, an ECA-controlled choke relay is used to more precisely control choke on time, Figure 7-69. The choke cover contains an electric heat element that applies heat to the choke spring. When the engine is started at low temperature, the ECA duty cycles the relay every 2.5 seconds. When the relay contacts are closed, battery voltage is applied to the choke heater element. When the engine reaches 27°C (80°F), the ECA holds the relay on continuously (100% duty cycle). This allows the choke to stay on while the engine is at low temperature but turns it off quickly as the engine begins to warm up and can function without it.

Temperature-compensated Accelerator Pump (TCP) Solenoid

The 2.8-liter engine equipped with a 2150A carburetor uses a TCP solenoid to control accelerator pump discharge. When the engine is below 35°C (95°F), the ECA does not ground the solenoid and the accelerator pump functions normally. When the engine coolant temperature reaches the criterion temperature and the engine needs less fuel as the throttle is opened, the TCP solenoid is energized and vacuum is routed to the accelerator pump. With vacuum applied the pump cannot deliver as much fuel to the accelerator pump discharge nozzle.

In 1987 the 4.9-liter engine was introduced with multi-point injection instead of a carburetor. Because the 4.9-liter engine is not a cross-flow engine (the exhaust valve is not across the combustion chamber from the intake valve) the injector is located just above the exhaust manifold. This exposes the injector and its connecting fuel line to a great deal of heat, especially during the heat soak that follows a hot engine shut off. Exposing the injector to this much heat could result in fuel vaporation in the injector and adjacent fuel line or injector seizing. Either condition could result in hard starting, rough running following start-up or no start at all. To offset this possibility, an injector cooling system was put on the 4.9 liter engine.

A tube that runs almost the length of the engine has been installed along the side of the

Chapter 7 Ford's Electronic Engine Control IV

FIGURE 7-69 Variable voltage choke circuit

engine, just above the injectors, Figure 7-70. Jets from the tube point to each of the first five injectors (the number six injector suffers less from exhaust manifold heat). A small blower motor, when turned on, blows air into the tube. The far end of the tube is closed off so the air is directed at the injectors. The fan is powered by an injector blower relay that is controlled by a module, Figure 7-71. The module is connected to a temperature-sensitive switch that is attached to the fuel rail. When the engine is turned off and fuel rail temperature reaches 170° F (77° C), the switch triggers the module, which then turns the fan on. The module will allow the fan to run for a maximum of 15 minutes. It will turn the fan off if the temperature switch opens or if the ignition is turned on before the 15 minutes has expired. This system is not part of, nor is it controlled by, the EEC system.

Chapter 7 Ford's Electronic Engine Control IV

FIGURE 7-70 4.9L EFI engine fuel injector cooling tube. *Courtesy of Ford Motor Company.*

Shift Light Indicator

Some EEC IV-controlled vehicles with manual transmissions feature a shift indicator light on the instrument panel to indicate to the operator when to shift into the next higher gear in order to obtain maximum fuel economy. The ECA uses information concerning engine speed, manifold vacuum, and engine temperature to control the light. When the ECA sees an engine speed greater than 1,800 RPM with light engine load, it grounds the indicator light circuit. The ECA will not turn on the light while coolant temperature is low. When the transmission goes into top gear, a switch in series with the ECA and the indicator light opens to prevent the light from coming on again.

VEHICLE SPEED CONTROL (VSC)

On limited applications in 1986, the cruise control system was integrated into the EEC IV system. In 1988, this integration became more universal for EEC IV systems. If the vehicle speed control system is activated, the ECA will compare vehicle speed to commanded speed and will control two solenoids that control vacuum to the speed control servo. The servo controls throttle position. These solenoids—speed control vacuum control (SCVAC) and speed control vacuum vent (SCVNT)—control vacuum to the servo just like the EGRC and EGRV solenoids (discussed earlier in this chapter) control vacuum to the EGR valve.

Top Speed Limiting

On some applications of the 5.0 liter engine beginning in 1988, the ECA is programmed to control top vehicle speed. This is done to limit top speed to a value that is within the rated limits of the vehicle's original equipment tires. It is achieved by reducing injector pulse width.

Programmed Ride Control (PRC). Beginning in 1987, selected Ford vehicles were equipped with a PRC system. A switch on the instrument panel allows the operator to select either a plush or firm ride. The shock absorbers have adjustable damping valves, which are adjusted by rotating a small shaft that extends up to the top of each shock absorber. A small electric motor mounted on the top of each shock absorber turns the shaft. The motor is controlled by a PRC module. If, while operating in the plush mode, certain operating conditions occur, the module will automatically adjust the shock absorbers to the firm position. Some of the information that identifies those

Chapter 7 Ford's Electronic Engine Control IV

FIGURE 7-71 Cooling fan control circuit. *Courtesy of Ford Motor Company.*

conditions is supplied to the PRC module by the EEC IV, ECA.

Information supplied directly to the PRC module from sensors is:

- hard braking—supplied by a pressure switch in the brake system
- hard cornering—supplied by a sensor on the steering shaft

Information supplied by the EEC IV, ECA is:

- more than 90% of WOT
- speeds above 83 miles per hour

SYSTEM DIAGNOSIS AND SERVICE

Malfunction Indicator Light (MIL)

Beginning in 1987, a limited number of Ford Motor Company vehicles were produced with a Malfunction Indicator Light. The MIL is

Chapter 7 Ford's Electronic Engine Control IV

in the instrument panel and works much the same as the Check Engine Light found on General Motors vehicles. When the ECA detects a fault in one of the circuits it monitors, it will turn the light on. In Self-Test the MIL will also blink out the service codes.

The first step for diagnosing an EEC IV system is to verify the driveability complaint, then go to the Diagnostic Routine index in the Diagnostic Routine section of the service manual and find the listing that most nearly identifies the driveability complaint. From that point the procedure provides a step-by-step approach to finding the cause of the problem.

Pay particular attention to the pretest inspection and test preparation steps. Malfunctions in any of the engine's non-EEC IV systems can prevent the EEC IV system from functioning properly and can cause inappropriate service codes to be set.

FIGURE 7-72 Self-Test connector. *Courtesy of Ford Motor Company.*

EEC IV Diagnostic Features

One of the outstanding features of the EEC IV system is its self-diagnostic capability. Like other systems, however, it is not totally self-diagnosing. The diagnostic features built into the ECA, combined with the diagnostic procedures in the service manual, will be very effective if the procedures are performed by a knowledgeable technician. The procedures seem confusing at first because of their extent and complexity, but studying and practicing the procedures make them a powerful and effective tool in diagnosing EEC IV system complaints.

Equipment Hookup. All EEC IV-equipped vehicles have a *Self-Test* connector somewhere in the engine compartment, Figure 7-72. This connector can be connected to an analog voltmeter; the *Star Tester*, which can be purchased through a Ford dealer, Figure 7-73; or

FIGURE 7-73 Voltmeter connection to Self-Test connector. *Courtesy of Ford Motor Company.*

to one of several other test instruments manufactured by test equipment manufacturers. The ECA displays service codes on the test instrument as a digital readout (or as needle sweeps on the analog voltmeter). A timing light should be hooked up at this time also. Consult the service manual or the test equipment manufacturer's instructions for specific directions on how to conduct the Self-Test.

Service Codes

The ECA monitors most of its input and output circuits. For any given set of driving conditions, it expects to see voltage values within a specified range on each monitored circuit. If the voltage is too high or too low, it will be interpreted as a fault (open or short) and a two-digit service code number representing that fault will be recorded in the ECA's memory. If the fault does not recur within the next twenty ignition cycles, the ECA will erase the recorded service code. The ECA reads the service code to the technician through a special test instrument or an analog voltmeter when it is put into the Self-Test mode.

Types of Codes. Six different types of service codes are stored by the ECA.

- *Fast codes:* These are codes that are useful to test equipment at the vehicle assembly plant and are used to test each vehicle on the assembly line. They contain the same information as the other service codes but occur much too fast for most test equipment to be able to read them. They can be seen as a slight needle quiver on an analog voltmeter or a flicker of the LED in the harness of the Star Tester.
- *On-demand codes* (sometimes referred to as Hard Codes): These codes refer to faults that exist at the time the Self-Test is being conducted.
- *Separator code:* This code (represented by the number 10) only occurs during the Key On/Engine Off segment of the Self-Test and indicates to the technician that the On-Demand codes are completed and that memory codes are coming next.
- *Memory codes* (sometimes referred to as intermittent codes or continuous codes): These codes refer to a fault or faults that the ECA detected during normal vehicle operation but that are not present during the Self-Test.
- *Dynamic response code:* This code, which only occurs during the engine running segment of the Self-Test and is also represented by the number 10 (the meaning of code 10 depends on the Self-Test segment in which it appears), tells the technician that he has fifteen seconds to quickly press the accelerator pedal to the floor and let it up. This allows the ECA to check the TPS, MAP, and the VAF for proper response. It is often referred to as the "Goose Test."
- *Engine identification codes:* These codes have no useful purpose for the technician. Like the fast codes, they are intended for testing in the assembly plant. They indicate to the factory test equipment how many cylinders the engine has. The number 20 means four cylinders, 30 means six cylinders, and 40 means eight cylinders.

Reading Service Codes with a Voltmeter. Reading service codes displayed on an analog voltmeter requires close attention because it is easy to miscount when higher numerical codes are presented. It involves counting needle sweeps and estimating the time lapse between sweeps. A code 21 (hard code) followed by a code 10 (separator code) and then an 11 (no memory code stored) appears as follows: the needle sweeps upscale and back, then immediately swings up and back again. It swings near full scale on a 12-volt scale. A two-second

Chapter 7 Ford's Electronic Engine Control IV

pause indicates that the first digit is completed and the two sweeps indicate that the digit is 2. After the two-second pause, one sweep occurs and is followed by a four-second pause. The single sweep indicates that the second digit is 1. The four-second pause indicates that the code is completed. After the four seconds, the 21 is repeated. A six- to nine-second pause precedes the separator code, which is shown as a single needle sweep. No needle sweeps occur for the second digit of a code if the second digit is zero. The separator code is not repeated. A second six- to nine-second pause precedes the memory codes. A single sweep followed by a two-second pause and another single sweep indicates a code 11. A four-second pause occurs and the 11 is repeated. The Engine Running segment of the Self-Test is presented in the same format except that a six- to twenty-second pause separates the engine identification code from the dynamic response code and a four- to fifteen-second pause separates the dynamic response code from the hard codes, Figure 7-74.

Quick-Test/Self-Test

The *Quick-Test* is a comprehensive set of procedures designed to help the technician identify problems within the EEC IV system. As presented in the Ford Service Manual, it includes preparation for testing (visual inspection and warming up the engine), Diagnostic Routines for symptoms without codes, the Self-Test, and Pinpoint Tests to locate the exact cause of faults that have been identified. The Self-Test is the portion of the Quick-Test in which the ECA tests itself and its related circuits and presents service codes for any faults it finds.

Key On/Engine Off. This is the first segment of the Self-Test. It is activated by connecting the single-lead portion of the Self-Test connector to pin 2 of the multiple-lead part of the Self-Test connector and turning the ignition on. If a voltmeter is being used to read codes, this will be done with a jumper lead; if a digital readout instrument such as the Star Tester is being used, this connection will be made by manipulating a button or switch as per

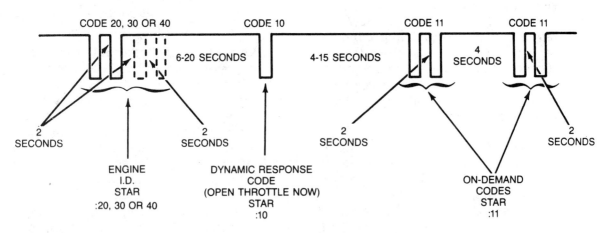

FIGURE 7-74 Typical service code display during Engine Running segment of Self-Test. *Courtesy of Ford Motor Company.*

test instrument instructions. In this segment the ECA tests first the input circuits. Any faults detected are reported as On-Demand Codes. A code 11 (system pass code) indicates that no input circuit faults have been found. Any Hard Codes that are identified should be repaired using the appropriate service manual before proceeding to the Computed Ignition Timing Check. After the repair is made, repeat this test segment to verify the repair.

Continuous Self-Test (still in Key On/Engine Off segment). The On-Demand Code or the code 11 are followed by the separator code, (10), which means that the On-Demand Codes are completed and memory codes (intermittent faults) are coming next. If no memory codes are stored, a code 11 will be displayed. If memory codes are displayed, they should be recorded by the technician; but no attempt to repair them should be made at this point.

Output Cycling Test (still in Key On/Engine Off segment). At this point, if the throttle is moved to WOT and released, the ECA will go into the Output Cycling Test in which it will turn most of the actuators on. If the throttle is moved to WOT and released again, it will turn the actuators off. This can be repeated as many times as desired. This step, however, is not normally used unless called for by one of the Pinpoint Routines, a later part of the Quick-Test. If the Output Cycling Test is called for in one of the Pinpoint Routines and to determine if one of the actuators is working properly, the Key On/Engine Off Test can be re-entered. It is only in this segment that the output cycling test can be initiated.

Computed Timing Check. This is the second segment of the Self-Test and is designed to test the ECA's ability to control ignition timing. It is initiated by having the ignition off for at least ten seconds, activating the Self-Test, and starting the engine. Any codes displayed at this time should be ignored. In this segment the ECA holds the timing at 20° above base timing for two minutes. If, for example, the base timing is 10° BTDC, the timing will read 30° BTDC.

FIGURE 7-75 In-line base timing connector. *Courtesy of Ford Motor Company.*

Check the timing with the timing light that was connected while the test equipment was being hooked up.

Be sure that the in-line base timing connector, Figure 7-75, is reconnected.

Engine Running Test. The third segment of the Self-Test tests actuators (output devices) and some sensor circuits for hard faults under dynamic conditions. Before going into this segment, the Self-Test should be deactivated and the engine started and run at 2,000 RPM for two minutes to warm the EGO sensor. Turn the engine off and wait ten seconds. To activate the Engine Running Segment, activate the Self-Test and restart the engine. The next display to appear is the Engine Identification Code followed by the Dynamic Response Code (10). The technician now has fifteen seconds to move the throttle to WOT and release it. The ECA next issues fast codes followed by the On-Demand Codes. If no hard faults are observed during the Engine Running segment, a code 11 will be issued.

Continuous Monitor Test. This test, which is the last of the Self-Test segments and is often called the *Wiggle Test,* is used to locate intermittent faults. It can be used in either of two modes: Engine Off or Engine Running. To initiate the Engine Off Mode, turn the ignition on with the test instrument connected but the Self-Test deactivated. Begin wiggling (pulling, twisting, etc.) the wires of the sensor circuit indicated by the intermittent fault code obtained during the Key On/Engine Off segment. Bump the sensor; manipulate any of its moving parts; apply heat from a heat gun to any thermistors involved, and so on. If this manipulation reproduces the intermittent open or short to which the ECA was responding when it set the code, the LED in the Star Tester harness will either flicker off or stay off. If a voltmeter is being used, the needle will pulse or swing upscale and stay there.

The Engine Running Mode of the Wiggle Test is automatically entered two minutes after the last code was displayed in the Engine Running segment of the Self-Test if the engine is not turned off.

Cylinder Balance Test. This is a feature of the Self-Test on the sequential injection engines. It is intended to help the technician identify a weak or noncontributing cylinder. By depressing and releasing the throttle within two minutes after the last code has been output in the engine running portion of the Self-Test, the system is put into Cylinder Balance Test mode.

The ECA fixes the duty cycle rate of the idle air bypass solenoid so that stabilized idle speed is established. Engine RPM is checked and stored in memory. Then the ECA stops pulsing the injector of the highest number cylinder. With the injector shut off, engine RPM is again measured and stored in memory. This is repeated for each cylinder with number 1 being the last tested. The ECA then selects the lowest RPM value that was obtained by shutting off injectors and subtracts that value from the original RPM with all the injectors being pulsed. The greatest RPM drop is then multiplied by a calibrated percentage. For example, if 125 RPM was the greatest RPM drop:

$$125 \text{ RPM} \times 65\% = 81 \text{ RPM}$$

If all cylinders dropped at least 81 RPM, a code 90 is output, indicating a pass.

During the 1987 model year, the Cylinder Balance Test strategy in the ECA was expanded to allow two additional levels of the test. Level 1 might require all cylinders to drop at least 65% of what the most efficient cylinder dropped in order to pass the test as shown in the example. Level 2 might require all cylinders to drop at least 43% of what the most efficient cylinder dropped in order to pass, while level 3 might require all cylinders to drop at least 20% of what the most efficient cylinder dropped in order to pass. As seen in Figure 7-76, if a cylinder fails to pass level 1 but passes level 2, it is not as efficient as it should be but is working. If it fails levels 1 and 2 but passes level 3, it is barely working. If it fails all three levels, it is very weak or dead. If all cylinders pass level 1, they are all working efficiently and no further testing is needed.

When level 1 has completed and code 90 for pass or a code representing a cylinder number has been presented, briefly pressing the throttle partway down within two minutes will put the system into level 2 of the Cylinder Balance Test. Pressing the throttle again within two minutes after level 2 has completed will initiate level 3. After level 3 has completed, pressing the throttle again within two minutes will cause level 3 to be repeated.

Pinpoint Procedures. This is the last part of the Quick-Test. It provides directed steps to locate the exact cause of faults that have resulted in codes being set. If the Quick-Test steps are being followed in the applicable service manual, a *Pinpoint Procedure* will be referred to for each service code requiring corrective action. The Pinpoint Procedures also provide directions that aid in finding the cause

Chapter 7 Ford's Electronic Engine Control IV

Car	Level 1 Code	Level 2 Code	Level 3 Code	Indication
A	40 cyl 4 did not pass	90 all cyl passed	level 3 not needed	cylinder 4 not as efficient as it should be but is working
B	50 cyl 5 did not pass	50 cyl 5 did not pass	90 all cyl passed	cylinder 5 barely working
C	70 cyl 7 did not pass	70 cyl 7 did not pass	70 cyl 7 did not pass	cylinder 7 very weak or dead
D	90 all cyl passed	level 2 not needed	level 3 not needed	all cylinders working efficiently

FIGURE 7-76 Cylinder bypass test

of driveability problems that do not have accompanying service codes. The symptoms of such problems are listed with suggested things to check.

Repairs should always be made in the order in which their codes are presented. After the repair the Self-Test should be repeated. The applicable service manual should be used for each vehicle application to be sure that all of the Quick-Test steps match the engine, vehicle, and model year. For instance, the 1983 1.6-liter EFI engine does not store memory codes.

Failure Mode Effects Management (FMEM). FMEM is a strategy programmed into the ECA, beginning in mid-1986, that is intended to maintain a good engine performance in the event of a failure of one of the most critical sensors. Prior to 1987, such a sensor failure would either stop the engine or put the system into Limited Operational Strategy in which the engine would run with no-spark advance and a fixed air/fuel mixture. With FMEM Strategy, the ECA will in most cases replace the failed sensor's input with a substitute value from another sensor. It will also set a code 98 in its Continuous Self-Test memory. Code 98, for 1987, indicates that a fault exists and that FMEM is operating.

The operator may not know that a fault exists, except for the following:

- idle speed may be fixed
- the cooling fan may be on all the time
- California vehicles are equipped with a Malfunction Indicator Light on the instrument panel that will come on when the fault occurs

All 1987 and later EEC IV vehicles feature FMEM Strategy. Figure 7-77 shows the sensors included in FMEM Strategy and the action taken in response to each one's failure.

Self-Test Improvements. Several improvements have been made in the Self-Test over the years. Some new codes have been added and some codes have been made more reliable. The Self-Test and the continuous monitor have the ability to check more functions. As a result, code numbers may not always have the same meaning from year to year.

Service Tips

The following suggestions should be adhered to when servicing an EEC IV system:

Chapter 7 Ford's Electronic Engine Control IV

Failed Sensor	Action Taken
MAP	Substitute TPS value. If TPS is faulty, use a fixed constant value. Disable Adaptive Learning. Disable EGR. Fixed idle speed.
TPS	Substitute fixed values during cranking, MAP value after engine starts.
ECT	Substitute ACT value during warm-up, fixed value after warm-up. Disable EGR. Disable Adaptive Learning. Fixed idle speed.
ACT	Substitute ECT value during warm-up, fixed value after warm-up. Disable EGR. Disable Adaptive Learning. Fixed idle speed.
EVP	No substitute. Disable EGR.

FIGURE 7-77 FMEM strategy

- Pinpoint Procedures shall not be used unless you are directed to do so by the Quick-Test. Each Pinpoint Procedure assumes that a fault has been detected in the circuit being considered, and to use them without being directed to by the Quick-Test can result in unnecessary repair procedures and replacement of nondefective parts. Do not replace parts unless specific test procedures indicate that they are defective.
- Do not measure voltage or resistance or connect a test light at the ECA harness connector unless directed to do so by a specific test procedure.
- Isolate both ends of a circuit and turn the ignition off before testing for shorts or continuity unless directed otherwise by a specific test procedure. Treat solenoids and switches the same way.
- An open is defined as any resistance that reads greater than 5 ohms, and a short is defined as any path to ground that reads less than 10,000 ohms, unless otherwise specified.

Adaptive Strategy and Driveability. If, for some reason, the continuous battery power to the ECA is removed, adaptive strategy learning is erased. The calibration values stored in the tables in the KAM revert to the original values stored in the ROM. Because the learned values were used to compensate for some engine or component inefficiency, the vehicle might demonstrate driveability problems such as surging, hesitation, or poor idle until the ECA has time to relearn.

On the other hand, if a component that the adaptive strategy has compensated for is repaired or replaced, the vehicle might demonstrate driveability problems until the ECA has an opportunity to learn the characteristics of the new or repaired part. To avoid this problem, battery power to the KAM should be interrupted for at least three minutes during or after such a repair.

MAP/BP Relocation. Beginning in 1987, MAP and/or BP sensors have to be repositioned so that they sit in a horizontal position with the vacuum line sloping downward to its manifold connection. This is to prevent fuel migration into the sensor. Fuel getting into these sensors has been responsible for many sensor failures.

REVIEW QUESTIONS

1. Name three types of fuel-metering devices used on the EEC IV system.
2. In what part of the ECA are the memory codes stored?
3. What type of signal does the MAP sensor produce?
4. Name at least three sensors used on EEC IV systems that are potentiometers.

5. Name at least three sensors used on EEC IV systems that are thermistors.
6. Name at least three sensors used on EEC IV systems that are switches.
7. Name at least one actuator on an EEC IV system that is duty cycled and one operated by pulse width.
8. What is the purpose of the IMS (inferred mileage sensor)?
9. What is the function of the PIP signal?
10. What is the function of the SPOUT signal?
11. Name three types of idle speed devices used on EEC IV systems.
12. What is the function of the TAD solenoid?
13. Although the EGR valve, Thermactor air, and the canister purge valve influence engine performance, they have a greater impact on _____ .
14. What is the function of the VVC (variable voltage choke)?
15. What is the function of the IAS (inlet air solenoid)?
16. What is the function of the TCP (temperature-compensated accelerator pump)?
17. List the order in which readable codes are presented during the Engine Running segment of the Self-Test.
18. Name two points during the Self-Test when the Wiggle Test can be initiated.
19. Name two different criteria for using the Pinpoint Test.

ASE-type Questions. (Actual ASE test questions will probably not be as product specific.)

20. Technician A is working on a vehicle equipped with an EEC IV system. He accidently reverses the polarity while hooking up a set of jumper cables. He fears that he has damaged the ECM. Technician B says that a diode in the power relay probably prevented damage from occurring to the ECA. Who is most likely correct?
   a. A only
   b. B only
   c. both A and B
   d. neither A nor B

21. Technician A says that if the PIP signal is lost on an injected EEC IV system, the engine will not run. Technician B says that if the SPOUT signal is lost, the engine will operate on base timing. Who is correct?
   a. A only
   b. B only
   c. both A and B
   d. neither A nor B

22. Of the four steps listed here (Diagnostic Routine, verify complaint, Pinpoint Procedure, Self-Test), technician A says they should be performed in the order in which they are listed. Technician B says they should be performed in the following order: verify complaint, Self-Test, Diagnostic Routine, and Pinpoint Procedure. Who is correct?
   a. A only
   b. B only
   c. both A and B
   d. neither A nor B

23. During the Key On/Engine Off portion of the Self-Test, technician A says that On-Demand Codes are presented followed by a separator code (10), which is followed by memory codes. Technician B says the On-Demand Codes represent faults currently present and that the memory codes represent intermittent faults. Who is correct?
   a. A only
   b. B only
   c. both A and B
   d. neither A nor B

24. Technician A says that of all the steps of the Self-Test, there are three that are intended to be used at the discretion of the

Chapter 7 Ford's Electronic Engine Control IV

technician or if directed to by one of the Pinpoint Procedures. Technician B says that she can only recall two: the Output Cycling Test and the Wiggle Test. Who is correct?
a. A only
b. B only
c. both A and B
d. neither A nor B

GLOSSARY

A4LD (Automatic Four-Speed Light-Duty) An overdrive transmission featuring an ECA-controlled torque converter clutch for rear-wheel-drive vehicles.

AXOD (Automatic Transaxle Overdrive) A four-speed automatic transmission featuring an ECA-controlled torque converter clutch used in some front-wheel-drive vehicles.

DIAGNOSTIC ROUTINE A section in the service manual that lists possible causes of a driveability complaint. It is the beginning point of the Diagnostic Procedure once the complaint has been verified.

E40D (Electronic Four-Speed Overdrive) An automatic transmission with some valve body functions performed by electronically controlled solenoids. It features an ECA-controlled torque converter clutch and is used primarily in trucks and vans.

PINPOINT PROCEDURE A service manual section that gives specific directions to find the solution of identified faults. This section should not be used until either the Diagnostic Routine or the Self-Test has identified it as the next step.

QUICK-TEST The whole set of procedures that make up the diagnostic procedure.

SELF-TEST The self-diagnostic portion of the quick-test in which the computer tests itself and its circuits for faults.

STAR TESTER A tester designed to test MCU and EEC IV systems that activates and deactivates the Self-Test procedure and displays a digital readout of the service codes.

THERMACTOR The name Ford Motor Company uses to identify its air injection pump for exhaust emissions reduction.

THICK FILM INTEGRATED (TFI) A term that identifies the type of chip used in the ignition module that is attached to the side of the distributor for EEC IV system applications.

FORD'S EEC I, EEC II, AND EEC III

Objectives

After studying this chapter, you will be able to:

- list the engine applications of the EEC I, II, and III systems.
- identify the controlled functions of the EEC I system.
- identify the controlled functions of the EEC II system.
- identify the controlled functions of the EEC III system.
- describe the self-diagnostic capability of the EEC I and II systems.

EEC I was introduced in 1978 and used on the 1978 and 1979 5.0-liter Lincoln Versailles. EEC II was introduced in 1979 but was limited to full-size Fords sold in California and Mercurys sold in all fifty states with 5.8-liter engines. EEC III was introduced in 1980 with expanded application but was still limited to V8 engines. In 1981 central fuel injection replaced the feedback carburetor as part of the EEC III system on some engine applications. In 1982 EEC III was introduced on some light truck applications.

The component parts of each of these earlier EEC systems are listed in this chapter. Many of them, however, are incorporated in the EEC IV system and were discussed in Chapter 6. Only the components or functions that are different from those found in the EEC IV system are discussed here.

ELECTRONIC ENGINE CONTROL (EEC) I

When considering computerized engine control systems, the most significant factor about the EEC I system is that it does not control air/fuel mixture. The engine is fitted with a variable venture (VV) carburetor, model 2700, which has no capability for electronic mixture control. No EGO sensor is used, and there is no closed-loop mode.

Electronic Control Assembly (ECA)

The ECA is similar to that used in the EEC IV system except that it:

- was designed to perform fewer functions.
- has no self-diagnostic capability.

Chapter 8 Ford's EEC I, EEC II, and EEC III

- has an externally mounted calibration assembly that calibrates the unit to specific vehicle values such as curb weight, axle ratio, high altitude use, and so forth, Figure 8-1.
- has a fuel octane adjustment switch that can be set to retard ignition timing either 3° or 6° if spark knock occurs (on EEC IV this feature was located on the distributor).
- provides a reference voltage of about 9 volts to its sensors.

Default Mode. If a failure occurs in the ECA, it stops sending commands to the three actuators it controls. The ignition timing stays at base timing, the thermactor system dumps its air into the atmosphere, and the EGR valve is left in the closed position.

As with the EEC IV system, the ECA is powered by a power relay that also protects it from reversed polarity. Both the ECA and the power relay are located on the passenger compartment side of the instrument panel, near the brake pedal.

Inputs

The ECA receives inputs from the following sensors:

- engine coolant temperature (ECT) sensor
- manifold absolute pressure (MAP) sensor
- barometric pressure (BP) sensor
- throttle angle position (TAP) sensor
- crankshaft position (CP) sensor
- inlet air temperature (IAT) sensor
- EGR valve position (EVP) sensor

The ECT, TAP, and EVP are essentially the same units as those used on the EEC IV system, although we find the word *angle* added to the name of the throttle position sensor. The IAT sensor is located in the air cleaner and is basically the same sensor that the EEC IV uses as either an ACT or a VAT.

MAP Sensor. The MAP sensor consists of an *aneroid* that moves the wiper of a potentiometer. The aneroid is an accordionlike capsule that contains a gas (so it can be compressed and expanded) at atmospheric pressure. As the pressure outside the aneroid increases, it compresses; as the outside pressure decreases, the aneroid expands. The aneroid is placed in a housing connected to manifold vacuum, Figure 8-2. As manifold vacuum changes, the expansion or contraction of the aneroid moves the wiper of the potentiometer to provide a signal to the ECA that is proportional to manifold pressure.

BP Sensor. The BP sensor works like the MAP sensor except that it is open to atmospheric pressure instead of being connected to manifold vacuum.

CP Sensor. The CP sensor consists of a four-lobe pulse ring on the back of the crankshaft and a probe with a permanent magnet and a coil of wire at its tip. Essentially the pick-

FIGURE 8-1 EEC I ECA and calibration assembly. *Courtesy of Ford Motor Company.*

Chapter 8 Ford's EEC I, EEC II, and EEC III

FIGURE 8-2 Aneroid manifold absolute pressure sensor. *Courtesy of Ford Motor Company.*

FIGURE 8-3 Crankshaft position sensor. *Courtesy of Ford Motor Company.*

up coil has been removed from the distributor and placed at the rear of the crankshaft, Figure 8-3. The probe is inserted into the rear of the block so that a small air gap exists between its tip and the pulse ring. As the crankshaft turns and as each lobe passes under the magnet in the probe tip, the air gap is suddenly made even smaller. This causes a sudden increase in the strength of the magnetic field being emitted from the pole of the magnet. The sudden change in the magnetic flux causes a voltage signal to be produced in the coil wound around the magnet. This steady stream of signals is transmitted to the ECA, and each one means that a piston is at 10° before top dead center.

Outputs

The EEC I system controls three functions:

- ignition timing
- Thermactor air control
- EGR flow control

Ignition Timing. To calculate ignition timing, the ECA evaluates coolant temperature, inlet air temperature, engine speed, throttle position, and engine load (intake manifold pressure). If temperatures are low, more spark advance will be provided; if they are high, less will be provided. For example, when inlet air temperature exceeds 90° F, spark advance decreases to avoid spark knock. If, however, the engine starts to overheat during idle, spark advance will be increased to raise engine speed and provide more coolant flow. As throttle opening increases, especially during rapid increases, spark advance is decreased. Spark advance is also decreased in response to an increase in manifold pressure even though throttle position does not change as might occur if the vehicle started up a hill and thus caused the engine to lose speed. The ECA's spark timing command is sent to the standard Duraspark II ignition module, which, upon receiving the signal, interrupts primary coil cur-

313

Chapter 8 Ford's EEC I, EEC II, and EEC III

rent flow to fire the appropriate spark plug. The ECA can provide as much as 30° of spark advance as opposed to the approximately 20° provided by pre-EEC systems.

A newly designed distributor that directs the spark from the coil to the appropriate cylinder is also used. The new distributor housing features a notch in its mounting flange that can index on a lug on the engine block so that once installed the distributor's base timing is set and cannot be changed, Figure 8-4. An enclosure for mounting a future miniaturized ignition module is also added. (This feature did not carry over to EEC III and was not used until the EEC IV system, when it was redesigned.)

To prevent crossfiring inside the distributor cap, the cap and rotor were given a unique design. (The increased spark advance, without the rotor being advanced as would occur if a centrifugal advance was used, increased the possibility of crossfire. This design was not carried over to the EEC IV system.) Alternate spark plug terminals in the cap extend down to a lower level. The rotor, Figure 8-5, is equipped with two electrode blades, a high one at one end and a low one at the other. The other end of each electrode blade forms a pickup arm that provides the same function as the spring contact found on most rotors. The center terminal, which receives input voltage from the coil, transmits the voltage to a center electrode plate instead of directly to the rotor, Figure 8-4.

FIGURE 8-4 EEC I distributor assembly. *Courtesy of Ford Motor Company.*

FIGURE 8-5 EEC I distributor rotor. *Courtesy of Ford Motor Company.*

Chapter 8 Ford's EEC I, EEC II, and EEC III

FIGURE 8-6 Cylinder firing pattern of EEC I distributor

FIGURE 8-7 EGR and Thermactor system control. *Courtesy of Ford Motor Company.*

From there the voltage is transmitted through a small air gap to the pickup arm, which is aligned with the center electrode plate.

By using this design, the next spark plug to be fired is almost directly across the cap (135°) from the one just fired instead of next to it. For example, let's assume that number one cylinder has just been fired from the lower electrode blade, Figure 8-6. The firing order is 1-5-4-2-6-3-7-8, and the distributor turns counterclockwise. The rotor turns 45° counterclockwise, and its lower electrode blade is pointed to the number three terminal. Its pickup arm, however, is between the spokes of the center electrode plate and does not receive the voltage surge from the coil. However, because the pickup arm for the upper electrode blade at the opposite end of the rotor is offset 45° from the rotor, it is aligned with the center electrode plate and picks up the coil voltage and transmits it to terminal number five through a second air gap. (The addition of a second rotor gap should add 2 millivolts to 3 millivolts to the required firing voltage.) Spark plug voltage continues to be delivered alternately from opposite ends of the rotor each 45° of rotor rotation.

Thermactor Air Control. The EEC I system uses a single-bed catalytic converter. The Thermactor system is therefore not designed to supply air to the converter during warm engine operation. Air is supplied to the exhaust manifolds during low-temperature operation and is dumped into the atmosphere after normal operating temperature is reached. The ECA controls these two operational modes with a solenoid-operated vacuum control valve (Thermactor control solenoid). This is a normally closed valve that controls manifold vacuum to the Thermactor bypass valve, Figure 8-7. The Thermactor pumps air to the bypass valve.

When underhood temperature is low (determined by air temperature in the air cleaner),

Chapter 8 Ford's EEC I, EEC II, and EEC III

FIGURE 8-8 Pressure-operated EGR valve. *Courtesy of Ford Motor Company.*

FIGURE 8-9 EGR cooler assembly. *Courtesy of Ford Motor Company.*

the ECA grounds the Thermactor control solenoid circuit. The solenoid opens the vacuum valve, and vacuum is routed to the bypass valve. The bypass valve then routes Thermactor air into the exhaust manifolds. When the engine warms up, the ECA de-energizes the solenoid. Vacuum is blocked from the bypass valve, which then dumps Thermactor air to get rid of it. The ECA also dumps during WOT operation.

EGR Flow. The EEC I EGR valve is opened by air pressure instead of vacuum, Figure 8-8. The pressure is supplied by the Thermactor from a port on the bypass valve, Figure 8-9. Exhaust gases are supplied to the EGR valve by an external tube that connects to an exhaust manifold. To improve driveability characteristics and to extend EGR valve life, the temperature of the EGR gases is reduced by routing the supply tube through a heat exchanger (EGR cooler). Coolant from a heater hose is circulated through the EGR cooler.

EGR flow is controlled by the ECA through two solenoid-operated valves, the normally closed pressure valve and the normally open vent valve, Figure 8-10. The ECA monitors sensor input information and determines what EGR flow rate is needed. For example, assume that an increased EGR flow is needed. The ECA energizes both solenoids. The normally closed pressure valve opens to allow pressure from the Thermactor system to be applied to the lower side of the EGR valve. The normally open vent valve closes to prevent the pressure from escaping to the atmosphere. The EGR valve opens wider until the EVP sensor indicates that the EGR valve has reached the desired position. When this occurs the ECA de-energizes the pressure solenoid. It closes and causes the existing pressure under the EGR valve diaphragm to be trapped and the EGR valve to be held in a fixed position. This condition is maintained until operating conditions change (throttle position changes, intake manifold pressure changes, and so forth) or until some of the trapped pressure leaks out and the EVP sensor indicates to the ECA that the EGR valve is no longer in the desired position. If the ECA wants to decrease EGR flow, it will de-energize both solenoids and thus close the pressure source and vent existing pressure. As barometric pressure decreases, EGR flow is reduced. EGR flow is cut off when coolant temperature is below 21° C (70° F) or above 110° C (230° F). Normally each of the two control solenoids is cycled at a rapid rate to maintain desired EGR valve position.

System Diagnosis and Service

The ECA used in the EEC I system has no self-diagnostic capability. Special test equipment and procedures have been developed and are listed in the service manual. To conduct tests special tester Rotunda No. T78L-50-EEC-I (or its equivalent) is placed in series in the ECA harness (a disconnect in the harness is provided for this). All signals going into or out of the ECA must now pass through the tester. A special DVOM (Rotunda No. T78-50) connects to the first tester and provides readings for each test in the test sequence.

The first of two test sequences is performed after the engine has been warmed up and with the engine ignition off. Eight test readings are taken to check the condition of the sensors, the power relay, the Thermactor control solenoid, and part of the ECA's circuitry. In the second test sequence, the technician checks the spark advance, EGR valve position, and Thermactor air mode with the engine running at 1,600 RPM in park with vacuum applied (from an external source) to the MAP sensor. This sequence checks the remaining ECA circuitry.

The results of each reading and check are recorded and compared to test result limits provided with the tester. If all results are within limits, the system is functioning properly; if not, diagnostic procedures are provided in the service manual for finding the fault within the identified circuit.

Rotor Alignment. Proper alignment of the distributor rotor is critical to proper engine performance. Having the rotor out of alignment with the distributor cap terminal does not affect timing but does shorten spark duration and

FIGURE 8-10 EGR control system. *Courtesy of Ford Motor Company.*

317

produce poor driveability. Notches in the upper electrode blade and the distributor body are provided so that a special tool, Figure 8-5 can be used to align the rotor. This is true for EEC I, EEC II, and EEC III systems even though the EEC III distributor uses a slightly different rotor design.

Rotor Alignment

Technicians who regularly service Ford products have learned that correct rotor alignment is critical to good engine performance. It is one of the first things they check in response to complaints of poor engine performance on vehicles with EEC I, II, or III systems.

EEC II

The EEC II system is similar to EEC I; however, it has some minor differences in its sensors and significantly more responsibility. The ECA and power relay are the same except for more internal circuitry in the ECA for its additional functions. It also provides a 9-volt reference voltage to its sensors. With the addition of an electronically controlled air/fuel mixture control device and an exhaust oxygen sensor, the system can operate in a closed-loop mode. The default mode is now called *limited operational strategy* (LOS), as we see in EEC IV. In this mode (caused by an electrical failure) all actuators are left in their deactivated mode.

Inputs

The sensors for EEC II include:

- engine coolant temperature (ECT) sensor
- manifold absolute pressure (MAP) sensor
- barometric pressure (BP) sensor
- throttle position (TP) sensor
- crankshaft position (CP) sensor
- exhaust gas oxygen (EGO) sensor
- EGR valve position (EVP) sensor

ECT. All temperature-related information is taken from the ECT, which has not changed; and the IAT is not used on EEC II.

MAP and BP Sensors. The MAP and BP sensors work like those on the EEC I system except that they are both located in one housing called a *B/MAP sensor,* Figures 8-11 and 8-12.

TP Sensor. The TP sensor is the same except that the word *angle* has been dropped from its name.

CP Sensor. The CP sensor is the same as that on the EEC I except that it has been moved to the front of the crankshaft and is shielded to protect it from electrical interference, Figure 8-13.

EGO Sensor. The EGO sensor first appeared on the EEC II systems and has been carried over to EEC III and EEC IV.

EVP Sensor. The EVP sensor has not changed.

FIGURE 8-11 BARO and MAP sensor. *Courtesy of Ford Motor Company.*

FIGURE 8-12 BARO and MAP sensor circuit. *Courtesy of Ford Motor Company.*

Outputs

The functions under the control of the ECA in the EEC II system are:

- air/fuel ratio
- ignition timing
- idle speed control
- EGR flow
- Thermactor air control
- canister purge

All of these functions are either new to the EEC system or are at least slightly different from their application on the EEC I system.

FIGURE 8-13 EEC II crankshaft position sensor. *Courtesy of Ford Motor Company.*

FIGURE 8-14 7200 feedback carburetor, back-suction system. *Courtesy of Ford Motor Company.*

Air/Fuel Ratio. The fuel bowl of a carburetor must be vented through either a balanced vent or an external vent to atmospheric pressure. The atmospheric pressure forces the fuel from the bowl through either the main metering or idle speed system and into the airstream moving through the carburetor throat. The Model 7200 VV carburetor used on EEC II applications is designed to control the pressure in the fuel bowl and thus to control the amount of fuel forced into the airstream. This is done by putting a restrictor in the bowl vent and then providing a vacuum passage that allows carburetor vacuum (called control vacuum in this case) to draw air (pressure) from the bowl area, Figure 8-14.

The amount of pressure drawn from the fuel bowl area is controlled by a pintle valve controlled by a stepper motor. As the stepper motor extends the pintle valve, the vacuum passage is restricted. This permits more pressure to develop in the fuel bowl, and the air/fuel mixture becomes richer. When the pintle valve is retracted, fuel bowl pressure is reduced and the air/fuel mixture becomes leaner.

The stepper motor contains four separate coils that can be selectively energized by the ECA to make it run forward or backward. The coils, however, do not get a steady voltage applied to them. Instead they get a series of short pulses. Each pulse rotates the armature a specific number of degrees. As the armature turns, the pintle valve is either extended or retracted depending on which way the armature turns. The ECA can then issue pulses to the stepper motor to achieve whatever air/fuel ratio it wants. The pintle has an extension range of 0.4 inch, and 120 pulses are requried to move it through its range.

Ignition Timing. The EEC II ignition system features the same coil and distributor as does EEC I and a new *Duraspark* III ignition module. The Duraspark III module was designed specifically to work with the ECA.

Idle Speed Control. The idle kicker solenoid and idle kicker actuator first appeared on EEC II and were carried over to EEC III and some applications on EEC IV. They are discussed in Chapter 6.

EGR Flow. The EGR control solenoids, EGR vent (EGRV), and EGR control (EGRC) function just like those in EEC I; however, the EGRP (P for pressure) solenoid used in EEC I was renamed. The word *pressure* was replaced by *control*. This name change occurred because the EGRC solenoid on the EEC II system is connected to manifold vacuum and the EGR valve is opened by vacuum rather than Thermactor system pressure. This system was carried over to EEC III and some EEC IV applications.

Thermactor Air Control. The EEC II system uses a dual-bed catalytic converter and therefore requires an additional Thermactor system mode: the ability to direct air to the converter in addition to being able to direct it to the exhaust manifold or to the atmosphere. To do this, two control solenoids and a combina-

FIGURE 8-15 EEC II Thermactor control system. *Courtesy of Ford Motor Company.*

tion air management valve are used: a Thermactor air bypass (TAB) solenoid, a Thermactor air divert (TAD) solenoid, and a combination TAB/TAD air control valve, Figure 8-15. This system was carried over to EEC III and EEC IV and is discussed in Chapter 6.

Canister Purge Solenoid. This is a new function for EEC II. It consists of a normally closed solenoid-controlled vacuum valve. The valve is placed in the purge hose that connects the canister to the intake manifold. When the ECA grounds the solenoid's circuit, it opens and allows the canister to be purged. This canister purge system was carried over to EEC III and some applications of EEC IV and is discussed in Chapter 6.

System Diagnosis and Service

Like EEC I the EEC II system has no self-diagnostic capability, and its diagnostic procedure is similar to that of EEC I. It requires the use of special test equipment, listed in the service manual, or its equivalent.

EEC III

On carbureted EEC III applications, we see the same components as we do on the EEC II. The air/fuel mixture control device on the carburetor and the ECA have, however, been modified. The ECA is becoming more sophisticated, and we see the addition of electronic fuel injection to some EEC III applications. The fuel injection system was the throttle body type that Ford most often called central fuel injection. In literature pertaining to EEC III systems, the terms *electronic fuel injection* and *central fuel injection* are used interchangeably. In literature pertaining to EEC IV systems, electronic fuel injection usually refers to multipoint fuel injection.

ECA

For EEC III the ECA is designed to operate in one of three operating strategies: the base engine strategy, the modulator strategy, and the limited operational strategy.

The base engine strategy covers normal operating conditions, which are divided into four modes:

- cranking
- closed throttle operation
- part throttle operation
- wide-open throttle operation

The ECA determines the operating conditions that are occurring from sensor input and selects the appropriate actuator commands according to its specific vehicle calibration programming.

The modulator strategy is used to compensate for conditions that require more extreme calibration commands. These conditions are:

- cold engine
- overheated engine
- high altitude

The limited operational strategy is like that of the EEC II system on carbureted vehicles. For CFI vehicles the LOS keeps the injectors operating but at a fixed pulse rate that produces a full, rich condition.

Inputs

The sensors for EEC III receive a 9-volt reference from the ECA. The only difference between the sensors for EEC III and those for EEC II is that the EEC III system equipped with CFI has an air charge temperature sensor screwed into the number seven intake manifold runner. The ACT sensor was carried over to some EEC IV applications and is described in Chapter 6.

Outputs

With the following exceptions, all of the actuators for the EEC III system are exactly like those used for EEC II.

EEC III 7200 VV Air/Fuel Mixture Control. The method of fuel mixture control is a little different on EEC III than on EEC II. A stepper motor and pintle valve are used in the same way (the stepper motor looks a little different), but the pintle valve does not control fuel bowl pressure. Instead the pintle valve controls the amount of air that is allowed to bleed into the main metering system, a variable main metering air bleed, Figure 8-16. Air is allowed to

FIGURE 8-16 7200 feedback carburetor, air control system. *Courtesy of Ford Motor Company.*

Chapter 8   Ford's EEC I, EEC II, and EEC III

FIGURE 8-17   EEC III second-generation distributor. *Courtesy of Ford Motor Company.*

enter a passage at the top of the carburetor from inside the air cleaner. The passage leads to an orifice that the stepper motor and pintle valve control. When the pintle is extended, less air is allowed to pass through the orifice; when it is retracted, more air is allowed to pass. After passing through the orifice, the air is channeled past the throats to the other side of the carburetor. Here it is bled into the main metering system. The more air that is bled into the main metering system, the less fuel it introduces into the carburetor throat.

CFI Injectors. The fuel delivery system and the operation of the CFI injector solenoids for EEC III were carried over to some EEC IV applications and are described in Chapter 6.

Second-generation Distributor. Later EEC III systems have a second-generation distributor, Figure 8-17. It has the upper- and lower-level electrode blades, but the rotor is cone shaped instead of rectangular. The rotor's pickup arms and the cap's center terminal are redesigned.

CFI Canister Purge. Four different canister purge systems have been used on EEC III CFI systems. A service manual for the specific vehicle should be consulted for information on how a specific purge system operates.

System Diagnosis and Service

The ECA for EEC III has self-diagnostic capability. It does not, however, have a memory that enables it to record and report faults that occur during normal driving. It can only report faults, in the form of two-digit code numbers, that are present when the self-test is being run. A quick-test procedure (similar to that for EEC IV) containing the self-test is outlined in the service manual. All necessary test equipment is listed in the service manual.

REVIEW QUESTIONS

1. What engine and vehicle use the EEC I system?
2. List the controlled functions of the EEC I system.
3. Where is the CP (crankshaft position) sensor located on the EEC I system?
4. What is the signal-producing device for the MAP and BP (manifold absolute pressure and barometric pressure) sensors? What operates the signal-producing device?
5. What are the controlled functions of the EEC II system?
6. What are the controlled functions of the EEC III system?
7. What is the reference voltage used for EEC I, II, and III systems?
8. What is unique about the EGR valve operation on the EEC I system?
9. What fuel-metering device is used on EEC I?
10. What method does the stepper motor and pintle valve of the EEC II system use to control the air/fuel ratio?
11. What method does the stepper motor and pintle valve of the EEC III system use to control the air/fuel ratio?
12. What type of service codes are presented by EEC I, EEC II, and EEC III?

ASE-type Questions. (Actual ASE test questions will probably not be as product specific.)

13. Technician A says that EEC III systems use the 7200 VV carburetor with a stepper motor for air/fuel mixture control. Technician B says that EEC III systems use CFI for air/fuel mixture control. Who is correct?
    a. A only
    b. B only
    c. both A and B
    d. neither A nor B
14. An EEC III-equipped car is brought into the shop with complaints of lost power and rough running. Technician A says that rotor misalignment can be the problem. Technician B says that rotor misalignment can-

Chapter 8 Ford's EEC I, EEC II, and EEC III

not be the problem because she has checked ignition timing and it is correct. Who is correct?
a. A only
b. B only
c. both A and B
d. neither A nor B

15. Technician A says that the EEC III system has self-diagnostic capability. Technician B says that the EEC III system has self-diagnostic capability but it does not have the ability to store codes for intermittent faults. Who is correct?

a. A only
b. B only
c. both A and B
d. neither A nor B

GLOSSARY

ANEROID An accordionlike capsule that expands or contracts in response to external pressure changes.

DURASPARK An electronic ignition system developed by Ford. Three generations of the system have evolved: Duraspark I, II, and III.

9

FORD'S MICROPROCESSOR CONTROL UNIT

Objectives

After studying this chapter, you will be able to:

- identify the applications of the MCU system.
- identify the operating modes of the MCU system.
- identify the operating modes in which open loop and closed loop are employed.
- describe the most commonly used sensing device in the MCU system.
- describe the self-diagnostic capability of the MCU system.

When it was designed, the microprocessor control unit (MCU) system's major purpose was to control air/fuel mixture and Thermactor air. That has not changed even though some additional functions have been added on some engine applications. The system uses a dual-bed catalytic converter first put into production in 1980 for California cars with 2.3-liter engines. In 1981 it was used on:

- passenger cars equipped with a 2.3-liter engine and Holley 6500, 2V carburetor (until 1983).
- California trucks with a 4.9-liter engine (until 1984).
- passenger cars equipped with a 4.2-liter engine (engine was discontinued after 1982).
- passenger cars equipped with a 5.0-liter engine (until 1984).
- passenger cars and trucks equipped with 5.8-liter engine (until 1984).

In 1982 and 1983, the MCU system was used on California passenger cars equipped with the 3.8-liter engine. In 1983 and 1984, it was used on 2.0- and 2.3-liter Ranger trucks. Its most recent applications have been on 1984, 1985, and 1986 police packages equipped with 5.8-liter engines.

There are two distinguishing characteristics of the MCU system: First, it does not control ignition timing; however, on some engine applications it has the ability to retard timing to prevent spark knock. Second, most of its sensors and actuators rely on engine vacuum. Figure 9-1 provides an overview of the MCU system.

MCU MODULE

The module for the MCU system is located under the hood and has no separate calibration unit, Figure 9-2. It is powered by the ignition

325

Chapter 9 Ford's Microprocessor Control Unit

MCU System									
Inputs					Outputs				
Engine					Engine				
2.3	4.9	3.8	8 Cyl.		2.3	4.9	3.8	8 Cyl.	
X	X	X	X	Exhaust Oxygen Sensor	X	X	X	X	Fuel Control Device
X	X	X	X	Tach Signal Engine Speed	X	X	X	X	Thermactor Bypass Sol.
X	X	X	X	Low Coolant Temp. Sw.	X	X	X	X	Thermactor Divert Sol.
		X	X	Warm Coolant Temp. Sw.		X	X	X	Canister Purge Sol.
		X	X	High Coolant Temp. Sw.		X	X	X	Spark Retard
X	X	X	X	Closed Throttle Sw.			X	X	Throttle Kicker
	X	X	X	Crowd/Mid. Throttle Sw.					
X	X	X	X	WOT Sw.					
		X	X	Knock Sensor					

FIGURE 9-1 MCU system overview

FIGURE 9-2 MCU module. *Courtesy of Ford Motor Company.*

switch (no power relay) when the switch is in the run position. No constant power supply is needed because the module has no keep-alive memory. The module operates in three fuel control modes.

Initialization Mode

When the ignition is turned to the run position, the module is activated and the initialization mode begins. The initialization mode provides a rich air/fuel ratio (between 12.8 and 13.2 to 1) for starting the engine. During warm- or moderate-temperature starts, this mode is maintained for just a few seconds after the engine is started and thus allows the induction system to stabilize. During colder starts (coolant temperature below about 10° C [50° F]) the initialization mode is maintained until the engine warms up to approximately 10° C coolant temperature.

Open-loop Mode

If after the initialization mode the engine and the EGO sensor are not warm enough to go into closed loop, the module will put the system into open loop. In open loop the air/fuel mixture is richer than 14.7 to 1 but is leaner than that provided in the initialization mode. In addition to the warm-up period, the system is in open loop during:

- idle.
- deceleration.

- heavy throttle operation (this can be defined as a throttle opening that produces a vacuum value of less than 10 inches Hg and includes high-speed cruise, acceleration, and WOT).

During open-loop operation, the MCU module sends a fixed command to the mixture control device. The command varies, however, from one open-loop operating condition to another (the command for an idle condition is different from that for a WOT or deceleration condition). During engine warm-up, the command changes as the engine coolant temperature goes up. In each of the open-loop operating conditions, the module commands an air/fuel mixture just rich enough to provide good driveability.

Closed-loop Mode

As the engine approaches normal operating temperature (measured by either a timer circuit in the module or a temperature switch, depending on module calibration), the module puts the system into closed loop. In order to go into closed loop, the coolant temperature must be at 53° C (125° F) or more, the exhaust gas oxygen sensor must be 346° C (650° F) or more, and the engine must be operating in a part throttle, light load condition. If brought to idle speed while in closed loop, the system will drop into open loop; however, some system will stay in closed loop for approximately three minutes before dropping into open loop.

INPUTS

Most of the MCU system's sensors are simple electric switches that when triggered by vacuum, temperature, or mechanical linkage provide an electrical signal to the module. For example, consider the low-temperature vacuum switch used on the 4.9-liter engine, Figure 9-3. Vacuum to this switch is controlled by a temperature-controlled, ported vacuum switch. When coolant temperature is below 35° C (95° F), the ported vacuum switch allows vacuum to the low-temperature switch and its contacts are held open, Figure 9-4. When coolant temperature exceeds 35° C, vacuum to the low-temperature switch is blocked and the contacts close.

The module applies 12 volts through an internal resistor to the insulated contact of the low coolant temperature switch. The other contact is grounded. When vacuum is applied to the diaphragm, the circuit is not complete to ground and there is no voltage drop across the resistor. The voltage signal to the sensing circuit stands at 12 volts. When vacuum is cut off, the points close and the completed circuit to ground causes a voltage drop across the resistor. This drop causes the voltage signal to the sensing circuit to drop to near zero. This tells the module that coolant temperature has reached 35° C.

Exhaust Gas Oxygen (EGO) Sensor

The EGO sensor is the same as that used on other Ford systems as well as by some other manufacturers.

FIGURE 9-3 Vacuum switch assembly (4.9 L California engine). *Courtesy of Ford Motor Company.*

Chapter 9 Ford's Microprocessor Control Unit

FIGURE 9-4 Vacuum-operated switch used as a low coolant temperature sensor

Tach Signal

The tach signal is obtained by running a wire from the distributor side of the ignition coil to the MCU module.

Low Coolant Temperature Switch

The low coolant temperature switch for the 4.9-liter engine was described in the example above. The low coolant temperature switch for another engine group, including the 2.3 liter, is similar except that it is mounted on top of the ported vacuum switch, Figure 9-5.

Mid- and Dual-temperature Switches

The 3.8-liter and eight-cylinder engine group use two switches that provide information about three different levels of coolant temperature, Figure 9-6. The dual-temperature switch (not vacuum operated) has one sliding and one stationary contact that are pushed together when the coolant reaches 13° C (55° F). The contacts are moved by a wax pellet in the nose of the switch. The pellet expands as the coolant temperature goes up. As the temperature continues to increase, the sliding contact continues to move. At 113° C (235° F) the contacts separate again.

The mid-temperature switch looks and functions like the low-temperature switch in

FIGURE 9-5 Low coolant temperature switch. *Courtesy of Ford Motor Company.*

Chapter 9 Ford's Microprocessor Control Unit

FIGURE 9-6 Mid- and dual-temperature switches. *Courtesy of Ford Motor Company.*

Figure 9-5 except that it closes at 53° C (128° F). When both switches are open, the module knows that coolant temperature is below 13° C. When the dual-temperature switch closes but the mid-temperature switch remains open, the module knows that the coolant temperature is between 13° C and 53° C. When the mid-temperature switch closes and the dual-temperature remains closed, the module knows that coolant temperature is between 53° C and 113° C. When the dual-temperature switch is open but the mid-temperature switch is closed, the module knows that coolant temperature is above 113° C.

Coolant temperature information affects all of the module's output decisions.

Throttle Position Switches

The 2.3- and 2.0-liter engine group uses a normally closed idle tracking switch that is mechanically opened by the throttle linkage when the throttle closes, Figure 9-7. The switch's opening and the resulting voltage rise in its circuit signal the module that the engine is at idle. This engine group also uses a vacuum-operated, normally closed WOT switch. The WOT switch gets manifold vacuum, but its vacuum signal is routed through a thermal vacuum switch in the air cleaner, Figure 9-8. If the engine is warm, vacuum gets to the WOT switch and holds it open. When the throttle is opened far enough for vacuum to drop to the switch's set point, it closes and the module knows to adjust the air/fuel mixture for WOT operating conditions.

The WOT switch also functions as a low-temperature indicator. When the engine is started cold and air temperature inside the air cleaner is low, the thermal vacuum switch in the air cleaner, which controls vacuum to the WOT switch, is closed and blocks vacuum to the WOT switch. Remember that the low coolant temperature switch used on the 2.3-liter engine group is closed at 35° C or more. As a result of the effectiveness of the heated air inlet system, air cleaner temperature reaches the thermal vacuum switch's set point before coolant temperature reaches 35° C. Therefore, if the WOT switch is closed and the low coolant temperature switch is still open, the module

FIGURE 9-7 Idle tracking switch. *Courtesy of Ford Motor Company.*

Chapter 9 Ford's Microprocessor Control Unit

FIGURE 9-8 Vacuum control to 2.3 L WOT switch

will know that engine temperature is below a value of about 10°C (50°F) as measured by coolant temperature. It can then command an air/fuel ratio that is appropriate for low-temperature starting and for the brief period following a cold start before the heated air inlet system raises inlet air temperature to the thermal vacuum switch's set point.

The throttle position switches for the 4.9-liter engine consist of three switches, Figure 9-3. They are the closed throttle switch, the *crowd* switch, and the WOT switch. The word *crowd* refers to a steady-state, heavy-throttle condition, but less than WOT, that produces 10 inches Hg or less of vacuum. These are part of a four-switch assembly; the fourth is the low coolant temperature switch discussed earlier. They are all normally closed.

The closed throttle switch is connected to ported vacuum. When the throttle is open, vac-

330

Chapter 9 Ford's Microprocessor Control Unit

KNOCK SENSOR (SOME 8-CYLINDER CALIBRATIONS)
(12A699)

FIGURE 9-10 Knock sensor. *Courtesy of Ford Motor Company.*

FIGURE 9-9 Zoned vacuum switch assembly (eight cylinder). *Courtesy of Ford Motor Company.*

uum holds the switch open. When the throttle closes, vacuum is cut off and the switch closes.

The crowd switch and the WOT switch are each connected to manifold vacuum. As the throttle is opened, vacuum drops. When vacuum drops to about 10 inches Hg, the crowd switch closes. When vacuum drops to around 2 to 3 inches Hg, the WOT switch closes.

The throttle position switches for the eight-cylinder engine group (called *zoned vacuum switches*) consist of a three-switch assembly, Figure 9-9. They provide basically the same information for the eight-cylinder engine module that the other three switches provide for the 4.9-liter engine module.

Knock Sensor

Only those engines in the eight-cylinder engine group are equipped with knock sensors, Figure 9-10. The sensor contains a piezoelectric crystal that produces a voltage signal in response to the vibration caused by spark knock.

OUTPUTS

Fuel Control Devices

Holley 6500 (Early 2.3-liter Engine). The Holley 6500 carburetor was modified slightly for use on MCU applications. Its main metering jet is reduced in size so that it alone cannot provide a rich enough air/fuel mixture. It relies on what resembles a power valve circuit to provide the additional fuel needed to produce a 14.7 to 1 mixture for closed-loop operation or a richer mixture for open-loop operation, Figure 9-11. This is, however, not a power valve; it is a feedback system that is part of the main metering circuit. When its metering rod is pushed down by the foot above it, additional fuel is allowed to flow past it into the main metering circuit. The farther it is pushed down, the more fuel that is admitted.

The foot that positions the metering rod is controlled by a diaphragm and spring. Vacuum applied above the diaphragm raises the foot.

Chapter 9 Ford's Microprocessor Control Unit

FIGURE 9-11 Feedback carburetor (6500 Holley). *Courtesy of Ford Motor Company.*

The metering rod is pushed up by its spring, and fuel flow is restricted. Vacuum above the diaphragm is controlled by a vacuum regulator solenoid, Figure 9-12. The solenoid is mounted away from the carburetor and in the right rear corner of the engine compartment with vacuum hoses going to its vacuum source and to the carburetor. One of the three hoses connected to the solenoid is vented to atmosphere. As the vacuum regulator solenoid is cycled, it alternately connects the vacuum chamber above the diaphragm to vacuum and to atmospheric pressure. The solenoid is cycled ten times per second by the MCU module, but the proportion of time its circuit is grounded or ungrounded within each cycle can be varied. In this way the module can control the amount of vacuum experienced above the diaphragm and the air/

FIGURE 9-12 Vacuum regulator solenoid. *Courtesy of Ford Motor Company.*

fuel mixture fed to the engine. A vacuum of 5 inches Hg produces a full lean mixture, and 0 Hg produces full rich.

Carter YFA IV (4.9-liter and Late 2.3-liter Engine). The fuel control solenoid on this carburetor, Figure 9-13, is cycled in the same way as the one just described. As it is stroked, however, it controls the amount of air that is allowed to pass through the air bleed passage into the main metering system. This control solenoid and carburetor was carried over to some applications of the EEC IV system.

Motorcraft 7200 VV (3.8-liter and Eight-cylinder Engine Group). This device was called a *feedback carburetor actuator* and used a stepper motor to control the air bleed system. It was carried over to EEC III and is discussed in Chapter 7.

Thermactor Control

Thermactor air control is achieved with the TAB and TAD solenoids and a Thermactor air control valve. It is essentially like the system used on the EEC II and EEC III systems discussed in Chapter 7. The module's programming for control of Thermactor air, however, is somewhat different. Thermactor air is delivered as indicated during the following operating conditions:

- prolonged initialization mode—atmosphere (bypass mode)
- open-loop warm-up—exhaust manifold
- WOT or crowd (open loop)—exhaust manifold
- extended idle or deceleration (open loop)—bypass
- closed loop—catalytic converter

Canister Purge

The canister purge function is controlled by a normally closed solenoid-controlled valve. It is located in the purge hose that runs between the charcoal canister and the manifold vacuum source, just as in the EEC II and carbureted EEC III systems. The module activates the solenoid and allows purging to occur:

FIGURE 9-13 Fuel control solenoid. *Courtesy of Ford Motor Company.*

- after a predetermined time has elapsed from when the engine was started.
- when the engine is within a predetermined RPM window.
- when the engine is within normal operating temperature limits.

Spark Retard

Two different spark retard systems are used on MCU systems.

4.9-liter Engine. This system does not have a knock sensor. The module is programmed to recognize conditions that are prone to produce spark knock. When it does, it eliminates vacuum advance by activating a solenoid-controlled valve, which then vents the vacuum advance hose to the distributor.

3.8-liter and Eight-cylinder Engine Group. Most of the engines in this group are equipped with a knock sensor and the universal ignition module. This module is one of three different ones used with various applications of the Duraspark II ignition system. It has the ability to retard ignition timing in response to a command from the MCU module.

Chapter 9 Ford's Microprocessor Control Unit

FIGURE 9-14 Throttle kicker control circuit

Throttle Kicker (TK)

The TK is similar to those used on other EEC systems. The TK solenoid controls vacuum to the actuator, Figure 9-14. When 12 volts is applied to the solenoid, vacuum is routed to the TK actuator. Vacuum causes the TK actuator to extend its plunger and slightly increase throttle opening when the engine is idling. Grounding the coil of the TK relay enables the relay to supply voltage to the solenoid. The relay can be grounded by either the A/C clutch being activated or by the MCU module. The TK actuator is actuated during idle conditions when:

- the engine is cold and warming up.
- the engine is overheated.
- the A/C compressor clutch is engaged.

SYSTEM DIAGNOSIS AND SERVICE

The MCU module has a self-diagnostic capability that is limited to faults observed by the module while it is in the diagnostic mode. It has no memory and therefore does not monitor and record service codes related to its sensor or actuator circuits during normal driving.

Functional Test

The Functional Test (occasionally referred to as Self-Test) is divided into three sections: Visual Inspection, Key On/Engine Off, and Engine Running.

Visual Inspection. The Visual Inspection is included as a part of the Functional Test to encourage the technician to check for obvious conditions such as loose wires, disconnected or broken vacuum hoses, faulty vacuum hose routing, low coolant temperature, faulty spark plugs or wires, and so on. Any of a number of problems such as these can cause the driveability problem for which the vehicle was brought in for service or can even cause the module to misread the results of the problem as a faulty sensor or actuator circuit. The mistaken service code produced by the module can likely result in needless testing by the technician and needless replacement of MCU system parts.

Key On/Engine Off. In this portion of the Functional Test, the module tests some of its sensor circuits electrically for proper operating condition. It is triggered by grounding the trigger terminal in the Self-Test connector, Figure 9-15, and then turning the ignition on. Service codes are read by connecting an analog voltmeter or a Star Tester to the output terminal and to ground.

Engine Running. In this portion of the Functional Test, the module tests its input and output circuits for proper operation under dynamic conditions. The procedure for entering this portion of the test varies slightly depending on engine application and model year. Be sure to follow test procedure steps and to use the appropriate service manual.

Service Codes. Service codes are presented by sweeps of the voltmeter needle or by a digital readout on the Star Tester just as they are for the EEC III and EEC IV systems. No special tools are required for testing the MCU system with the possible exception of the Star Tester as an alternative to the voltmeter.

Service Procedures

Service procedures for circuits that have been identified by a service code are covered by **Subroutines** found in the service manual. Driveability problems with no associated service code are covered by the **Diagnostic Routine** section of the service manual.

REVIEW QUESTIONS

1. What are the three operating modes used by the MCU system?
2. What is the most common sensing device used by the MCU system?
3. What is the reference voltage used by the MCU system?
4. Indicate where Thermactor air should be going during the following operating conditions:
   warm-up
   WOT or crowd
   extended idle
   closed loop
5. What is the source of the tach signal?
6. What method is used to anticipate spark knock on the 4.9-liter engine?
7. What method is used to retard timing on the 4.9-liter engine?

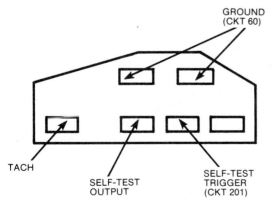

FIGURE 9-15 MCU Self-Test connector

Chapter 9 Ford's Microprocessor Control Unit

8. Which MCU engines use a knock sensor?
9. What are the two segments of MCU's self-diagnostic mode?
10. What type of memory does the MCU module have?

ASE-type Questions. (Actual ASE test questions will probably not be as product specific.)

11. Technician A says that service codes obtained from an MCU system always refer to a problem that the computer perceives as currently existing. Technician B says that service codes obtained from an MCU system can refer to a problem the computer recognized while the car was being driven. Who is correct?
    a. A only
    b. B only
    c. both A and B
    d. neither A nor B
12. The Functional Test of the MCU system consists of three segments. In what order do these segments occur?
    a. Visual Inspection, Engine Running Test, and Key On/Engine Off
    b. Engine Running Test, Key On/Engine Off, and Visual Inspection
    c. Visual Inspection, Key On/Engine Off, and Engine Running Test
    d. Key On/Engine Off, Visual Inspection, and Engine Running Test
13. Technician A says that the Key On/Engine Off segment of the MCU system's Functional Test tests the sensor circuits electrically. Technician B says that the Engine Running Test segment of the MCU system's Functional Test tests the actuator circuits electrically. Who is correct?
    a. A only
    b. B only
    c. both A and B
    d. neither A nor B

GLOSSARY

CROWD A steady-state, heavy-throttle condition that is less than full throttle.

10

CHRYSLER CORPORATION'S SINGLE-POINT AND MULTIPOINT FUEL INJECTION

Objectives

After studying this chapter, you will be able to:

- identify the operating modes of Chrysler's single-point and multipoint fuel injection systems.
- describe the function of the power module and how it interacts with the logic module.
- describe the two methods that each of these systems uses to control idle speed.
- describe the method of spark knock control used on the multipoint (turbocharged) system.
- describe the three methods of turbocharger boost control.
- identify malfunctions within the system that cause the engine to stop.
- identify the test modes used to diagnose driveability problems.

In the early 1980s, Chrysler introduced an electronic fuel injection system on the Chrysler Imperial. Few of those systems were produced; therefore, they are not discussed in this text.

The first current fuel-injected systems by Chrysler Motors were introduced in 1984 on 2.2-liter engines. They were the single-point system—referred to by Chrysler as Electronic Fuel Injection (EFI)—and the multipoint system—referred to by Chrysler as Multi-Point Injection (MPI). The 2.2 liter MPI system is often called the "Turbo" system because it is equipped with a turbocharger. Since then, other engines have been equipped with either the EFI or the MPI system. The two systems are quite similar and, for the sake of time and space, will be treated in the same chapter. Variations between the systems will be pointed out.

Prior to 1986 the single-point system was a high-pressure system operating at a fuel pressure of about 36 PSI. In 1986 the single-point, high-pressure-style injector was replaced by a

337

Chapter 10 Chrysler's Single-Point and Multipoint Fuel Injection

low-pressure-style injector, and the fuel pressure was reduced to 14.5 PSI. The pintle-style valve was replaced by a ball-shaped valve similar to those used by other low-pressure, single-point systems.

LOGIC MODULE/ POWER MODULE

Chrysler Motor's electronic fuel injection systems from 1984 through 1987-88 were controlled by two separate modules. On these vehicles, the logic module, which contains the central processor, is inside the passenger compartment, Figure 10-1. It receives most of its power from and works with the power module, which is located in the engine compartment. The logic module sends a 5-volt reference signal to its sensors and controls all system calibrations.

The power module drives by ground side switching the ignition coil, the injector/injectors, and the automatic shutdown relay, Figure 10-2. Beginning in 1985 it controlled the charging system by providing ground for the alternator's field coil. These are the circuits that have relatively heavy current flows and have been separated from the logic module to protect it from electrical interference and heat buildup. The power module is located near the battery in such a way as to receive a bath of air to help cool it, Figure 10-3.

The power module is instructed by the logic module when to open and close the primary ignition circuit and when to turn on and off the injector/injectors. The power module also provides a constant 8 volts to the logic module and the distributor pickup unit.

When the ignition switch is in the run position, it powers the power module by way of a fused circuit. The circuit that supplies this power is commonly referred to by Chrysler literature as J2. The printed circuit board of the power module has a rail that acts as a fuse and protects the module against reversed polarity. If a reversed polarity condition occurs, the fuse will blow. A reserve circuit with a diode continues to supply J2 power to the power module but with a slight voltage drop resulting from the resistance of the diode.

Single-Module Engine Control (SMEC). In mid-1987 on some 3.0-liter engine applications, the Logic Module and the Power Module were placed as a logic board and power board into a single unit called a SMEC module. The SMEC is located behind the battery in the duct that supplies air to the air cleaner where the Power Module had been. In 1988, the SMEC was used on most 1988 electronic fuel injection applications.

Because much of the discussion in the remaining portion of this chapter could apply to the earlier separate module systems or the later SMEC systems, references from this point on will be made to either the logic board or the power board unless the reference applies only

FIGURE 10-1 Logic module. *Courtesy of Chrysler Motors.*

Chapter 10 Chrysler's Single-Point and Multipoint Fuel Injection

FIGURE 10-2 2.2 L turbo EFI system. *Courtesy of Chrysler Motors.*

FIGURE 10-3 Power module. *Courtesy of Chrysler Motors.*

to the earlier systems in which case logic module or power module will be used. If a reference applies to only the later systems, SMEC will be used.

The eight-volt regulator on the power board has been changed to supply 9.2 to 9.4 volts to the SMEC. This is due to the use of a different type of transistor (Metal Oxide Semiconductor Field Effect Transistor) on the power board that requires the higher voltage. The battery temperature sensor that was formerly in the Power Module has been moved to the logic board.

Features

Automatic Shutdown (ASD) Relay. The ASD relay is powered by the ignition switch

Chapter 10 Chrysler's Single-Point and Multipoint Fuel Injection

FIGURE 10-4 Typical ASD relay location. *Courtesy of Chrysler Motors.*

when the ignition is turned to the run position. When the power module grounds the ASD coil, the ASD provides power for the electric fuel pump, the ignition coil, and the injector(s). The fuel pump has a permanent ground and comes on automatically. The ignition coil and injector(s) wait for their circuits to be grounded by the power module. For 1985 through 1987 models, the ASD relay was located in the power module. In 1984 and after 1987, it is outside the module, Figure 10-4. Mounting it outside saves space in the module and improves module reliability.

On-board Diagnostics or Self-diagnosis. The logic board monitors its most vital input and output circuits for faults and records fault codes in its memory when it perceives an inappropriate voltage value.

Adaptive Memory. The logic board has the ability to modify some of its programmed calibrations concerning fuel metering to compensate for production tolerances and changes in barometric pressure.

Operating Modes

Starting. While the engine is being cranked, the injector pulse frequency is doubled. The single-point injector is normally pulsed twice per revolution, and the multipoint system normally pulses its injectors in pairs, with each pair being pulsed once each revolution. The double pulsing only occurs for a preprogrammed time interval (to avoid flooding). The pulse width is influenced only by coolant temperature. After the engine has started and if it is below a criterion temperature, the logic board will provide a fuel enrichment program based on coolant temperature and manifold pressure. The start-up enrichment program decays to base enrichment over a programmed time span.

Primer Function. When the 3.0-liter engine's ignition is turned on, all six injectors will be pulsed to spray fuel into the intake ports. This is intended to improve starting. It will occur regardless of engine coolant temperature, but if the engine is warm, less fuel will be injected; if the engine is cold, more fuel will be introduced.

Open Loop. Open loop occurs during the following conditions:

- coolant temperature is below normal operating temperature
- oxygen (O_2) sensor is below minimum operating temperature (315°C [600°F])
- O_2 sensor is not switching across a preset switch point of about 0.45 volt
- wide-open throttle (WOT) or acceleration
- when the multipoint system reaches 1 PSI or more of boost
- deceleration
- when the single-point system is at idle

Closed Loop. The following conditions must be met before the system will go into closed loop:

- coolant temperature must reach a preset value
- O_2 sensor must be producing a valid signal
- predetermined amount of time must have elapsed since engine was started

Limp-in Mode. If the logic board through its on-board diagnostics circuit receives a voltage signal (or no signal) from one of its most vital sensors and the signal does not fit within an acceptable range or does not change under specific conditions, it will consider the sensor inoperative and will put the system into limp-in mode. In this mode the logic board will substitute one or more other sensor values for the failed sensor and will turn on the power loss lamp on the instrument panel. Although the engine may not operate efficiently, it will continue to operate. Those sensors for which this will occur include the following (the indicated codes will be stored in memory):

- *manifold absolute pressure*—Throttle position and engine speed are used to create a simulated MAP value. (13 and/or 14)
- *throttle position*—The MAP is used to produce rough estimates of throttle position. (24)
- *coolant temperature*—Charge temperature (manifold air/fuel mixture temperature) is used as a substitute. (22)
- *charge temperature*—The engine usually functions effectively without this input. The power loss lamp is not turned on. (23)

Loss of the distributor reference causes the engine to stop.

INPUTS

Manifold Absolute Pressure (MAP) Sensor

This is a piezoresistive-type pressure sensor. The logic board uses it as a barometric pressure sensor during periods of ignition on/engine off or during certain other conditions on some models as described in the **Outputs** section of this chapter, Figure 10-5. The MAP is usually located on the logic module.

FIGURE 10-5 MAP sensor. *Courtesy of Chrysler Motors.*

Some of the earlier applications had a small bleed hole in the vacuum line to the MAP to prevent condensation from collecting in the line. The MAP sensor was calibrated to allow for the vacuum leak.

Throttle Position Sensor (TPS)

The TPS is a rotary potentiometer attached to the end of the throttle shaft on the side of the throttle body, Figure 10-6.

FIGURE 10-6 Throttle position sensor. *Courtesy of Chrysler Motors.*

Chapter 10 Chrysler's Single-Point and Multipoint Fuel Injection

Oxygen (O₂) Sensor

This is a single-wire O_2 sensor. Some applications use a three-wire sensor, and the additional wires power an internal heat element.

Coolant Temperature Sensor (CTS)

The CTS is a single-element thermistor that screws into the thermostat housing, Figure 10-7. It has a resistance from 11,000 ohms at −20°C (−4°F) to 800 ohms at 90.5°C (195°F); 195°F is considered full operating temperature for these systems.

Charge Temperature Sensor (Charge Temp)

The charge temp sensor is a single-element thermistor screwed into a runner of the intake manifold. It has about the same resistance range as and looks similar to the CTS. On earlier models its input is used during cold enrichment calculations; otherwise it serves as a backup to the CTS. On later model multipoint systems, its input is used in the control of air/fuel mixture (especially cold) and boost control.

Throttle Body Temperature Sensor

The low-pressure, single-point system no longer uses a charge temperature sensor; rather, it uses a throttle body temperature sensor that measures fuel temperature in the throttle body, (Figure 10-8). This information is used to calculate air/fuel mixture during hot restarts.

Distributor Reference Pickup (REF Pickup)

A Hall-effect switch in the distributor generates the REF pickup signal from which the logic module determines engine speed and crankshaft position. On the single-point system, this same signal is supplied to the power board as well. Without this signal it does not keep the ASD relay grounded, which in turn removes power from the ignition coil, the injector, and the fuel pump. On the multipoint system, this signal is sent only to the logic board; but in response to this signal, the logic board

FIGURE 10-7 Charge sensor and coolant sensor. *Courtesy of Chrysler Motors.*

FIGURE 10-8 Low-pressure, single-point throttle body with throttle body temperature sensor. *Courtesy of Chrysler Motors.*

relays to the power board exactly when to turn on the injectors.

Distributor Sync Pickup (SYNC Pickup)

Most multipoint system has a second Hall-effect switch assembly in its distributor and beneath the REF pickup, Figure 10-9. This

Chapter 10 Chrysler's Single-Point and Multipoint Fuel Injection

FIGURE 10-9 Turbo distributor with reference pickup and sync pickup Hall-effect switches. *Courtesy of Chrysler Motors.*

FIGURE 10-10 Optical-sensing device in distributor

Hall-effect switch has a shutter wheel with only one vane instead of four. Its one vane, however, occupies 180°, or one-half of the shutter wheel. The leading and trailing edges of the vane each provide a signal to both the logic board and the power board. The power board uses this input to determine which pair of injectors (1 and 2 or 3 and 4) to turn on when the on signal from the logic board arrives. The logic board uses the SYNC pickup signal along with other sensor information to help determine pulse width.

Optical Distributor. A 3.0-liter V-6 engine was introduced in 1987 with multipoint fuel injection. This was Chrysler's first naturally aspirated MPI-equipped engine. The computer uses a speed-density formula (does not use a vane airflow meter or a mass airflow meter) to calculate fuel quantity and is similar to the system used on the 2.2-liter multipoint injection system. Rather than using Hall-effect switches for the reference and sync pick-up signals, this engine uses a pair of optical sensors (Figure 10-10). A disc rotates with the distributor shaft. The disc has two sets of slits. The first set, closest to the center of the disc, consists of six slits. These are called "low data rate" slits. The second set, "high data rate," are near the outside circumference. These slits are spaced every two crankshaft degrees with the exception of one small, blank spot where there are no slits. This one blank section alerts the SMEC where number one cylinder is at in the engine cycle. The high data rate of this sensor allows the SMEC to monitor crankshaft acceleration and deceleration.

Two light-emitting diodes (LED) are placed on one side of the disc so that their light beams can pass through the slits, one for each set of slits. When one of the slits aligns with its respective LED, the light beam strikes a photo diode, of which there is one for each set of slits. When the light from one of the LEDs strikes one of the photo diodes, the photo diode generates a small voltage that is applied to the base of a

Chapter 10 Chrysler's Single-Point and Multipoint Fuel Injection

FIGURE 10-11 Reference optical sensor

transistor (Figure 10-11). Let's assume that we are talking about the photo diode that triggers the reference circuit. The transistor acts as a ground switch in the reference circuit that extends from the SMEC. The SMEC applies a constant 5 volts to the reference circuit. When the reference circuit transistor is turned on, current flows across resistor A in the SMEC through the transistor to ground. Voltage drop across resistor A causes the voltage at point B to be near zero. When the disc blocks the light beam from the LED, the switching transistor in the reference circuit is turned off. The reference has no path to ground and, there is no current flow across resistor A, so the voltage at point B is 5 volts. Each time the voltage at point B drops to near zero, the SMEC knows that the next piston in the firing order has reached a specified position in its cycle. The sync circuit works the same way with each signal meaning that 2 degrees of crankshaft travel have occurred.

At engine speeds below 1200 RPM, the SMEC will use the sync signal (high data rate)

Chapter 10 Chrysler's Single-Point and Multipoint Fuel Injection

FIGURE 10-12   3.3-liter direct ignition system. *Courtesy of Chrysler Motors.*

Chapter 10 Chrysler's Single-Point and Multipoint Fuel Injection

for calculating ignition timing and injector pulsing. At speeds above 1200 RPM, it will use the reference signal for these calculations. This is because at lower engine speeds there is a significant, momentary RPM gain each time a spark plug is fired. However, due to variations in compression, spark plug condition, and so on, the RPM change may not be the same for each cylinder. The high data rate signal allows the SMEC to monitor these RPM changes and adjust ignition timing and injector pulse width accordingly. At higher engine speeds, there is much less variation in RPM between power impulses and such adjustments are not as necessary.

Direct Ignition System (DIS). In 1990 Chrysler introduced a 3.3-liter, 60-degree V/6 for some passenger cars and vans. This engine features a distributorless ignition system which Chrysler calls Direct Ignition System, Figure 10-12. Similar to General Motors and Ford distributorless systems, the Chrysler DIS system uses Hall-effect sensors to monitor camshaft and crankshaft position, an ignition module and a coil pack that contains a separate coil to fire each pair of companion spark plugs. The coil pack is mounted on the ignition module.

The camshaft sensor mounts in the timing cover and triggers off slots on the camshaft timing gear. The slots on the timing gear are coded so that the signals provided by the sensor identify for the ignition module which pair of companion cylinders to fire next in response to the computer's command to fire. Which pair of injectors that the computer will pulse next is also determined from this signal. Because the camshaft sensor can identify each cylinder, the computer can begin firing spark plugs and pulsing injectors within one crankshaft revolution during cranking.

The crankshaft sensor mounts on the transaxle bell housing and triggers off slots on the torque converter drive plate. There are four slots per pair of companion cylinders. These signals are used by the computer to determine crankshaft position and engine speed. This information assists in determining ignition timing, injector timing, and injector pulse width.

Vehicle Speed (Speed Sensor)

The speed sensor is a small switching device that receives a steady 5-volt signal from the logic board and breaks into eight on/off signals for each speedometer cable revolution. It is located on the transmission where the speedometer cable attaches, Figure 10-13.

Detonation (Knock) Sensor

The knock sensor is the piezoelectric type. It is screwed into the intake manifold.

Battery Temperature Sensor

A thermistor in the power board, which is near the battery, senses the battery's ambient temperature and sends the information to the logic board.

Charging Circuit Voltage

The logic board senses the voltage that the charging system is maintaining from the fuel pump power circuit.

Switch Inputs

Park/Neutral (P/N). The P/N switch tells the logic board if the transmission is in gear or not. Its input influences idle speed, and normal spark advance is not issued when the transmission is not in gear.

Electric Backlite (Heated Rear Window). If the heated rear window switch is on while at idle, the logic board will increase the throttle blade opening to compensate for the additional load that the alternator puts on the engine.

Brake. The brake switch is used as a backup in the event that the TPS fails to signal the logic board that the throttle has closed.

Air-conditioning (A/C). If the A/C is on, a signal will be provided from the A/C control circuit to the logic board so that it can increase throttle opening to compensate for the additional load on the engine during idle.

Chapter 10 Chrysler's Single-Point and Multipoint Fuel Injection

FIGURE 10-13 Vehicle speed sensor. *Courtesy of Chrysler Motors.*

Air-conditioning Clutch. As the A/C system cycles the compressor clutch on and off during idle, the A/C clutch switch informs the logic board of the clutch's status so that the throttle blade can be adjusted accordingly to maintain a steady engine idle speed.

The two A/C system inputs were combined into one signal on later models.

Battery. One terminal of the logic board receives power directly from the battery. This input is absolutely necessary to sustain power to the keep-alive memory (KAM) when the ignition is turned off.

OUTPUTS

Injector/Injectors

Figures 10-14 and 10-15 show a comparison of the low-pressure and high-pressure injectors. They are each solenoid-operated injectors.

During normal operation the single-point system pulses its injector twice during each engine revolution. In the middle and late 1980s all four-cylinder multipoint systems pulsed two

FIGURE 10-14 Low-pressure injector. *Courtesy of Chrysler Motors.*

injectors simultaneously; they pulsed one pair during the first revolution of the engine cycle and the other pair during the second revolution of the cycle. During cranking, this schedule was doubled as discussed earlier in this chapter.

Chapter 10 Chrysler's Single-Point and Multipoint Fuel Injection

FIGURE 10-15 High-pressure injector. *Courtesy of Chrysler Motors.*

FIGURE 10-16 Fuel pump. *Courtesy of Chrysler Motors.*

Later 2.2-liter Turbo IV and 2.5-liter Turbo 1 systems featured sequential injection strategy where each injector is pulsed individually in the cylinder firing order. The 3.0- and 3.3-liter six-cylinder multipoint systems pulse injectors in pairs of two. The pulse width is determined by the logic module based on input from the:

- CTS.
- MAP.
- RPM.
- TPS.
- O_2—in closed loop.
- charge temp—during cold enrichment.
- speed sensor—during deceleration.
- fuel enrichment and enleanment multipliers programmed into the logic board.

In contrast to the most common practice, the single-point system from 1984 to 1987 used a fixed ground for the injector, and the power module controlled it by switching power on and off (positive side switching). From 1987 the single-point system used ground side switching to be compatible with the multipoint systems.

Electric Fuel Pump

The fuel pump is a DC electric motor with a permanent magnet field, Figure 10-16. It is submerged in fuel at the bottom of the fuel tank. It contains two check valves. One acts as a pressure relief valve; the other, near the outlet port, prevents fuel from flowing in either direction when the pump is not running. This helps maintain the fuel line full of fuel and reduces the occurrence of vapor in the fuel line.

When the ignition is turned on, the power board grounds the ASD relay, which powers the fuel pump. If a distributor pickup signal is not received within two seconds, the ASD ground will be removed and the pump turned off. It will remain off until the pickup signal arrives or until the ignition is turned off and on again.

Some applications such as the Shelby turbocharged units use two fuel pumps, one in the tank and one nearer to the engine.

Fuel Pressure Regulator. The fuel pressure regulator for the pre-1986 single-point system maintains a constant fuel pressure at the injector of 36 PSI; the low-pressure system maintains 14.5 PSI. The multipoint system maintains a pressure differential across the injector tip of 55 PSI, Figure 10-17. The pressure differential is maintained by connecting manifold vacuum/pressure to the pressure regulator as explained in Chapter 2. Fuel pressure as measured in the injector line varies from 42 to 62 PSI.

On the 3.0-liter engine, the fuel pressure was raised to 46 PSI to avoid starting problems during hot restarts due to fuel vaporizing in the fuel lines and injectors.

Ignition Timing

Spark advance is calculated by the logic board and using information from the:

Chapter 10 Chrysler's Single-Point and Multipoint Fuel Injection

FIGURE 10-17 Turbo multipoint system fuel pressure regulator. *Courtesy of Chrysler Motors.*

- CTS.
- REF pickup.
- MAP.
- barometric pressure (taken from MAP).

During warm idle the logic board uses spark advance as its first choice to control normal idle speed fluctuation. If a predetermined amount of spark timing change does not bring idle RPM within acceptable limits, the logic board will use the automatic idle speed motor.

Beginning in 1985 the multipoint system used a BARO-read solenoid that the logic board used to briefly vent the vacuum line to the MAP sensor. While the line is flooded with atmospheric pressure, the MAP sensor updates the BARO reading. This vent cycle occurs at least once each time the throttle closes and the RPM is below a specific value and can occur as often as every thirty seconds while conditions prevail. Using a turbocharger makes ignition timing especially critical, and barometric pressure significantly influences allowable spark advance.

Detonation Control. Beginning in 1985 the multipoint system has had the unique ability to selectively retard the timing on just the cylinder that is knocking. When the knock signal occurs, the logic board checks which cylinder has just been fired. The spark advance for that cylinder is then retarded slightly. If the knock continues, the logic board will begin to lower boost pressure. In 1986 a feature was added that provides a ten-second delay from the time the throttle is moved to WOT before the logic board responds to the knock sensor.

Chrysler's Detonation Control

Chrysler has put a strong emphasis on performance in their turbocharged engine control strategy. This is reflected in their method of detonation control. There are really only two ways to stop detonation on a turbocharged engine during boost conditions, and to allow detonation to go unchecked on turbocharged engines destroys them. To reduce the boost pressure everytime a cylinder detonates hampers performance. To retard timing during boost conditions raises exhaust temperature, which is already critically high. To retard the timing on only the offending cylinder raises exhaust temperature only slightly, if at all, and still allows the other cylinders to produce full power.

Wastegate Control Solenoid

Prior to 1985 turbocharged units had three types of boost control. The first was the wastegate as discussed in the **Outputs** section of Chapter 3. The wastegate actuator, Figure 10-18, opens the wastegate at a boost pressure of 7.2 PSI. Secondly, boost pressure is sensed by the MAP. If the boost pressure rises above 7.2 PSI, as it will momentarily in response to snap acceleration, the logic board will recognize the high pressure and will stop issuing injection pulse commands until boost pressure falls below 7.2 PSI. The third control is in response to engine RPM. If RPM goes above

Chapter 10 Chrysler's Single-Point and Multipoint Fuel Injection

1. Turbine Housing (Hot Side)
2. Turbine Wheel (Hot Side)
3. Wastegate
4. Shaft Wheel Assembly
5. Water Passage
6. Oil Passage
7. Compressor Wheel
8. Compressor Housing

FIGURE 10-18 Turbocharger. *Courtesy of Chrysler Motors.*

6,650, the logic board will again stop fuel injection until the RPM drops to 6,100.

In 1985 a fourth feature was added. The logic board controls a boost control solenoid that when activated vents the line that pipes manifold pressure to the wastegate actuator. If operating conditions are favorable, the logic board will pulse the boost control to bleed off some of the pressure to the actuator. In this way boost pressure can actually go to 10 PSI before the actuator opens the wastegate. The inputs that the logic board considers for this decision are:

- barometric pressure.
- RPM.
- TPS.
- CTS.
- charge temp.
- detonation.

FIGURE 10-19 Multipoint throttle body with AIS motor. *Courtesy of Chrysler Motors.*

Additionally the logic board considers detonation history and how long the engine has been in boost.

Revised Boost Control

Beginning in 1988, the actuator for turbo boost control on the Turbo I engine gets its pressure signal directly from the turbocharger rather than from the intake manifold. This allows the wastegate to be opened during part throttle operation and eliminates any possible boost until WOT. Without boost at part throttle:

- exhaust back pressure is reduced
- the intake charge is cooler
- tendency toward spark knock is reduced
- part throttle fuel economy is improved

Automatic Idle Speed (AIS) Motor

The AIS motor, Figure 10-19, is a reversible electric motor controlled by the logic board based on information from the:

- TPS.
- CTS.
- speed sensor.
- P/N switch.
- brake switch.

The AIS motor moves a valve to control airflow through a channel in the throttle body to bypass the throttle blade. With the valve in its most closed position, enough air passes through the throttle body to support the engine with no load at a minimum idle speed. The logic board commands the AIS motor to move the valve to provide a preprogrammed idle speed for all operating conditions. During deceleration the valve opens to prevent engine stalling and to compensate for the fuel that begins to evaporate from the intake port walls.

EGR Control Solenoid

This solenoid controls the ported vacuum to the EGR valve. When it is energized by the

FIGURE 10-20 EGR valve and transducer. *Courtesy of Chrysler Motors.*

logic board, vacuum to the EGR valve is blocked. The solenoid is energized under the following conditions:

- coolant temperature below 21°C (70°F)
- engine speed below 1,200 RPM
- WOT

When the solenoid is de-energized, the EGR valve is additionally controlled by a back-pressure transducer, Figure 10-20. Some applications have a bleed in the vacuum line between the transducer and the EGR valve. Its purpose is to prevent a pressure buildup in the EGR valve.

EGR Valve Temperature Sensor

In 1988 a thermistor-type temperature sensor was introduced on the California edition of the 1.5-liter multipoint injection engine used in the Colt. It is used to detect an EGR valve that fails to open when commanded to do so. When the EGR valve opens and exhaust gas flows through it, the temperature of valve's lower body goes up sharply. If the SMEC commands the EGR valve to be opened and does not then see a signal from the EGR valve temperature sensor indicating a temperature rise, it will store a fault code and turn on the malfunction indicator light. The EGR valve was eliminated on the turbo 2.2-liter engine in 1988.

Chapter 10 Chrysler's Single-Point and Multipoint Fuel Injection

FIGURE 10-21 EGR, canister purge, wastegate control, and BARO-read solenoids. *Courtesy of Chrysler Motors.*

FIGURE 10-22 Radiator fan relay. *Courtesy of Chrysler Motors.*

Canister Purge Solenoid

This solenoid is in the purge line that runs from the throttle body to the charcoal canister. When coolant temperature is below 82°C (180°F), the logic board energizes the solenoid. This blocks the purge line, Figure 10-21.

Radiator Fan Relay

The radiator fan relay, Figure 10-22, supplies power for both the cooling fan motor and the A/C compressor clutch. The normally open relay can be activated (ground the relay coil) by the logic board or when the A/C is turned on (the A/C and blower motor are controlled by ground side switching). The logic board activates the relay when coolant temperature reaches a specified value.

Charging Circuit Control

Beginning in 1985 the logic board, working through the power board, assumed control of the alternator's output; they replaced the voltage regulator. The logic board's decisions are based on output voltage and battery temperature. When a criterion output voltage as sensed by the logic board is reached, it duty-cycles a switching transistor inside the power board. This transistor is used to ground the alternator's field, Figure 10-23. The duty cycling varies as necessary to control output voltage at the desired level. The criterion voltage varies based on battery temperature.

A/C Cut-out Relay

The normally open A/C cut-out relay is in the ground side of the A/C compressor clutch circuit and between the compressor clutch and the A/C switch. When open it breaks the A/C clutch circuit. It, too, is controlled by the logic board. If the A/C is turned on, the logic module will keep the A/C cut-out relay energized (closed) whenever the engine is running except:

- when the engine is operated at WOT.
- when engine speed is below 500 RPM.
- when the engine is cranked with the A/C on. It remains de-energized for ten to fifteen seconds after the engine is started.

Chapter 10 Chrysler's Single-Point and Multipoint Fuel Injection

FIGURE 10-23 Charging circuit control. *Courtesy of Chrysler Motors.*

Lock-up Converter

Beginning in 1988, the Torqueflite transmission was equipped with a SMEC-controlled lock-up converter for some vehicle applications. The converter clutch is hydraulically applied and is controlled by a computer-actuated solenoid. The solenoid is mounted on the valve body transfer plate and is fed third clutch oil when the transmission goes into third gear. If the solenoid is activated (grounded by the computer) while third clutch oil pressure is present, the clutch is applied. If it is not, the non-applied solenoid allows the oil to be bled off, and the clutch is released. Its operation is similar to the systems used by General Motors and Ford. The SMEC will not apply the clutch unless:

- the coolant temperature is above 150° F
- park/neutral switch indicates the transmission is in gear
- brake switch indicates that the brakes are not applied
- TPS is about a minimum value

Power Loss Lamp

This is a lamp on the instrument panel and that the logic board controls. (On some early applications, it was on the logic board and could be seen through a hole in the module housing.) It is turned on as a bulb check for three seconds each time the ignition is turned on. If the logic board sees a fault in any of its most vital circuits, it will turn the lamp on and put the system in limp-in mode. If the fault is

353

somehow cleared up, the lamp will remain on and the system will stay in the limp-in mode until the ignition is turned off. When the engine is started again, the lamp will be off and the system will no longer be in limp-in; but the code will remain in memory until the ignition has been cycled thirty times without the problem reappearing.

The Limp-in strategy was changed somewhat for 1988. On those and later applications, if a sensor fails, the computer will use a substitute value for the failed sensor and the Power Loss Lamp (referred to as a Check Engine Light on some newer applications) will come on as in the past. But if the sensor's signal comes back into proper range, the computer will go back to normal function without the ignition having to be cycled as on past models.

Top Speed Limit

Beginning in 1988, vehicles equipped with 3.0-liter or 2.2-liter Turbo I engines have a speed limit feature programmed into the SMEC. If the vehicle exceeds 118 MPH, the computer stops fuel injection to prevent speeds that exceed the speed rating of the original equipment tires.

SMEC-Controlled Cruise Control

Many SMEC-controlled systems incorporate cruise control as one of the functions that the engine control computer is responsible for. The system operates similarly to the Ford system described earlier in the **Outputs** section of Chapter 7. One feature it has in addition to those on the Ford system is a third solenoid called a dump solenoid. The ignition switch powers the dump solenoid through a fuse, the cruise control on/off switch, and the brake switch. The circuit has a fixed ground. Turning the on/off switch off or stepping on the brake will interrupt the power to the dump solenoid and it will immediately bleed off vacuum to the servo, allowing the throttle to close.

SYSTEM DIAGNOSIS AND SERVICE

A thorough diagnostic procedure is published by Chrysler Corporation for each model year and application. This *driveability test procedure* should be in the service manual you are using.

On-board Diagnostics

The logic board monitors specific input and output circuits for faults (voltage values that are too high or too low or that do not change appropriately). If a fault is detected, a two-digit code number will be stored in the module's diagnostic memory. The codes and other diagnostic information can be obtained by putting the system in one of its diagnostic modes. This can be done with the diagnostic readout box (tool C-4805) or its equivalent. The fault codes can also be presented and the input switches tested with just the power loss lamp.

Test Modes

Diagnostic Test Mode. Connect the readout box to the diagnostic test terminal under the hood. Activate the test by turning the ignition on-off-on-off-on within five seconds. The readout box displays any stored codes. After the codes are completed, a 55 is displayed.

Switch Test Mode. If the above test is completed and the 55 appears (be sure that all system input switches are turned off). turning on any of the input switches will cause the displayed 55 to change if the logic board receives the signal. Turning the switch off causes the 55 to reappear.

Circuit Actuator Test Mode (ATM). If after the diagnostic test mode is completed and the 55 appears the ATM button on the readout box is pressed and held until the desired ATM code number is displayed and then released, the ATM will be initiated. The logic board now cycles the selected actuator on and off at two-second intervals for five minutes or until the ignition is turned off. This allows the technician to verify that the actuator and circuit are operating properly. The ATM code numbers are two-digit numbers that represent each of the actuator circuits that this mode can test. They should be listed in the service manual along with the fault codes.

Sensor Test Mode. Added to system capability in 1986, this test allows the technician to select individual sensors, and the readout box will display the voltage signal that the selected sensor is sending to the logic board.

Engine Running Test Mode. This test allows you to check O_2 sensor switching, idle speed motor operation and, on turbo applications only, knock sensor performance. Connect the readout box or its equivalent to the diagnostic test connector. Start the engine and observe the display on the readout box. If the O_2 sensor is hot enough, the display will switch between 0 and 1 (0 = lean and 1 = rich). The readout box can be used to command the AIS motor to increase idle speed, and on turbo applications, striking the intake manifold near the knock sensor as directed in the service manual will cause an 8 to appear on the display along with the switching 0–1 (the 8 means that timing is being retarded).

Test Mode 10. The random access memory (RAM) of the SMEC module is less volatile than that of the Logic Module. Therefore, removing battery power from it—the procedure used to clear the memory for Logic Module-equipped vehicles—may not erase stored fault codes for several minutes. Because of this, a new sensor test mode has been added. When sensor test mode 10 is selected, fault codes are erased.

There are also some additional fault codes and some revisions in diagnostic procedures. For example, the harness connector on the SMEC is hard to probe with test instruments. As a result, diagnostic procedures may have you disconnect the harness and make open circuit tests rather than closed circuit tests.

Testing without the Readout Box. Cycling the ignition switch on and off as described earlier without the readout box connected triggers the logic module to read out the codes through the power loss lamp. It flashes on and off to indicate the code number. For example, two flashes separated by a short pause of about one-half second followed by a pause of about two seconds and then followed by three flashes separated by a short pause again indicates a code 23. Once the codes are presented, the system is ready for the switch test. If it is receiving the signals, the logic module will turn the power loss lamp on and off in response to its input switches being turned on and off. None of the other test modes can be performed without the readout box or its equivalent.

Driveability Test Procedure

The driveability test procedure is designed to use the test modes just described and to help diagnose engine performance complaints that are not caused by parts of the computer system. It should be used whenever driveability complaints are being diagnosed. The first step is to verify the complaint. This probably requires that the vehicle be operated under the same conditions in which the owner experiences the problem.

The second step is the visual inspection. The driveability test procedure provides a guide for this step. If the problem is not located during the visual inspection, the most appropriate driveability test section should be selected and used. If the engine does not start, select the no start section. If the problem occurs only with a cold engine, use driveability test cold section, and so forth. Once a driveability test section is selected, begin with the first step of that section.

PERSONAL SAFETY: Because of the high fuel pressure in many of these systems, it is very important to relieve the pressure before attempting to open the fuel system. A serious safety hazard is created if a fitting is loosened while the system is still under pressure. Only the correct part number fuel system hose and hose connections should be used when making repairs.

REVIEW QUESTIONS

1. Name four different operating modes of the single-point and/or multipoint system.
2. What is the logic board's power supply?
3. Name at least three things that the power board controls with instructions from the logic board.
4. What is the function of the automatic shutdown relay?
5. What purpose does the adaptive memory serve?
6. What is the purpose of the bleed hole in the MAP sensor line used on some applications?
7. What two signals does the logic board receive from the distributor on turbocharged applications?
8. Name at least four inputs that affect ignition timing.
9. Name two functions that the fan control relay controls.
10. Describe how spark knock is controlled on turbocharged applications.
11. Describe four strategies used to control boost pressure.
12. Name four inputs whose failure puts the system into limp-in.
13. Name an input that causes the engine to stop.
14. What causes the power loss lamp to come on?
15. Name four test modes used for driveability problem diagnosis.
16. What are the first two steps of the diagnostic procedure?

ASE-type Questions. (Actual ASE test questions will probably not be as product specific.)

17. Technician A says that the logic board prefers to use the AIS motor to control idle speed. Technician B says that the logic board prefers to use spark timing to control idle speed. Who is correct?
    a. A only
    b. B only
    c. both A and B
    d. neither A nor B
18. Technician A says that on late-model Chrysler fuel-injected engines the logic and power boards control charging circuit voltage. Technician B says that the charging system on those vehicles no longer has its own voltage regulator. Who is correct?
    a. A only
    b. B only
    c. both A and B
    d. neither A nor B
19. Technician A says that the engine running test mode of the Chrysler diagnostic sequence allows the technician to test individual actuator circuits for proper operation. Technician B says that technician A is wrong. Who is correct?
    a. A only
    b. B only
    c. both A and B
    d. neither A nor B
20. A car equipped with Chrysler's EFI system is brought into the shop with a complaint of poor idle during warm-up. What driveability test section should be used to diagnose the problem?
    a. driveability test cold
    b. driveability test warm
    c. both A and B
    d. neither A nor B

11
CHRYSLER'S OXYGEN FEEDBACK SYSTEM

Objectives

After studying this chapter, you will be able to:

- identify the functions controlled by the O₂ feedback system.
- identify the system's operating modes.
- identify the conditions during which the system stays in closed loop.
- describe the three different air/fuel mixture control solenoids and their methods of control.
- describe the diagnostic procedure used to locate the cause of driveability problems on models without on-board diagnostics.
- describe the diagnostic procedure used to locate the cause of driveability problems on models with on-board diagnostics.

The oxygen feedback (O₂ feedback) system was Chrysler's first comprehensive computerized engine control system, and it has continued to be the system used for carbureted vehicles. The first system, introduced in California in 1979 on the 3.7-liter slant 6, was little more than the existing spark control computer system with the addition of an oxygen sensor and a mixture control solenoid. In 1980 the system was in limited production on four-, six-, and eight-cylinder engines. By 1981 it was in use nationwide and was beginning to take on additional responsibilities such as limited control of idle speed, EGR, and air injection switching.

As development of the system continued, some applications came to control such functions as:

- shift indicator light.
- coolant fan.
- a solenoid that controlled vacuum to the vacuum-operated secondary throttle blade for altitude engines.

Figure 11-1 presents an overview of the system.

Chapter 11 Chrysler's Oxygen Feedback System

Not all inputs or outputs are used on all Oxygen Sensor Feedback applications.

FIGURE 11-1 O_2 feedback system overview

COMBUSTION CONTROL COMPUTER

On the first applications, the combustion control computer was located in the air cleaner and looked very similar to that of the lean burn system of the seventies. On later applications it can be located in other areas of the engine compartment. It is often referred to as the spark control computer. In 1985 a self-diagnostic function and memory were added to the computer to facilitate the diagnostic process.

Open Loop

In open loop the computer does not respond to the oxygen sensor. The system is in open loop during any of the following conditions:

- coolant below required operating temperature (49° C [120° F])
- oxygen sensor below its required operating temperature (315° C [600° F])
- idle
- acceleration (low manifold vacuum)
- deceleration (high manifold vacuum)
- hot engine restart
- oxygen sensor failure

Closed Loop

In closed loop the computer uses input from the oxygen sensor as well as other sensors to determine fuel-metering calibrations. The system goes into closed loop after:

- the coolant temperature reaches a criterion temperature (about 49° C [120° F]).
- the oxygen sensor starts producing voltage (must be around 315° C [600° F]).
- a predetermined amount of time has elapsed since the engine was started.

INPUTS

Coolant Temperature

Three different types of coolant temperature-sensing devices have been used on differ-

Chapter 11 Chrysler's Oxygen Feedback System

FIGURE 11-2 Temperature sensors/switches. *Courtesy of Chrysler Motors.*

ent O_2 feedback applications: a thermistor, a dual thermistor, and a switch, Figure 11-2.

Thermistors. The single-element thermistor was first used on six- and eight-cylinder engines and later used on four-cylinder engines. Still later, about 1985, some of the transverse engine applications were equipped with a dual-element thermistor. One element is

FIGURE 11-3 Coolant sensors

359

Chapter 11 Chrysler's Oxygen Feedback System

used for fuel, spark, EGR, and canister purge control; the other is used for radiator fan relay control, Figure 11-3. Some O_2 feedback systems used a separate coolant temperature sensor in the radiator receiving tank for control of the radiator fan relay.

Switch. Most four-cylinder engines through 1983 were equipped with a temperature-sensing switch instead of a thermistor. By closing at a specific temperature, the switch tells the computer whether the coolant temperature is above or below the criterion temperature.

Charge Temperature

On some six- and eight-cylinder applications, a temperature sensor is mounted in the intake manifold to sense air/fuel mixture temperature. Keep in mind that the heated air inlet system brings incoming air temperature up to normal operating temperature before the coolant reaches normal temperature. The computer's operating strategy on some O_2 feedback systems requires that it knows both air/fuel mixture temperature and coolant temperature.

Distributor Pickup

Engine speed and crankshaft position information are provided by a Hall-effect switch located in the distributor on four-cylinder engines, Figure 11-4, and by the dual pickup coils in the distributor of most six- and eight-cylinder engines, Figure 11-5.

Vacuum Transducer

Vacuum is sensed by a vacuum *transducer*. The transducer is mounted on the computer housing and contains a diaphragm that is attached to a movable metal core, Figure 11-6. Either manifold or ported vacuum is connected to the transducer. As the vacuum changes, the diaphragm and the core move. The computer passes an AC current through a coil that surrounds the core. The alternating magnetic field

FIGURE 11-4 Hall-effect switch. *Courtesy of Chrysler Motors.*

Chapter 11 Chrysler's Oxygen Feedback System

FIGURE 11-5 Dual pickup units. *Courtesy of Chrysler Motors.*

produced by the AC current produces an *inductance,* an electromotive force produced by current variation within the coil. The inductance changes as the core's position in the coil changes; during low vacuum the core is extended farther into the core and the inductance is reduced. The inductance impedes the AC current flow through the coil. This particular type of *impedance,* or resistance, is called *reactance* and is expressed in ohms. The computer measures volt drop across the coil and can therefore determine engine vacuum.

Oxygen Sensor

The O_2 feedback system uses a single-wire oxygen sensor.

Throttle Stop Switch

The O_2 feedback system does not use a throttle position sensor. It does, however, use the throttle linkage as a switch to indicate to the computer when the throttle is closed, Figure 11-7. As the throttle closes, a contact on the throttle lever contacts the throttle stop, which acts as a ground contact. The path to ground is completed, and a voltage drop occurs across the circuit's resistor inside the computer. This alerts the computer that the throttle is closed.

Note: The 1979 system used a type of throttle position sensor called a throttle position transducer.

FIGURE 11-6 Vacuum transducer

FIGURE 11-7 Throttle stop switch. *Courtesy of Chrysler Motors.*

361

Chapter 11 Chrysler's Oxygen Feedback System

Speed Sensor

Vehicle speed information is provided on many 1985 and later systems by a pulse generator mounted on and driven by the transmission.

Detonation Sensor

A piezoelectric knock sensor is used on high-performance four-cylinder and eight-cylinder engines equipped with four-barrel carburetors.

Heated Rear Window

Some vehicles with a heated rear window provide an input signal to the computer when the window heater is turned on. This allows the computer to increase throttle opening during idle to support the additional load that the heavy current load of the heat element causes the alternator to put on the engine.

OUTPUTS

Mixture Control

The O_2 feedback system has used three different mixture control solenoids, which Chrysler calls O_2 feedback solenoids, on four different carburetors.

Holley Model 6145. This carburetor was used in 1979 and 1980 on California applications of the 3.7-liter engine, Figure 11-8. The O_2 feedback system controls idle mixture by way of the idle air bleed and the main metering mixture by way of fuel metering, similar to the operation of the solenoid used on the Holley 6149 discussed in the **Outputs** section of Chapter 7. Air for the idle and low-speed system air bleed is supplied through a fixed orifice plus a passage controlled by a metering rod. The metering rod is controlled by a vacuum diaphragm. For main metering control, a second diaphragm controls a spring-loaded rod and foot; the spring wants to drive the rod down. If the diaphragm allows the rod's foot to come down, it will force open a valve that when open allows additional fuel to enter the main metering circuit.

FIGURE 11-8 Holley 6145 mixture control circuit. *Courtesy of Chrysler Motors.*

Vacuum to the two diaphragms is controlled by a remote-mounted, computer-controlled solenoid, Figure 11-9. The lower portion of the unit is a simple vacuum regulator valve. It receives manifold vacuum and reduces it to 5 inches of mercury (Hg). The 5 inches of Hg is then supplied to a valve seat higher in the assembly. This valve seat is closed by a conical tip on the lower end of the solenoid armature when the solenoid is not energized by the computer. When the solenoid armature is in this position, the assembly's vacuum output port is exposed internally to the vent port, and

FIGURE 11-9 Solenoid-operated vacuum regulator. *Courtesy of Chrysler Motors.*

vacuum at the output port is zero. When the computer energizes the solenoid, the armature moves up and a second conical tip on the top of the armature closes off the vent exposure. The lower passage is now open, and the 5 inches of Hg is exposed to the output port. The computer duty cycles the solenoid at whatever rate is necessary to maintain the desired output vacuum (between 0 and 5 Hg) at the two diaphragms in the carburetor. High vacuum produces a leaner mixture in both idle and main metering circuits; low vacuum produces a rich mixture for either circuit.

Holley 6520 and Later Applications of the Holley 6145. In 1980 the two-barrel Holley 6520 was equipped with an O_2 feedback solenoid for four-cylinder engines, and in 1981 the same type solenoid was applied to the Holley 6145 for six-cylinder engines. This solenoid is essentially the same as the one used in the Holley carburetor used on some General Motors' CCC applications discussed in the **Outputs** section of Chapter 3.

Carter BBD and Thermo-Quad. The BBD two-barrel and Thermo-Quad carburetors were equipped with an O_2 feedback solenoid in 1980 for eight-cylinder applications. This solenoid controls the air bleed for both the idle and main metering circuits.

At idle air is brought through a passage from above the venturi area to the O_2 feedback solenoid valve, Figure 11-10. When the solenoid is energized, air is allowed into the idle circuit and the mixture is lean. When the solenoid is off, air is blocked from the idle circuit and the mixture is rich. The main metering circuit air bleed draws air from the same passage under the same O_2 feedback solenoid control, Figure 11-11. The circuit being metered depends on the circuit from which fuel is being drawn.

Ignition Timing

Ignition timing in the O_2 feedback system is still based on the strategy of the earlier lean burn system. The negative side of the ignition coil is switched to ground by the computer. During cranking, the distributor pickup signal triggers the computer to open the primary circuit. The spark plugs are fired at base timing.

Once the engine starts and if it is cold, the computer will begin to add spark advance

FIGURE 11-10 Carter carburetor mixture control idle circuit. *Courtesy of Chrysler Motors.*

Chapter 11 Chrysler's Oxygen Feedback System

FIGURE 11-11 Carter carburetor mixture control main metering circuit. *Courtesy of Chrysler Motors.*

based on coolant temperature, engine speed, and engine vacuum. Once the coolant reaches a criterion temperature, advance is dependent on vacuum and RPM. If the throttle stop switch closes, the amount of advance issued in response to engine vacuum will be removed.

EGR Solenoid

Only some of the later O_2 feedback applications use the computer to control EGR valve operation. On these applications the computer controls a solenoid that either allows or denies vacuum to the EGR valve. Vacuum is denied during:

- cold engine operation.
- light throttle operation.
- the first minute or so following hot engine restarts.
- WOT operation.

Air Switching Solenoid

On O_2 feedback systems that control the air injection system, a computer-controlled solenoid controls vacuum to the switch relief valve, Figure 11-12. During cold engine operation, the computer commands that vacuum be applied to the valve's diaphragm and air pump air is directed upstream to the exhaust manifold. When the engine is hot enough to go into closed loop, vacuum to the valve's diaphragm is cut off and the air is directed downstream to the catalytic converter. For a brief period after each engine start-up, the computer commands vacuum to be applied so that injection air is sent upstream. In this system there is no provision for pump air to be dumped into the atmosphere except as a pressure relief function.

On systems where the computer does not control the air injection system, a similar strategy is achieved by using a coolant temperature-controlled vacuum valve in place of the computer-controlled solenoid.

Vacuum Secondary

The Holley 6520 is a progressive two-barrel carburetor with a vacuum-operated secondary throttle blade. Some later applications of this carburetor feature a solenoid-operated vacuum valve in the vacuum line that supplies venturi vacuum to the diaphragm that opens the secondary throttle blade. This computer-controlled solenoid vents the line and does not allow the secondary throttle blade to open until:

- coolant temperature reaches 60° C (140° F).
- vacuum drops below a criterion value.
- engine speed is above a specified RPM.

FIGURE 11-12 Air injection switch/relief valve. *Courtesy of Chrysler Motors.*

Radiator Fan Relay

The radiator fan relay is used on later model vehicles with transverse engines and electric cooling fans. It functions like the one discussed in the **Outputs** section of Chapter 10.

Shift Light Indicator

On 1985 and 1986 models with manual transmissions, the computer controls a light that is on the instrument panel and that indicates to the driver the optimum time to make the next upshift.

Tach Drive

On some 1986 models, the computer operates a tachometer on the instrument panel. The tach drive signal comes from the distributor side of the ignition coil.

Other features often included on engines with O_2 feedback systems but not controlled by the computer are:

- a vacuum-powered throttle kicker.
- a choke heater.
- a carburetor bowl vent solenoid.

These units are usually self-grounded and are activated by turning on the ignition switch.

SYSTEM DIAGNOSIS AND SERVICE

Diagnostic Procedure 1979-1984

Prior to 1985 the O_2 feedback system computer had no self-diagnostic capability. Diagnosis of the cause of a no-start condition or driveability complaint is performed with the aid of the driveability test procedures booklet or its equivalent. Chrysler has published such a booklet for each system application.

Diagnosis is divided into three parts: verify the complaint, do a visual inspection, and perform the appropriate test steps. The driveability test procedure outlines the visual inspection and the test steps. The test procedures are also divided into three parts: no start, driveability test cold, and driveability test warm. If the engine does not start and the visual inspection is completed, begin with the first test of the no start section. If the complaint concerns cold driveability, begin with the first test of the driveability test cold section and likewise for a driveability complaint that only occurs when the engine is warm.

If you begin with the first test of the appropriate section and then go to the next test indicated by the results of the test just completed, the test procedure will take you through all of the components and systems that could cause the driveability complaint, including systems not controlled by the computer.

Diagnostic Procedure, 1985 and Later

With the addition of self-diagnosis, the diagnostic test procedure has been modified to make use of the fault codes and other self-diagnostic features. The fault codes can be recorded for the most vital sensors and actuators.

The system is put into one of its diagnostic modes with the diagnostic readout box (tool C-4805 or its equivalent). The 1985 system has the following three different diagnostic modes:

- *diagnostic test mode*—The computer reads out any stored fault codes.
- *circuit actuation test mode (ATM test)*—The technician selects an actuator circuit and activates the test mode. The computer then turns that circuit on and off at quarter-second intervals for five minutes while allowing the technician to inspect the actuator for proper operation.
- *switch test mode*—This allows the technician to verify that input switch signals are being received by the computer.

Assuming that the no start test procedures are not required and the visual inspection is completed, the diagnostic test mode is the first test of the driveability test procedure. The latter two test modes are most useful if used when called for by one of the remaining driveability test procedures.

The 1986 system was upgraded still more by adding a sensor test mode to the other three self-diagnostic test modes. In this mode the computer reads out the sensor input voltage values through the readout box.

Reading O_2 Solenoid Calibration

A voltmeter can be connected across the O_2 solenoid to read the calibration commands being sent to the solenoid to maintain the desired air/fuel ratio. A voltage reading closer to battery voltage indicates a consistently lean command, probably in response to a rich condition. A voltage reading nearer to zero volts indicates a rich command either in response to a lean condition or an operating condition that requires a rich mixture.

Service Tips

Overcompensation. An overrich carburetor condition or excessive fuel introduced by the charcoal canister or the PCV system can cause the system to overcompensate and go too lean while in closed loop. On the other hand, a vacuum leak can cause the system to go overrich.

REVIEW QUESTIONS

1. What year did the O_2 feedback system come into wide-scale use?
2. List the operating modes of the O_2 feedback system.
3. List seven operating conditions that keep the system in open loop.

4. What device is used to monitor engine load?
5. List at least five engine support functions that the combustion control computer controls.
6. List the three types of mixture control solenoid used on different O_2 feedback engine applications by their carburetor application.
7. For each of the mixture control solenoids identified by carburetor application, identify the medium (air or fuel) used to control the mixture during idle and part throttle operation.

Carburetor	Idle	Off Idle
_____	_____	_____
_____	_____	_____
_____	_____	_____

8. When the engine is cold, what inputs are used to determine ignition timing?
9. What inputs are used to determine ignition timing when the engine is warm?
10. During what operating conditions is the EGR valve turned off?
11. What is the function of the vacuum secondary solenoid?
12. List the three parts of the driveability test procedure for vehicles without on-board diagnostics.
13. List the four parts of the driveability test procedure vehicles with on-board diagnostics.

ASE-type Questions. (Actual ASE test questions will probably not be as product specific.)

14. Technician A says that O_2 feedback systems do not use a throttle position sensor. Technician B says that during one model year in the early development of O_2 feedback, a type of throttle position sensor was used. Who is correct?
   a. A only
   b. B only
   c. both A and B
   d. neither A nor B
15. Technician A says that there is no easy way to determine if the O_2 feedback system is tending to run rich or lean other than by checking for oxygen sensor codes on later models. Technician B says that a voltmeter can be connected to the O_2 solenoid to determine if the system is tending to run rich or lean. Who is correct?
   a. A only
   b. B only
   c. both A and B
   d. neither A nor B
16. A 1981 O_2 feedback system is brought into the shop with a driveability problem. Technician A starts to perform the driveability test warm procedure because the customer says the problem occurs when the engine is warm. Technician B says that technician A should not start with the driveability test warm procedure. Who is correct?
   a. A only
   b. B only
   c. both A and B
   d. neither A nor B

GLOSSARY

IMPEDANCE The total resistance to current flow within an AC circuit.

INDUCTANCE The phenomenon by which electromotive force is induced as a result of a variation of current flow within a circuit.

REACTANCE Resistance, expressed in ohms, within an AC circuit that is a result of inductance or capacitance.

TRANSDUCER A device that receives energy from one system and transfers it to another system, usually in a different energy form.

ROBERT BOSCH CORPORATION'S MOTRONIC

Objectives

After studying this chapter, you will be able to:

- identify the operating conditions that invoke different operational modes.
- describe how the control unit determines engine load.
- identify the major input sources.
- identify the controlled functions that are standard features of Motronic.
- identify the two major steps of the diagnostic procedure.

The Robert Bosch Corporation builds engine support components and control systems that any vehicle manufacturer can purchase to use on their vehicles. The Bosch Corporation is well known for its computer-controlled fuel injection systems. Although later versions of such systems as the L-Jetronic and K-Jetronic feature an oxygen sensor, they are still single-function systems in that they only control fuel metering and therefore are not discussed in this text. The Motronic is Bosch's first multi-function engine control system. Motronic's primary concern is the control of air/fuel mixture, ignition dwell, and ignition timing. Its other function control capabilities such as idle speed control, EGR control, and turbo boost control are offered as options to vehicle manufacturers.

CONTROL UNIT

Motronic grew out of the L-Jetronic system and like its predecessor is a multipoint injection system with fuel metering controlled by pulse width. The L-Jetronic's analog computer was replaced by a digital unit with additional function control capability such as spark timing, idle speed control, and turbo boost control. An overview of the Motronic system is provided in Figure 12-1. The control unit's

---

Bosch and Imports

Of the imported European cars equipped with a comprehensive computerized engine control system, most use the Bosch system. It first appeared on selected BMW models. Most of the systems used on Japanese cars for export to the United States are built using Bosch patents. In fact all of the gasoline engine, multipoint fuel injection systems used on U.S.-built vehicles use some Bosch components.

mount brackets serve to dissipate heat from the transistors that drive the ignition and injectors, Figure 12-2. A thirty-five-pin connector connects to the vehicle's system harness.

Main Relay

A main relay, Figure 12-3, similar in function to the power relay used on the Ford EEC systems (Chapters 7 and 8), powers the control unit when the ignition is turned on. This relay also contains a diode to protect the control unit against accidental reversed polarity.

On-board Diagnostics

The first few model year applications of Motronic do not feature any form of self-diagnosis except that if a fault is detected in the closed-loop operating circuit, a calculated pulse width aimed at maintaining an air/fuel mixture near stoichiometric will be used.

OPERATING MODES

The control unit is programmed with different operational strategies as operating conditions change. Most of the different operational strategies are conducted as part of the open-loop mode.

Cranking

Two fuel-metering programs can be employed during cranking. One is based on crank-

Inputs	Control Unit	Outputs
Airflow Meter		Fuel Injection
Engine Speed		Fuel Pump Relay
Crankshaft Position		Ignition Timing
Throttle Position		Dwell Control
Coolant Temperature		RPM Limit
Air Temperature		Peak Coil Current Cutoff
Lambda Sensor (oxygen sensor)		
Knock Sensor	Main Relay	The following functions are
Altitude Sensor		system options (available to the
Starter Signal		vehicle manufacturer):
Battery Voltage		
		Rotary Idle Adjuster
		Turbo Boost Control
		EGR Control
		Canister Purge
		Transmission Control
		Start-Stop Control

Noncomputer-Controlled Functions: Fuel Pressure Regulation, Fuel Pressure Pulsation Damper, Cold-Start Injector and Thermo-Time Switch, and Auxiliary Air Device (used in place of Rotary Idle Adjuster)

FIGURE 12-1 Overview of Motronic system

Chapter 12 Robert Bosch Corporation's Motronic

1 Additional program memory
2 Analog-digital converter
3 Microcomputer for standard program and data
4 Integrated circuit for engine-speed and reference-mark signal processing
5 Ignition output stage
6 Fuel-injection output stage

FIGURE 12-2 Control unit. *Courtesy of Robert Bosch.*

Chapter 12 Robert Bosch Corporation's Motronic

FIGURE 12-3   Main relay. *Courtesy of Robert Bosch.*

ing speed, the other on temperature. At lower speeds the quantity of fuel introduced is constant regardless of fluctuations in airflow (airflow sensor input is not reliable under these conditions). At higher cranking speeds, air intake diminishes slightly as a result of lower volumetric efficiency, so fuel quantity is reduced. During cold cranking the control unit superimposes an enrichment program over the speed-dependent program. Some applications provide enrichment from a cold-start injector in the intake manifold, Figure 12-4, and some use additional pulses from the port injectors. Ignition timing is also adjusted as a function of cranking speed and temperature. The cold-start injector system is the same as that described in the **Outputs** section of Chapter 5 (an example of a Bosch subsystem being used in a U.S.-built fuel injection system).

On applications that do not use a cold-start injector, the control unit pulses the port injectors several times during each crankshaft revolution as opposed to the one pulse per engine cycle used during normal operation. This is to enhance fuel evaporation and to avoid fouling the spark plugs with wet fuel. The enrichment decays to zero over a specific number of engine revolutions. The number depends on engine temperature when cranking begins. If RPM reaches a criterion value before the specified number of revolutions has occurred, the cold-cranking enrichment program is stopped. The criterion RPM also varies with the engine temperature that exists when cranking begins.

Post-start

When the engine first starts, the intake port and combustion chamber surfaces are still cool and thus cause fuel to condense. During the brief time it takes the surfaces to begin heating up, the control unit initiates a post-start enrichment program to improve idle quality and throttle response. Once initiated this enrichment decays within a brief time to zero. The actual time depends on engine temperature. In this mode the ignition timing is advanced in relation to engine temperature to provide the most efficiency with the least possible amount of fuel.

Warm-up

After the post-start period has expired, any additional warm-up enrichment necessary is provided based on engine temperature and load. An idle speed increase is also provided to improve driveability and to hasten warm-up. The idle speed increase is initiated during the post-start mode and continues into the warm-up mode. As an added means of bringing the oxygen sensor and catalytic converter up to operating temperature as quickly as possible, some applications of Motronic retard ignition timing during warm-up.

Acceleration

If throttle opening suddenly increases from a prior fixed position, a brief fuel enrichment is commanded by the control unit. This function is similar in effect to the performance of an accelerator pump in a carburetor. The base acceleration enrichment program is modified during low engine temperature.

Full Load (WOT)

During wide-open throttle operation, the control unit commands a fixed, enriched air/fuel mixture. To avoid fuel quantity calculation errors that could occur as a result of fluctuations in the airflow sensor, fuel quantity is calculated based on engine speed. The objectives during this operational mode are to produce maximum torque and to avoid detonation.

Overrun (Deceleration)

During deceleration fuel injection is stopped but over a short time period (tapered off), and ignition timing is retarded along with the fuel cut-off. The retarded timing provides

The Motronic system.
1. Fuel tank
2. Electric fuel pump
3. Fuel filter
4. Fuel distributor
5. Pressure regulator
6. Pulsation damper
7. Control unit
8. Ignition coil
9. High-tension distributor
10. Spark plug
11. Injection valve
12. Cold-start valve
13. Idle-speed adjusting screw
14. Throttle valve
15. Throttle-valve switch
16. Air-flow sensor
17. Air-temperature sensor
18. Lambda sensor
19. Thermo-time switch
20. Engine-temperature sensor
21. Auxiliary-air device
22. Idle-mixture adjusting screw
23. Reference-mark sensor
24. Engine-speed sensor
25. Battery
26. Ignition-starting switch
27. Main relay
28. Pump relay

FIGURE 12-4 The Motronic system. *Courtesy of Robert Bosch.*

better engine braking and reduces HC production in the event that fuel has been evaporated from the intake port walls. When engine speed falls below a programmed RPM or the throttle is opened again, fuel injection is restored to its appropriate level. This occurs over a programmed number of engine revolutions. Ignition advance is restored in the same way for a smooth transition. If the fuel cut-off is not used during deceleration, a fuel-enrichment program is employed to avoid engine bucking and high HC production (a result of lean misfires).

Closed Loop

The Motronic system goes into closed loop when the engine and lambda sensor reach normal operating temperature. It maintains a closed-loop mode as long as operating temperature of the lambda sensor is maintained and engine load does not require an air/fuel mixture richer than stoichiometric.

INJECTORS

The injectors are like those used in other U.S.-built multipoint injection systems. In fact all American manufacturers have historically used Bosch injectors for their multipoint injections.

INPUTS

Airflow Meter

The airflow meter is a vane type similar to that used on some of the Ford multipoint fuel injection systems (Chapter 7). Its vane shaft moves the wiper of a built-in potentiometer designed to provide accurate voltage signals to the control unit in spite of aging and rapid changes in temperature, Figure 12-5. By comparing crankshaft position input with airflow meter data, the control unit can determine the air intake for each cylinder's intake stroke. This determination is the basis for pulse width calculation.

Engine Speed Sensor

The engine speed sensor is a pulse-generating device that uses the flywheel teeth as its triggering device, Figure 12-6.

Reference Mark (Crankshaft Position) Sensor

The reference mark sensor is like the engine speed sensor except that its trigger is a reference mark on the flywheel, Figure 12-6. It provides the control unit with information about crankshaft position.

Throttle Valve Switch

The throttle valve switch contains two sets of electrical contact points: one for idle and one for WOT, Figure 12-7. Each set of contacts has its own circuit to and receives a reference voltage from the control unit. This enables the control unit to determine if each of the switches is open or closed. When the throttle blade is closed, the idle contacts are closed; during part throttle operation, both sets are open; and at WOT the full-load contacts are closed. The control unit can therefore recognize these three driving conditions. The switch is attached to the end of the throttle shaft.

Coolant Temperature Sensor

The coolant temperature sensor is a thermistor that screws into the water jacket near the thermostat housing.

Air Temperature Sensor

The air temperature sensor is a thermistor mounted in the intake opening of the airflow sensor, Figure 12-8. It measures income air temperature.

Lambda Sensor (Oxygen Sensor)

Bosch's term for oxygen sensor is *lambda sensor*.

Chapter 12 Robert Bosch Corporation's Motronic

34) Air-flow sensor (above).
1 Ring gear for spring preload
2 Return spring
3 Potentiometer
4 Sliding contact

Thick-film potentiometer of the air-flow sensor. The resistances can be identified as dark rectangular surfaces in the upper half of the Figure. The resistive material is a ceramic-metal mixture which is burnt into the ceramic plate at a high temperature.

FIGURE 12-5 Airflow meter. *Courtesy of Robert Bosch.*

1 Permanent magnet
2 Housing
3 Engine block
4 Soft iron core
5 Winding
6 Flywheel ring gear
7 Reference mark

FIGURE 12-6 Engine speed and reference mark sensors. *Courtesy of Robert Bosch.*

1 Full-load contact
2 Switch guide
3 Throttle shaft
4 Idle contact
5 Electrical connection

FIGURE 12-7 Throttle valve switch. *Courtesy of Robert Bosch.*

Chapter 12 Robert Bosch Corporation's Motronic

1 Electrical connection, 2 insulation tube, 3 connector, 4 NTC resistor, 5 housing, 6 rivet pin, 7 securing flange. Arrow denotes the direction of intake air.

FIGURE 12-8 Air temperature sensor. *Courtesy of Robert Bosch.*

Determining Air/Fuel Ratio

Lambda (λ) is one of the Greek letters and is used here to represent a relationship that defines the air/fuel ratio. The relationship is:

Lambda = actual inducted air quantity ÷ the theoretical air quantity requirement

The theoretical air requirement is 14.7 units (by weight). If the actual air quantity used was 14.7, then lambda = 1. Using this method of expression, an air/fuel ratio richer than 14.7 is expressed, for example, as 0.9; 0.9 is equivalent to 13.23 to 1 (13.23/14.7 = 0.9). Conversely, an air/fuel ratio leaner than 14.7 is expressed, for example, as 1.03; 1.03 is equivalent to 15.14 to 1 (15.14/14.7 = 1.03).

Knock Sensor

The knock sensor is the piezoelectric type, Figure 12-9. This sensor produces a voltage signal in response to most engine vibrations, but the signal's strength is proportional to the vibration frequency. The signal produced in response to a vibration frequency of 5 to 10

FIGURE 12-9 Knock sensor. *Courtesy of Robert Bosch.*

376

kilohertz is recognized as spark knock. The knock sensor is located on the block between two cylinders. On some applications two sensors are used.

Altitude Sensor

The altitude-sensing device or barometric pressure sensor is an aneroid unit attached to the wiper of a potentiometer similar to the one described in the **Inputs** section of Chapter 8. It is usually located in the air cleaner.

Starter Signal

The cranking control circuit provides a signal to the control unit when the starter is being cranked.

Battery Voltage

A system voltage input from the vehicle's electrical system is provided to the control unit so that it can monitor battery voltage.

OUTPUTS

Of the functions controlled by the control unit, injector operation, engine speed limit, fuel pump relay control, ignition timing, dwell control, and peak coil-current cut-off are standard features of Motronic. Other features are optional to the vehicle manufacturer.

Injectors

The injectors are solenoid-operated, pintle-style valves, as used in all of the other multi-point injection systems presented in this text. In open-loop operation, the control unit calculates pulse width based on engine load and engine speed; engine load is derived by computing airflow and engine speed. Figure 12-10 presents a three-dimensional graph or map that represents the relationship between load, speed, and air/fuel ratio for a given engine application; air/fuel ratio is expressed as lambda. This map is stored in the control unit's read-only memory and serves as the look-up table to determine pulse width. The arrows indicate increased load, speed, and richness. Each line intersection represents a relative value of air/fuel ratio.

The control unit can, however, modify the pulse width value represented by each line intersection in response to the following input information:

P Load
n Engine speed
λ Air ratio

FIGURE 12-10 Pulse width control map. *Courtesy of Robert Bosch.*

Chapter 12 Robert Bosch Corporation's Motronic

1. Electric fuel pump
2. Fuel filter
3. Fuel distributor tube
4. Injection valve
5. Cold-start valve
6. Pressure regulator
7. Pulsation damper

FIGURE 12-11 Fuel supply system. *Courtesy of Robert Bosch.*

- engine temperature.
- throttle position. Beyond the influences of engine temperature and throttle position already discussed under **Operating Modes,** the control unit adjusts air/fuel mixture (pulse width) at warm WOT in an attempt to avoid detonation.
- air density. The airflow sensor input needs to be corrected at higher elevations. This correction is made with input from the air temperature sensor plus, on some applications, the altitude sensor.
- battery voltage. System voltage affects the injector's response time (how fast it opens) and therefore how fuel actually flows during a pulse width. A correction factor is programmed into the control unit to modify the pulse width accordingly.

Engine Speed Limit. The control unit has a programmed engine RPM limit for each engine application. If engine RPM exceeds that limit by more than 80 RPM, the control unit will stop fuel injection until engine speed is 80 RPM below the limit. This prevents engine damage resulting from overrevving and avoids dumping excessive HC into the exhaust system as occurs if engine speed is controlled by stopping spark delivery.

Fuel Pump Relay

The fuel pump is controlled on the Motronic system like it is on most other electronic fuel injection systems. The pump is powered by an external relay that has power available to it when the ignition is turned on. The control unit grounds the pump relay's control coil, and the pump is powered. If the engine is not turning at a criterion speed, the control unit will turn off the relay. The fuel pump is mounted inside the fuel tank, Figure 12-11.

Fuel Pressure Regulator. The fuel pressure regulator works like those on other electronic

Chapter 12   Robert Bosch Corporation's Motronic

Motronic ignition subsystem.
1 Ignition switch
2 Ignition coil
3 High-tension distributor
4 Spark-plug leads
5 Plug connectors
6 Spark-plugs
7 Control unit
8 Battery

FIGURE 12-12   Ignition circuit. *Courtesy of Robert Bosch.*

fuel injection systems. It is connected to manifold vacuum so that pressure differential across the injectors is constant Figure 12-11.

Fuel Pressure Pulsation Damper. The pulsation damper, Figure 12-11, eliminates pulsation noise. It is designed much like the pressure regulator; however, it is not connected to manifold vacuum and acts as an accumulator to absorb the pressure pulses produced by the opening and closing of the injectors. It is in the fuel return line and downstream from the pressure regulator.

Ignition Timing

The control unit controls ignition by providing ground side switching for the ignition primary winding, Figure 12-12. Like pulse width, ignition timing is based on engine load and speed. Figure 12-13 presents a three-

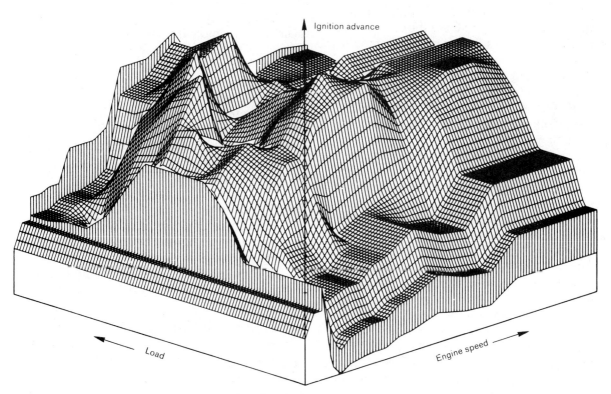

FIGURE 12-13   Ignition timing map. *Courtesy of Robert Bosch.*

379

dimensional map of the spark advance programming for a given engine application. The arrows point in the direction of increased load, speed, and advance. The line intersections represent the spark advance for all combinations of engine speed and load. These values are stored as the look-up table in the read-only memory of the control unit. Input data is reviewed and spark timing is recalculated between each cylinder firing. The control unit can modify the values on the map in response to inputs from:

- coolant temperature sensor.
- intake air temperature sensor.
- throttle position switch. During WOT operation timing is calculated for maximum torque with additional consideration for detonation only.
- altitude sensor (if used).

Temperature of either the coolant or incoming air most readily causes the control unit to modify the spark advance map. The modification depends on the operating condition. During cranking timing depends on cranking speed and temperature. At low cranking speeds and low temperature, timing is near TDC. At higher cranking speeds and with a warm engine, timing is still near TDC; but with high cranking speed and low temperature, timing is advanced. During a cold post-start period, timing is advanced significantly for a short time to partially reduce the need for fuel enrichment. Once the engine is warm enough to reduce the likelihood of engine stalling, timing is retarded (on some applications), at the cost of optimum torque, to hasten engine warm-up.

During warm idle, ignition timing is used to control idle speed. With no additional load, transmission in neutral, air-conditioning off, and so forth, the timing can be retarded slightly and still maintain a desired idle speed. This keeps combustion chamber surfaces warmer to reduce HC production. If a load is applied such as the transmission being pulled into gear, the idle speed will start to drop. The control unit will respond by advancing ignition timing enough to obtain more torque and bring idle speed back to the desired value. The throttle opening need not be increased, and greater fuel economy is realized. On applications with the rotary idle adjuster, idle air is increased if spark timing manipulation is not enough to maintain desired idle speed.

During WOT operation normal or higher engine and/or incoming air temperature cause the control unit to reduce spark timing at those load ranges where spark knock has the greatest probability of occurring. If a specified value of load increase occurs (acceleration), short of WOT, the control unit will reduce the existing spark advance and will then gradually begin to restore it. This tends to reduce spark knock and NO_x production.

Spark Knock Control

On earlier systems the spark knock control system was remote from the Motronic unit. On later applications it is incorporated into Motronic's control unit. Its application allows a more aggressive spark advance schedule. When the knock sensor produces a signal that the knock control module or control unit recognizes as spark knock, it reduces spark advance until the knock disappears. It then gradually readvances the timing to its original value or until the knock recurs.

Dwell Control

The control unit controls dwell in a manner similar to that of General Motors' HEI module. As engine speed increases, dwell angle is increased. In this case, however, the amount of increase is also affected by battery voltage.

Peak Coil-Current Cut-off

Allowing the ignition to remain in the on position with the engine not running can cause both the ignition coil and the coil-driving transistor in the control unit to overheat. To avoid this the control automatically turns the ignition primary circuit off anytime engine RPM drops below a programmed speed of about 30 RPM.

Idle Speed

Two different methods are used to control idle speed.

Auxiliary Air Device. On some applications warm idle speed is controlled by an adjustable screw in the idle air bypass passage, Figure 12-14. On a cold start and during warm-up, idle speed is increased by an auxiliary air valve, Figure 12-14. A bimetal strip inside the valve holds the valve open at low temperature, Figure 12-15. When the valve is open, additional air is allowed to bypass the throttle blade. This air has, however, passed through the airflow sensor so that the appropriate amount of fuel is injected to mix with it. When the ignition is turned on, voltage is applied to a heat element wrapped around the bimetal strip. As the strip warms up, it gradually closes the valve

1 Electrical connection
2 Electric heating
3 Bimetal
4 Bimetal
5 Slider valve

FIGURE 12-15 Auxiliary air device schematic. *Courtesy of Robert Bosch.*

and blocks the flow of the auxiliary air. When the engine is warmed up, its heat is enough to keep the valve in the closed position.

Rotary Idle Adjuster. Other applications use the rotary idle adjuster, which is controlled by the computer, Figure 12-16. It maintains idle

FIGURE 12-14 Auxiliary air device. *Courtesy of Robert Bosch.*

FIGURE 12-16 Rotary idle adjuster circuit. *Courtesy of Robert Bosch.*

speed on a cold or warm engine. It controls air through a throttle bypass passage by rotating a spool valve attached to the end of the armature shaft, Figure 12-17. The motor has a permanent magnet field and two armature windings. By controlling the voltage applied to the two windings, the control unit can rotate the armature and therefore position the spool valve to control idle air. The control unit has a programmed idle speed that it wants to maintain. It first uses ignition timing as a means to adjust idle speed; if that is not enough, it will use the rotary idle adjuster to bring idle speed to the desired value.

Turbo Boost Control

To control spark knock on turbocharged engine applications, Motronic uses a combination of spark retard and boost pressure reduction. Bosch feels that just reducing boost pressure puts unnecessary limits on engine performance. They also feel that just using spark retard to control spark knock is a poor choice because retarding ignition timing raises exhaust temperature. This rise in temperature can be threatening to the already high-temperature exhaust-driven turbine. Using both measures proves quite effective. When spark knock is detected, timing is retarded and boost pressure is reduced. Spark retardation is immediately effective to stop spark knock, but the drop in boost pressure takes a little longer (a few seconds because the air in the intake manifold must cool slightly) to take effect. As the drop in boost pressure takes effect, spark timing is moved back to its optimum value. Boost pressure is controlled with a wastegate control diaphragm similar to that of other turbocharged systems.

EGR Valve Control

Most European-built vehicles do not use an EGR valve to control NO_x. They rely instead on valve overlay to provide an EGR function and on the reducing function of the three-way

1 Electrical connection
2 Housing
3 Permanent magnet
4 Armature
5 Air passage as bypass to throttle valve
6 Rotary slider

FIGURE 12-17 Rotary idle adjuster motor. *Courtesy of Robert Bosch.*

catalytic converter. Predictions are, however, that EGR valves will become increasingly used on European-built vehicles as well. The Motronic control unit is capable of providing that function similar to the way it is done on most U.S.-built systems.

Evaporative Emissions Control

Motronic-equipped vehicles sold in the United States or Japan feature an evaporative emission control system. Vapors from the fuel tank are stored in a charcoal canister, Figure 12-18. When the system is in closed loop, the control unit activates a solenoid, which opens a valve in the purge line that connects the canister to manifold vacuum.

Electronic Transmission Control

By adding sensing devices to provide to the control unit information concerning transmission output speed (a vehicle speed sensor), gear selector position, transmission gear position, and transmission kickdown, the control unit can control the transmission's hydraulic pressure and shift points (solenoid valves replace traditional shift valves), Figure 12-19. Shift point look-up charts (one for economy, one for performance, and one for manual shifting) stored in the control unit's read-only memory provide optimum transmission performance according to the programming (look-up chart) selected by the operator.

Smooth shifts with reduced clutch slippage and wear are achieved, especially during WOT operation, by momentarily retarding ignition timing to reduce torque output during the shift. Motronic transmission control improves fuel economy, shift quality, transmission torque capacity, and expected transmission life.

Start-Stop Control

Another Motronic feature available for manual transmission applications conserves fuel by automatically stopping the engine during idle periods and restarting it when the throttle is depressed. It requires the use of a vehicle speed sensor, a clutch pedal position–sensing switch, and a separate module.

When vehicle speed falls below about 1 MPH with the clutch pedal depressed, the mod-

1 Fuel tank
2 Fuel-vapor canister with active charcoal
3 Solenoid valve
4 Control unit
5 Intake manifold
6 Lambda sensor
7 Engine-temperature sensor
8 Air-flow meter
9 Engine-speed sensor

FIGURE 12-18 Evaporative emissions control system. *Courtesy of Robert Bosch.*

Chapter 12 Robert Bosch Corporation's Motronic

FIGURE 12–19 Electronic transmission control. *Courtesy of Robert Bosch.*

ule signals the control unit to stop fuel injection. When the driver depresses the accelerator pedal with the clutch pedal still depressed, the module activates the cranking circuit and fuel injection is restored. This function does not occur until the engine is warm.

SYSTEM DIAGNOSIS AND SERVICE

The vehicle manufacturer establishes the diagnostic procedure for each vehicle model; therefore, diagnostic procedures can vary from one make of vehicle to another. The diagnostic procedure presented here is selected as a representative sample of what should be found in the service manual.

Troubleshooting Guide

The troubleshooting guide consists of a list of driveability complaints such as: Will Not

---

Variable Displacement System

Another feature that Bosch has considered for the Motronic system is a variable displacement system that cancels cylinders during periods of light load and is similar to Cadillac's modulated displacement system. Plans for this system include:

- deactivating cylinders by shutting off the injectors to selected cylinders.
- having the intake and exhaust valves of the deactivated cylinders continue to function.
- providing intake and exhaust manifold valving that allows the exhaust from the working cylinders to be passed through the deactivated cylinders and thus keep them up to normal operating temperature.

Start Cold, Erratic Idle During Warm-up, Backfires, and so forth. Each complaint is accompanied by a list of possible causes. The possible causes should be checked out in the order of the easiest to check or in the order of the most likely cause and based on the technician's experience with the problem and knowledge of the specific vehicle. It is important to verify **that all of the components or systems identified in the Possible Cause list for the identified complaint that are not a part of the Motronic system are working properly before proceeding to test procedures of Motronic system components.**

Motronic Test Section

The service manual is likely to have a section that contains test procedures for each component of the Motronic system. The tests in this section should not be conducted until the engine is at normal operating temperature. In testing components and their circuits, remember that connections cause more problems than components.

REVIEW QUESTIONS

1. Name seven operating conditions that cause Motronic to select a specific control strategy for fuel and/or ignition timing control.
2. How does the Motronic system protect its control unit against reversed polarity?
3. What is the function of the airflow sensor?
4. What are the sources of engine speed and crankshaft position input?
5. List seven control unit-controlled functions that are standard features of Motronic.
6. Name at least six optional control unit-controlled features.
7. What does the term *lambda* mean as used in automotive application?
8. Name at least two functions affected by battery voltage.
9. What two measures are used to control spark knock on turbocharged applications?
10. What does the troubleshooting guide do?
11. What does the test section do?

ASE-type Questions. (Actual ASE test questions will probably not be as product specific.)

12. Technician A says that Motronic's pulse width during open-loop operation is determined based on data concerning engine speed and throttle position. Technician B says that Motronic's pulse width determination in open loop is based on engine speed and load. Who is correct?
    a. A only
    b. B only
    c. both A and B
    d. neither A nor B
13. Technician A says that Motronic determines engine load from engine speed and throttle position. Technician B says that Motronic determines engine load from engine speed and airflow. Who is correct?
    a. A only
    b. B only
    c. both A and B
    d. neither A nor B
14. Technician A says that Motronic controls idle speed with ignition timing and idle airflow. Technician B says that technician A is correct but there are two different methods of controlling idle airflow. Who is correct?
    a. A only
    b. B only
    c. both A and B
    d. neither A nor B
15. Technician A says that Motronic limits maximum engine speed by shutting off the injectors if the criterion speed is exceeded. Technician B says that technician A is correct but that some Motronic engine appli-

cations limit engine speed by shutting off the ignition instead of the injectors. Who is correct?
a. A only
b. B only
c. both A and B
d. neither A nor B

GLOSSARY

LAMBDA (λ) A Greek letter used to indicate how far the actual air/fuel mixture deviates from the ideal air/fuel mixture. Lambda equals actual inducted air quantity divided by the theoretical air required (14.7).

APPENDIX: 1991 AUTOMOBILES

This appendix is included as a means of informing the reader about significant changes in electronic engine control systems and new engines plus engineering improvements that were introduced on 1991 models.

GENERAL MOTORS

An all-new 3.4L double overhead cam 24-valve V-6 engine is the big news on the General Motors performance front in 1991. Delivering an estimated 210 horsepower and 215 pound-feet of torque in manual transmission form, the twin dual cam V-6 is standard in the GM cars covered in the following paragraphs.

Developed from the oil pan up to the cylinder heads, the twin dual cam V-6 is a step up from the 2.3L four-cylinder or 3.IL V-6. General Motors engineers defined the design as a naturally aspirated engine with multiport fuel injection, six cylinders, direct ignition system, and a single accessory belt drive. GM engineers wanted to team the engine with an electronically controlled four-speed automatic or five-speed manual transmission. When the performance goals and fuel economy requirements were taken into consideration, the design was frozen as a 3.4L, dual overhead cam V-6 with four valves per cylinder. Engineers claim that this is the highest displacement multivalve V-6 available from any carmaker—domestic or foreign.

The new 3.4L engine has the same bore spacing as the 3.1L, allowing it to be machined on the same line. However, the block is not identical to that of the 3.1L. To achieve its larger displacement, the cylinder bore of the new block has been increased from 89.0 to 92.0 mm. The head bolt bosses have been reinforced for less strain and lower bore deformation. Deck height, head bolt location, crankshaft location, and stroke remain identical to those of the 3.1L engine.

The aluminum heads of this engine feature four valves per cylinder for optimal breathing. Four valves per cylinder allow the use of smaller and therefore lighter valves, which have a higher rpm limit as well as a higher total flow capacity. The higher rpm limit means that the inertia of the valve train at high speeds will not overcome the valve spring force and allow the valves to float/bounce. The 3.4L's smaller valves are among the largest used for its cylinder bore size, when compared to other four-valve-per-cylinder engines. The diameters measure 36.5 mm for the intake and 32 mm for the exhaust.

The cylinder head and camshaft carrier are separate assemblies. The assemblies are split into upper (cam carrier) and lower (cylinder head) sections at the lifters. The alloy cast-iron cams run directly in the bearing webs of the aluminum cam carriers without additional bearing inserts. The cam bearings are cut in a single tunnel boring operation. This simplifies the design and eliminates the need for several separate bearing caps and split inserts. Cylinder head gaskets are of a nonasbestos composition material similar to those of the 3.1L.

Appendix: 1991 Automobiles

The heads have a classic pent-roof design, with a centrally located spark plug. A four-valve combustion chamber allows larger total valve flow area for better breathing, yet the lighter individual valves permit the engine to rev higher without encountering valve float. This four-valve cylinder design achieves high-end power thanks to its good volumetric efficiency, and high rpm due to its light, direct-acting valve train and optimum combustion chamber shape. The twin dual cam V-6 engine has a tach redline at 7000 rpm.

The twin dual cam V-6 uses lightweight cast aluminum Mahle pistons with elliptical skirts. Steel expansion struts embedded in the piston control their shape as they reach operating temperature. While a sump in the piston crown forms the 3.1L's combustion chamber, the 3.4L, with its pent-roof chamber and higher compression ratio, has flat-topped pistons with machined "eyebrows" for valve clearance. Piston ring tension is higher due to the increased piston diameter.

Like the 3.1L, the 3.4L features the speed density engine control system. This system calculates through the ECM and several sensors the amount of air and fuel entering the engine in order to maintain the proper ratio during closed-loop (warm engine) operating conditions. The speed density system eliminates the need for a mass airflow (MAF) sensor and a cold start injector. The exhaust manifolds for the 3.4L are made of high silicon molybdenum ferritic nodular iron. The material was chosen for the manifolds because it can withstand higher temperatures with less expansion. Although the log design of the manifolds is similar to that used on the 3.1L, the flow characteristics have been improved due to better routing of the exhaust gas at the crossover pipe and from accommodating the 3.4L's larger port openings. The 3.4L's ignition coils and module are identical to those used on the 3.1L, except for the mounting. The thermostat housing and power steering pump are also the same.

The 3.1L 60-degree V-6 is still found in A, J, L, and W of General Motors car lines. Only the 1980 3.1L has been dropped from the line.

There are a number of other interesting high-output GM power plants. For example, the high-output Quad 4 in the Chevy Beretta GTZ and Olds Cutlass Calais generates 180 horsepower at 6200 rpm. Features include four valves per cylinder (16 total), twin overhead camshafts, an integrated direct-fire ignition system, and a single-belt accessory drive system. The legendary Camaro Z28 is powered by a standard 5.0L V-8 with tuned-port fuel injection. An optional 5.7L V-8 with TPI (that is right, the "350") is a 245 horsepower powerhouse.

GM has smaller engines, too. For instance, in Chevrolet's international Geo line, all tracker models are powered by a Suzuki 1.6L, electronically fuel-injected, four-cylinder engine, rated at 80 horsepower at 5400 rpm. The Geo Storm hatchback and coupe both are powered by a 95 horsepower, multiport fuel-injected (MPI) 1.6L, 12-valve, single overhead camshaft (SOHC), four-cylinder engine. A 130 horsepower, 16-valve, dual overhead cam (DOHC) is also available in Storm models. The Geo Prizm offers the same range of engines. A water-cooled, three-cylinder, 1.0L, single overhead camshaft (SOHC) electronically fuel-injected engine is standard on all Metro models.

The 1991 Grand Am SE is the first Pontiac to carry GM's new ABS VI antilock brake system as standard equipment. ABS VI is based on an operating system different from that of all other ABS brakes on the market today. The system will also be offered on Buick, Oldsmobile, and Saturn cars, and eventually in high volume across GM's car lines. Conventional ABS systems use solenoids to control brake modulation. When a conventional system senses wheel lockup, it continuously releases brake pressure to that wheel and then reapplies it through opening and closing solenoids. ABS VI, on the other hand, actively adjusts the pressure. If conventional ABS systems can be considered digital, either on or off, then ABS VI is analog with continuous modulation of brake line pressures controlled by the system. The new system provides excellent control of brake circuit pressure.

The ABS VI does not have a separate pump and accumulator like other systems. Instead, it uses electric motors to modulate displacement pistons. The ABS system is piggybacked onto a conventional brake master cylinder and vacuum booster unit. A modulator contains three electric motors that drive pistons to modulate brake line pressure. This integrated approach makes ABS VI more compact than competing systems.

Each motor drive gear turns a screw, which in turn drives a nut, to which is attached the pressure modulating piston. The electric motors can rapidly drive the pistons up to increase pressure or down to decrease it. When the system senses impending lockup through speed sensors located at each wheel, a solenoid closes the hydraulic path from the master cylinder. The electric motor for that brake circuit then cycles the piston up and down to modulate braking force and prevent wheel lockup. Because the hydraulic path that leads directly to the driver's foot is closed, the ABS VI is much smoother in operation than other systems currently on the market. In use, drivers will experience less pedal pulsation than in earlier ABS systems. ABS VI also has fewer parts, which implies greater reliability.

The first electronically controlled automatic transmission ever offered in a GM luxury sedan is standard in Buick Motor Division's 1991 Park Avenue and Park Avenue Ultra models, introduced in mid-summer 1990. It is now standard on the 1991 Buick Riviera, Reatta, the Olds Cutlass Supreme, and several other GM vehicles featuring the Hydramatic 4T60-E. Featuring a computerized shift control, this new front-wheel-drive four-speed automatic delivers precise, barely detectable upshifts and downshifts during normal driving, smooth operation at highway speeds, and excellent fuel economy. The Hydramatic transmission has freewheeling or one-way clutches to smooth shifts between forward gears. This eliminates any potential for band-to-clutch overlap, which is the cause of more noticeable shifts in automatic transmissions. Shift pattern control is accomplished electronically by a computer known as the power train control model (PCM). Hydraulic complexity is minimized and valves are used only to control the smoothness of the shifts, not their timing.

Superior shift quality in the new transmission is also achieved with separate trim valves for each of the shifting clutches. Two electronically activated shift solenoids inside the transmission control the shift pattern according to instructions from the PCM. Two other electronic solenoids are used in the transmission. One, the torque converter clutch enable solenoid, is used to activate the lockup feature of the torque converter, thereby improving fuel efficiency. The other solenoid, a pulse-width modulated solenoid, is used to vary the rate at which the torque converter is applied.

The biggest thing the driver should notice is an absence of "busyness" when the computer is making shift decisions. When the driver demands power, during passing maneuvers, the shifts between gears have a positive character. That is because the delay between engagement is short, and full-throttle upshifts and downshifts are crisp and precise. The engagement and disengagement of the torque converter clutch at highway speeds is smooth. The precision built into this new transmission translates directly into customer benefits. Fuel economy is optimized because calibrating the engine and transmission combination into a fully integrated power train is much more exact. Precision also means power is available when the driver needs it. With electronic control, there is no delay in downshifting, and the transmission can shift down from third gear to first without going into second—so the car feels very responsive.

The Cadillac is still the crown jewel of the General Motors line. Its 1991 V-8 engine is more powerful and more refined due to a displacement increase from 4.5 to 4.9 liters and numerous detail improvements. This smoother, quieter V-8 increases horsepower by more than 10 percent and is standard on DeVille, Fleetwood, Eldorado, and Seville models without any loss in fuel efficiency.

Since its introduction as a 4.1 liters engine, it has grown in displacement and horsepower. In 1988, engine size was increased to 4.5 liters and horsepower jumped from 130 to 155 bhp. In 1990, the 4.5 engine was equipped with port fuel injection and a higher compression ratio to produce 180 bhp. Zero-to-sixty acceleration times have decreased nearly two seconds since 1988. The new 4.9 V-8 offers increased horsepower and torque, resulting in both better acceleration off the line and improved power for highway passing. Zero-to-sixty acceleration times for each model fitted with the new engine will be reduced by about one second from 1990. Highway fuel economy ratings increase to 26 mpg, while city ratings remain at 16 mpg. Horsepower increases from 180 to 200, and torque is now 275 foot-pounds, up from last year's 245 foot-pounds. Unlike some engines that concentrate their horsepower at higher engine speeds, Cadillac's 4.9L V-8 has substantial torque output across its entire operating range. The resulting engine matches perfectly with Cadillac's new electronically controlled four-speed automatic transmission, providing spirited performance in stop-and-go traffic and relaxed highway cruising.

The displacement increase from 4.5 to 4.9 liters is accomplished by lengthening the stroke of the engine. Both bore and stroke are now 92 mm. Major mechanical changes are limited to a new crankshaft and the connecting rods. The compression ratio remains 9.5:1.

Improvements in emission control systems also help increase performance and reduce the complexity of the engine. A smooth 2.75-inch intermediate pipe and a wide oval catalytic converter reduce restriction by approximately 38 percent on DeVille and Fleetwood models. Air injection reaction (AIR), an emission control system previously used on certain models, has been eliminated with the addition of a new catalyst. This reduces engine noise, eliminates power loss from the pump, and simplifies the underhood layout. New, larger fuel injector nozzles have been incorporated into the standard port fuel injection system.

For 1991, Cadillac DeVille, Fleetwood, Seville, and Eldorado models feature a new advance in automatic transmission technology, an electronically controlled four-speed automatic. Called the 4T60-E, the new standard transmission combines computerized shift control with advanced mechanical features.

Based on the proven four-speed automatic currently used by Cadillac, the 4T60-E adds computer control of shift timing, precisely coordinating engine, transmission, and cruise control. The transmission now also has one-way clutches on all forward speeds for smoother gear changes. At the heart of the four-speed automatic are four electronic solenoids. These solenoids allow the power train control module (PCM), the computer that controls the 4.9L V-8, to also determine the timing of each forward shift, including the operation of the viscous converter clutch. Because the entire power train is now controlled by a single computer, the transmission, converter clutch, and cruise control can work in harmony. For example, the transmission shift pattern can be altered when cruise control is engaged, minimizing the hunting between third and fourth gear that can occur on long mountain upgrades.

The computer eases off the power at the moment of the shift, then gently reapplies it after the next gear is engaged. Called torque management, this electronic system provides almost imperceptible gear changes during normal driving and significantly reduces wear and tear on the drivetrain. Shift quality is further enhanced by use of a roller clutch for the 2 to 3 gear set. All forward shifts are now handled by one-way clutches, eliminating band-to-clutch overlap that occasionally causes shifting to lurch.

The key to this precise shift control is the set of four solenoids in the valve body of the 4T60-E. Responding to signals from the power train control module, these small electric valves control the timing of each forward shift, as well as the application of the viscous converter clutch. One pair of solenoids controls the four forward speeds in the transmission. By using binary logic—the same type of logic used in computers—these two simple solenoids can provide four gear choices with a minimum of hardware. A second pair of solenoids controls the viscous converter clutch, which provides a firm link between the engine and drive wheels at highway speeds to increase fuel economy. One solenoid in the pair applies the converter clutch. The second solenoid, called a *pulse-width modulated solenoid,* varies the rate at which the converter clutch is applied. When teamed with the viscous damper in the clutch itself, these solenoids make converter clutch operation almost imperceptible to the driver.

In addition to computer control of shift timing, this four-speed automatic transmission also benefits from a significant mechanical upgrade. With the addition in 1991 of a roller clutch for the 2 to 3 gear set, all forward shifts are now handled by one-way clutches. These clutches automatically release one gear as the next one is applied, providing smooth shifts under all operating conditions. In a traditional automatic, shifts are handled by friction elements, commonly known as bands or clutches. Under ideal circumstances, the transmission shifts smoothly as one friction element releases exactly as the second applies. But normal wear, temperature changes, and other variations in driving conditions can upset this strategy. If a band is released too soon, the engine speeds up until the clutch grips and the engine flares. If a band is released too late, the transmission momentarily binds, since it is trying to be in two gears at once. In both instances, the driver is aware of the shift. But with one-way mechanical clutches, any potential overlap is eliminated. The result is an automatic transmission that consistently shifts smoothly no matter what the driving conditions.

Engine functions are monitored and controlled by a GMP4 power train control module (PCM). This 64-kilobyte on-board computer senses vital operating conditions, including engine rpm, intake manifold pressure and air temperature, outside air temperature, coolant temperature, exhaust gas oxygen content, throttle position, vehicle speed, accessory load level, and brake operation. The PCM works in microseconds to constantly maintain the optimum balance between the fuel injection, ignition, exhaust gas recirculation, and idle speed. It also stores data for service technicians to diagnose malfunctions and controls the shift points of the new electronic transmission.

Another on-board computer—body computer module (BCM)—monitors and controls such functions as air-conditioning, twilight sentinel headlight operations, theft alarm, and electric cooling fans.

Computer command ride (CCR) is standard on the Fleetwood, Eldorado, and Seville, and optional on the DeVille. An enhanced speed dependent damping system, which shares major technology with CCR, has been improved for 1991 and will continue as standard equipment on the Allanté, where it was first introduced in 1990.

Suspension dampers, commonly called *shock absorbers* or *struts,* control wheel and body motions by forcing hydraulic fluid back and forth through valves in the damper. Without this damping, the car would bounce and wallow out of control. With traditional dampers, the rate of fluid flow is essentially fixed by the design of the valves. Though engineers have developed more sophisticated valves, including those used in deflected disc struts, the result is still somewhat of a compromise. Gentle damping, which gives a smooth, quiet ride at low speeds, can be overwhelmed when the car crosses a swell in the highway or when the driver makes a sudden maneuver at higher speeds. But the firmer damping that provides good control at higher speeds tends to make the suspension harsh and noisy around town.

CCR can provide optimum damping at both extremes by changing the valving in response to commands from an on-board computer. Damper settings are changed by a selector valve, which is driven by a small DC motor inside each strut. Using different size orifices, the selector valve restricts fluid flow through the main valve in the damper. The selector valves, one in each strut, are operated by a small, powerful computer. This computer processes information on vehicle speed, acceleration and braking rates, and lateral acceleration and alters damping to optimize comfort and control. Sophisticated electronics allow the computer to change damping rates in as little as 1/10 of a second.

At speeds up to 40 mph, the computer generally selects the *comfort* setting. This light damping reduces the amount of road shock transmitted to the body. As a result, far less noise and vibration reach the driver. As vehicle speed increases, however, the car hits bumps and swells harder, feeding more energy into the suspension system. So at speeds from 40 to 60 mph, *normal* damping is selected, which provides a firmer ride and better sense of control. At speeds greater than 60 mph, the *firm* setting provides still more damping. The computer also selects the *firm* setting during heavy acceleration and hard braking to reduce body pitch and enhance driver control. A lateral acceleration sensor also signals the damping to switch to *firm* during hard cornering, reducing roll motions of the body. All of these changes occur without any input from the driver. In many cases, the driver is not aware that a change in damping even occurred, since the extra damping essentially matches the increased demands being placed on the suspension.

For 1991, Allanté retains its 4.5L, transverse-mounted V-8 engine and traction control system, which established new technology standards in 1990. The aluminum V-8 is capable of powering the Allanté from 0-to-60 mph in 8.0 seconds and a top speed of more than 130 mph. The Allanté engine has a compression ratio of 9.0:1 and delivers 270 foot-pounds of torque at 3200 rpm, and 200 horsepower at 4400 rpm. Bore measurement is 92 mm and stroke is 84 mm. The Allanté engine also retains its tuned port fuel injection system for 1991.

Allanté's engine is controlled by a computerized power train management system. Based on inputs from engine, transaxle, and cruise control, the system automatically makes adjustments to provide the most coordinated operation possible. To ensure a quality air/fuel ratio, particularly during start-ups, cold-weather operation, and transient throttle changes, the engine control system employs the latest in state-of-the-art port fuel injection and dual exhaust gas oxygen sensors. The system provides improved combustion efficiency and reduced emissions, which allows engineers to remove the power-robbing air injection pump and exhaust gas recirculation. This reduces engine complexity and parasitic power losses.

The power train management system can also retard the engine timing during shifts to reduce stress without hampering performance. Under extreme conditions that could harm the transaxle, the system can automatically shut off fuel delivery to selected injectors. To ensure that engine operating temperatures remain constant, even at high-speed cruising, the engine control system can retard the timing and provide an enriched, cooler air/fuel mixture to the cylinders. This prevents detonation and preignition.

An idle speed control system maintains the idle rpm at a nearly constant level and adjusts engine power to cope with the changing engine loads due to the power steering, airconditioning compressor, alternator, transaxle, and the internal functions of the engine. Additionally, this power train management system includes sophisticated diagnostic subroutines to recognize faults, substitute values, and inform the driver. With any fault, the system tries to work around the problem without inconveniencing the driver. For service technicians, the system has a memory and can provide a snapshot of key operating variables from its memory bank to help isolate particular faults.

Two of the Allanté's electronic systems coordinate operation to make traction control possible—the vehicle's sophisticated Bosch III antilock brake system (ABS) and the engine control module. By linking the vehicle's two systems together, the right balance of power is delivered to the wheels of Allanté based on the traction available. When either of the front wheels begins to spin, the speed sensor in the wheel signals the ABS system to gradually apply the brake to the spinning wheel to limit slipping until it regains traction. The computer continuously monitors wheel spin and adjusts the brakes to maintain about 10 percent slippage at each drive wheel to optimize traction. If wheel spin continues, due to extremely slippery surfaces, or if the driver applies too much power, the engine computer slowly will reduce the power to the vehicle.

When traction control is in operation for more than 2 seconds, fuel delivery to one cylinder is cut off, reducing engine power. If slippage continues, engine power can be cut to up to four cylinders. This minimizes stress, reduces the possibility of overrevving the engine, and signals the driver that traction is extremely limited. Also, a small indicator light on the dashboard illuminates after the traction control system has been in operation for more than three seconds. The delayed signal helps prevent the light from flickering when traction is briefly lost, as can happen when the vehicle accelerates over pavement stripes at a wet intersection. The Allanté traction control system operation is completely automatic; there are no switches to activate or deactivate the system regardless of vehicle speed. And, the system does not need to be turned off due to brake or power train overheating after prolonged use.

CHRYSLER

Chrysler's cars enter 1991 with technological refinements and additional features to further distinguish them in the automotive marketplace. All models have improved engines, front suspensions and steering, antilock brakes either standard or optional, and styling. But as far as new electronic controls, like many of the vehicle manufacturers, Chrysler is taking a break for 1991.

However, the electronic vehicle information center (EVIC) system for 1991 Chrysler Motors' passenger cars has been revised slightly to make operation easier and availability has been expanded to include Chrysler Imperial, New Yorker Fifth Avenue, and LeBaron coupe and convertible. Use of centralized electronic body computer controls has also been expanded to include the Chrysler Imperial, New Yorker Fifth Avenue, LeBaron coupe and convertible, and Dodge Daytona.

A refined electronic speed control system is being made available on most 1991 Chrysler passenger cars. It features new steering wheel-mounted switches for easier operation of the speed control system by the driver.

The popular ACCEL feature of 1990 has been continued in 1991 models. When speed control is engaged, pressing and holding the RESUME/ACCEL button increases the vehicle speed at a constant rate. When the button is released, the vehicle's current speed will be the new set speed. Pressing the button and then releasing it in less than 1 second will produce a 2 mph increase in SET speed.

Appendix: 1991 Automobiles

A new DECEL feature also has been added. It is similar to the coast feature in the previous system, but with better control. When the speed control is engaged, pressing and holding the SET/DECEL button will allow the vehicle to decelerate. When the button is released, a new SET speed is established.

Additional improvements have been made in the system's performance for 1991. For example, after a passing maneuver, the vehicle speed will return quickly to the SET speed as a result of new software. The software also includes additional diagnostic features to increase ability to identify the source of a system malfunction.

Embarking on its second year of a six-year plan to refine its power train offerings, Chrysler has introduced the following series of engine improvements for its 1991 model lineup to boost engine performance characteristics:

- *2.2L DOHC, 16-Valve Engine.* Dodge Spirit R/T for 1991 is powered by a dual overhead cam, 16-valve, turbocharged 2.2L engine with intercooler and twin balance shafts. It generates the highest specific output—more than 100 horsepower per liter—ever achieved by a Chrysler production engine. This engine delivers 224 horsepower at 6000 rpm and 217 pound-feet of torque at 2800 rpm. The dual overhead cams and lightweight valves allow a redline of 6500 rpm. A relatively broad and flat torque curve, steady at the near maximum 217 pound-feet between 2800 rpm and 5000 rpm, assures superior reserve acceleration through upper gear ranges.

  Chrysler engineering tests in the Spirit R/T sedan with the new double overhead cam 2.2 turbo engine and five-speed manual transmission with Getrag gear set achieved 0 to 60 mph speeds in 6.8 seconds and a quarter-mile time of 15.4 seconds. Without gear changes, the 2.2 DOHC turbo propelled the Spirit R/T from 40 to 60 mph in 3.9 seconds and from 50 to 70 mph in 8.4 seconds.

- *2.5L High-Torque Turbo Engine.* Chrysler's 2.5L turbocharged Turbo I engine has had its peak torque rating increased 16 percent from the 1990 level of 180 pound-feet to 210 pound-feet and is featured in selected Chrysler, Dodge, and Plymouth models. As a result, acceleration during the initial stages of launch is significantly increased. The increased rating is the result of a revised calibration of the turbocharger waste gate control for more low-speed boost and revised fuel and ignition timing calibrations. No hardware has been changed. The horsepower rating is unchanged.

  The revised Turbo I is optional on the Dodge Daytona (standard on Daytona Shelby) and Chrysler LeBaron convertible with either manual or automatic transaxle. It is available with automatic transmission only on Dodge Spirit ES, Shadow, and Shadow convertible.

- *3.8L V-6 Engine.* A 3.8L multiport, fuel-injected V-6 engine was introduced on selected 1991 Chrysler models. It provides added power and performance, and is the standard engine for the Chrysler Imperial and optionally available on the Chrysler Fifth Avenue.

  Compared to the 3.3L engine on which it is based, the 3.8L engine offers more torque at all speeds below 4800 rpm—up to 15 percent additional. Estimated power ratings are 150 horsepower at 4400 rpm and 204 pound-feet of torque at 3200 rpm. EPA city fuel economy estimates for the cars equipped with this engine are expected to be slightly lower, but highway estimates are the same as for the same vehicles with the 3.3L engine.

  The 3.8L engine has a larger bore and longer stroke than the 3.3L engine

(3.78 inch and 3.42 inch, respectively). The compression ratio is 9.0:1 compared to 8.9:1 for the 3.3L engine. Parts that differ from the 3.3L engine are the crankshaft, connecting rods, pistons, upper intake manifold, and the vibration damper. The upper intake manifold carries a unique 3.8 logo. The 3.8L engine also uses a larger torque converter.

- *4.0L High-Output I-6 Engine.* Horsepower for the 4.0L, six-cylinder engine in Cherokee and Comanche applications has been increased to 190 at 4750 rpm, while torque moves up to 225 pound-feet at 4000 rpm. Comparable figures for 1990 were 177 horsepower at 4750 rpm with a torque rating of 220 pound-feet at 4000 rpm.

  In the Wrangler, this same engine generates 180 horsepower at 4750 rpm and torque of 220 pound-feet at 4000 rpm. These lower ratings are primarily due to differences in the Wrangler's induction and exhaust systems. A new progressive throttle body is used with the sequential fuel injection system, and it offers dual 60 mm diameter throttle bores rather than the previous 52 mm diameter bores. An improved intake manifold, improved intake and exhaust ports in the cylinder head, along with improved port-to-manifold alignment all contribute to increased horsepower. A tubular steel, exhaust manifold design also helps scavage exhaust gases more effectively.

- *2.5L Engine.* Jeep's four-cylinder engine for 1991—the base engine in all three vehicle models—will receive a more efficient and technically sophisticated, multipoint fuel injection system (MPI) to replace the previous throttle body injection system. Performance characteristics from this and several other technical improvements will raise the 2.5L engine's output—across the board—to 126 horsepower at 5250 rpm, along with torque of 148 pound-feet at 3000 rpm.

The 2.5L engines for Cherokee and Comanche vehicles offer approximately a 5 horsepower increase for 1991 compared to 1990. Sequential multipoint fuel injection provides increased horsepower (approximately a 7 percent increase for Cherokee/Comanche and approximately 11 percent in Wrangler) and torque compared to single point injection. Additionally, driveability for Wrangler, Cherokee, and Comanche vehicles is improved. A revised intake manifold with longer and larger runners helps increase engine torque and power. The intake and exhaust ports in the cylinder head and port to manifold alignment are modified to improve flow characteristics.

A progressive throttle body is used with the multipoint fuel injection system. It is equipped with a 52 mm diameter throttle bore, rather than the 44 mm diameter bore of the throttle body used in the previous single point injection system. An aluminum valve cover and crankcase ventilation system also are introduced on the 2.5L engine for 1991. A serpentine accessory drive belt system is used for the first time on the 2.5L engine in Wrangler models. It replaces the multiple V-belt accessory drive system used through 1990 and offers increased durability, quieter operation, and reduced maintenance.

Chrysler is using several imports in its 1991 engines; for instance, TC by Maserati, which is powered by an Italian 3.0L V-6 multiport, fuel-injected engine that is mated to Chrysler's Ultradrive four-speed fully adaptive electronic automatic transmission.

Engine features include Chrysler Motors' first direct (distributorless) ignition system, multipoint electronic fuel injection, a tuned in-

Appendix: 1991 Automobiles

take manifold, premium Inconel exhaust valves, some of the lightest pistons in the industry and a self-adjusting serpentine belt accessory drive system. It also includes many other premium features to make the engine industry competitive for smoothness, NVH, reliability, and durability.

Availability of Chrysler's Ultradrive electronic four-speed automatic transaxle will be expanded for 1991. One design refinement that has been implemented is sequential closed-throttle downshifts. As the car decelerates, the transmission will downshift sequentially through all gears rather than bypassing the third to second gear shift. The downshifts are also smoother and second gear is immediately usable without kick down, below 15 mph.

The transmission control computer logic has also been modified to assure that, when climbing a grade at highway speeds, downshifts and upshifts will not occur cyclically. After a downshift occurs, the logic observes the torque used to maintain the present speed and/or acceleration level, and only allows an upshift if the torque required to maintain speed is also available in fourth. This logic also includes the ability to compensate for the loss of engine power, which can occur at high altitude as a result of lower air density. When speed control is engaged, the downshifts will occur somewhat earlier than before to assure that speed is maintained. Rapid throttle changes will also result in earlier downshifts to provide better vehicle response and more rapid acceleration on demand.

A major engineering program has noticeably improved front suspension and steering systems on all 1991 Chrysler Motors American-made front-wheel-drive passenger cars and minivans. These include Dodge Daytona, Shadow, Spirit, and Dynasty; Plymouth Sundance and Acclaim; Chrysler LeBaron Sedan, Coupe, and Convertible, New Yorker Fifth Avenue, New Yorker Salon, and Imperial; and Dodge Caravan, Plymouth Voyager, and Chrysler Town & Country minivans.

Extensive refinements have improved directional stability, steering returnability, and steering response linearity to be as quick on center as at larger steering angles. These changes are covered in the following paragraphs.

The roll center has been raised approximately 3 inches, reducing body roll or lean during cornering. The vehicle responds more quickly to steering inputs because less roll means stable tire loads are established sooner in the cornering process. Reducing roll through a geometry change also means the low body roll angle gradient associated with sporty vehicles can be achieved with a lighter sway bar. This improves steering feel and straight-line stability, that is, the ability to continue on a straight course when cross winds attempt to deflect it. It also improves response and increases the lateral force provided by the outside tire during cornering by reducing positive camber.

Steering geometry has been revised to improve steering returnability and reduce tire scrub at low speeds. Computer-aided design was used to increase the Ackerman effect from the 40 to 45 percent of the theoretical ideal on prior models to 53 to 74 percent. (Ackerman effect is the ability of the inside wheel to turn more sharply than the outside wheel when cornering.) The steering geometry toe change during suspension movement has been reduced by approximately 50 percent to provide quicker steering response and increase high-speed and directional stability when driving on bumpy or uneven roads.

Power steering gear valving has been revised to increase steering response to small angular inputs from the steering wheel. The steering ratio for the Chrysler LeBaron coupe and convertible, Dodge Daytona, Dodge Spirit, Dodge Shadow, Plymouth Acclaim, and Plymouth Sundance has been commonized at 16:1. Most of these vehicles had a steering ratio of 14:1 in 1990, but the overall enhancement of response made the 16:1 ratio more appropriate in the overall vehicle context and reduced steering effort.

The higher steering ratio also reduces the amount of steering wheel movement resulting from wheel-to-wheel traction variations during rapid acceleration.

Availability of four-wheel antilock brakes has been extended to a broader range of Chrysler vehicles for 1991, bringing the number of Chrysler models to offer this added security system to twenty. Three ABS systems will be used: Antilock 10 on Caravan, Voyager, Premier, and Monaco; Antilock 6 for the LeBaron, Daytona, Spirit, and Acclaim; and all-new antilock systems on the Talon and Laser. Caravan, Grand Caravan, Voyager, Grand Voyager, and Town & Country models with ABS represent the first Chrysler application of this technology in conjunction with drum-type rear brakes. Application of ABS to the all-wheel-drive versions of Caravan and Voyager models includes custom computer software to meet the performance requirements in the AWD setting.

This system, identified as Antilock 10, was introduced in the mid-1990 model year on the Chrysler Imperial, Fifth Avenue, New Yorker, and Dodge Dynasty. It includes an integrated hydraulic assembly incorporating a hydraulic booster, master cylinder, and a 10-valve hydraulic modulator. A remote pump/motor provides hydraulic power to a bladder accumulator. Four wheel-speed sensors feed speed information to an electronic speed controller. In the event of excessive wheel slip, the controller signals the modulator to precisely control brake pressure in the affected wheels. To accomplish this, the modulator has four isolation valves, three build valves, and three decay valves—a total of ten valves, hence the name Antilock 10. Upon detection of incipient wheel lock, the isolation and decay valves for the affected wheel are energized to reduce brake pressure. When antilock action is not needed or desired on any of these vehicles, the brake pedal feel is the same as a non-ABS system.

During ABS action, a low level of pedal pulsation occurs to alert the driver that the system is operational.

The optional antilock braking system for Eagle Talon and Plymouth Laser is similar in operation to that of Chrysler's Antilock 6. It is an add-on or closed-type system that uses a conventional master cylinder and vacuum booster. The system modulates brake pressure in any or all wheels using only displaced fluid from the master cylinder.

FORD/LINCOLN-MERCURY

Like the other domestic car manufacturers, power train and other functional improvements have been made on Ford car lines such as the Thunderbird, Taurus, Mustang, and Tempo. The 1991 Festiva and Crown Victoria also have exterior styling changes.

The addition of a 5.0L V-8 engine in the Ford Thunderbird and Mustang GT, a new twin-plug version of Ford's 2.3L engine for Mustang, which increases horsepower from 86 to 105, a new electronically controlled transaxle for the Taurus, and a 15 percent horsepower improvement for the Taurus 2.5L engine are just a few of the functional changes Ford has made in its 1991 car lineup.

The 1991 Escort offers a much larger interior package than previous Escorts. Interior features and enhancements include an ergonomically designed, wraparound instrument panel, five-passenger seating capacity, and improved shoulder and hip room. Power train changes include the addition of sequential electronic fuel injection to Ford's 1.9L engine.

The two-door Probe is available in three series—GL, LX, and GT. The Probe GL continues with a peppy 2.2L, 12-valve, electronically fuel-injected four-cylinder engine.

The Probe LX continues to offer a standard 3.0L, electronically fuel-injected V-6 engine. An antilock brake system can be added to LX's four-wheel power disc brakes. The 1991 Probe GT maintains its high performance with a 2.2L electronically fuel-injected, turbocharged I-4 engine, which can be coupled with an optional four-speed automatic transmission. GT's handling suspension features special sport-tuned performance components, plus a computerized automatic adjusting suspension.

The five-passenger Tempo is available in four series—GL (two-door and four-door sedans), GLS (two-door and four-door sedans), LX (four-door sedan), and the four-door Tempo four-wheel-drive sedan. Four-wheel-drive has made it possible for a family-type sedan such as Tempo to operate with greater traction during adverse weather conditions. The four-wheel-drive feature can be activated at the touch of a conveniently located switch—even while the car is in motion.

The addition of sequential multiport electronic fuel injection and other modifications on the Taurus 2.5L four-cylinder engine have increased engine horsepower, as already mentioned, by 15 percent. The 3.0L V-6 also gets sequential fuel injection. The Taurus SHO has a special 3.0L, dual overhead cam, 24-valve V-6 engine that generates up to 200 horsepower. In addition, a new electronically controlled four-speed automatic overdrive transaxle is now standard on all models except the Taurus SHO. This electronically controlled transaxle helps drivers achieve optimum fuel economy and performance along with smooth, consistent shifts. An antilock brake system also is offered for the first time on 1991 Taurus station wagon models.

The Crown Victoria has a 5.0L V-8 engine with sequential fuel injection and standard automatic overdrive transmission that automatically shifts into an overdrive fourth gear when the vehicle exceeds approximately 45 mph and the driver eases up on the accelerator pedal. In overdrive, engine speed is reduced by about one-third for improved fuel economy and reduced engine wear.

In 1991, six out of nine Lincoln-Mercury car lines will have new or improved power trains. For instance, the Lincoln Town Car adds an all-new 4.6L V-8 as standard equipment in 1991. The new engine is the first of a family of all-new engines that will go into various Ford and Lincoln-Mercury vehicles in the 1990s. The new engine represents an improvement in performance and fuel economy, as well as a reduction in noise and vibration. Despite 40 horsepower more than the previous Town Car engine, the 4.6L makes the car even quieter than the previous model.

The Continental also gets a boost in 1991. Engineers have added a more precise airflow sensor and improved the airflow into the combustion chamber of the 3.8L sequential electronic fuel injection (SEFI) V-6 to give it a 15 horsepower increase for improved performance. Additionally, they have paired that power plant with Ford's all-new electronic automatic overdrive transaxle (AXOD-E), an electronically assisted automatic overdrive, for greater smoothness and efficiency.

The 1991 Mercury Cougar has a new look on both sides of the hood. The sporty XR7 carries, as standard equipment, a 5.0L HO electronic fuel injection (EFI) V-8 engine, and the LS features the same engine as an option. A four-speed automatic overdrive transmission is standard on both models. On the outside, both Cougar models feature revised body side moldings, front grille, fascia, hood, headlights, and taillights.

The 3.0L Mercury Sable engine has been refined to include sequential, multiport fuel injection, which delivers more precise fuel distribution and more efficient combustion. Sable also receives the electronically assisted automatic overdrive transaxle for 1991. Mercury also offers an antilock braking system (ABS) as an option on the Sable station wagon for 1991. ABS already is offered as an option on the Sable sedan. Additionally, a moonroof is now available on the station wagon as well.

The 1991 Tracer is an all-new front-wheel-drive car that is available in three models: a

Ford/Lincoln-Mercury

standard four-door notchback, a station wagon, and the sporty four-door Tracer LTS. The Mercury Tracer and Tracer station wagon feature a new electronically fuel-injected 1.9L engine. Tracer LTS features a powerful new electronically fuel-injected 1.8L dual overhead camshaft engine. All Tracer models offer a standard five-speed manual or optional four-speed automatic transmission.

The Mercury Capri series combines a new electronically fuel-injected 1.6L double overhead cam engine with either a five-speed manual or optional four-speed automatic transaxle.

APPENDIX: 1992 AUTOMOBILES

This appendix is included as a means of informing the reader about significant changes in electronic engine control systems and engine modifications for 1992.

GENERAL MOTORS

Coolant Flow

The 1992 Corvette 5.7 liter engine, now designated the LT1 instead of the L-98, has undergone several changes that have resulted in a 20% increase in horsepower, from 240 @ 4000 RPM to 300 @ 5000 RPM. Borrowing a trick from racers, engineers have reversed coolant flow so that the cooled coolant is pumped into the heads first. It then flows down through the block, back to the pump, and back to the cross-flow radiator. This results in better control of combustion camber temperatures and a more predictable combustion process, which leads to more power, better fuel economy, and lower emissions.

The water pump is shaft driven instead of belt driven. The shaft extends forward out of the timing cover to engage the rear of the water pump and is driven by a gear that is meshed to the camshaft timing gear.

Electric A.I.R. Pump

The A.I.R. system on the LT1 features an electrically driven air pump instead of the typical belt driven pump. The pump is controlled by the ECM by way of a relay, Figure A-1. To turn the pump on, the ECM grounds the control coil of the pump relay, and the relay supplies power to the pump motor. The pump is used only during engine warm-up or following an engine start-up and then only after the engine has reached a temperature of 15°C (59°F). Once the pump is turned on, the ECM allows it to run until the system goes into closed loop or for a maximum of 240 seconds. The system usually goes into closed loop within 80 seconds.

When the ECM turns the pump on, it also grounds the divert valve. The divert valve is similar to other GM divert valves. When energized, vacuum is applied to a diaphragm that in turn moves a directional valve. The directional valve directs air to the exhaust manifolds. If the divert valve is not energized, air is dumped to atmosphere. The electrically driven pump, however, eliminates the need to continue pumping air when it is not needed.

Opti-Spark Ignition

Corvette's Opti-Spark ignition system features a distributor mounted on the front of the engine, tucked in between and behind the water pump and the harmonic balancer. It is driven by a short shaft that is splined to the center of the camshaft timing gear and extends forward into the back of the distributor. The shaft drives the distributor sensor disk and the rotor, Figure A-2. All parts of the assembly, including the cap and rotor, are designed for long service life.

Rather than the magnetic pulse generator of the traditional GM HEI distributor, the Opti-Spark distributor uses optical diodes in a manner similar to those described under

Appendix: 1992 Automobiles

FIGURE A-1 Electric A.I.R. pump control circuit. *Courtesy of Service Technology Group, General Motors Corporation.*

FIGURE A-2 Opti-Spark distributor. *Courtesy of Service Technology Group, General Motors Corporation.*

General Motors

Optical Distributor in the Inputs section of Chapter 10. The sensor disk has two sets of slots that GM calls tracks, Figure A-3. The outer track, which consists of 360 equally spaced slots, one slot per degree of camshaft rotation, is called the high-resolution track. Each time one of these slots lines up with the high-resolution LED, the light passes through the slot and strikes the photo diode, and a voltage is generated in the photo diode. As the slot passes by and the light is blocked, the voltage signal in the photo diode falls back to zero, Figure A-4. Therefore, two voltage transitions are produced per degree of camshaft rotation for a total of 720 switch point signals per engine cycle.

The inner or low-resolution track has eight slots, one per cylinder. The leading edges of each of these slots are spaced 45° apart, but their widths vary. The slots for every other cylinder in the firing order, cylinders 1, 4, 6, and 7, are each 2 degrees wide, but the slots for cylinders 2, 8, 3 and 5 are 22, 7, 12, and 17 degrees wide respectively. Each time one of these slots lines up with the low-resolution LED, a voltage signal is produced and the

FIGURE A-3 Opti-Spark sensor disk.

Appendix: 1992 Automobiles

FIGURE A-4 High-resolution signals, low-resolution signals; dwell and spark commands.

signal stays high as long as the slot allows the light to pass through the disc.

Both the high- and low-resolution signals are sent to the ECM. By counting the number of high-resolution signals that occur while each low-resolution signal is high, the ECM can identify which cylinder the low-resolution signal represents. The ECM also identifies the 2° signal following the 22° signal as the firing order starting over again. Each leading edge (voltage rise) of the low-resolution signal represents TDC for that cylinder.

The ECM uses the low-resolution signals in much the same way as the distributor reference is used in other GM systems. Based on these signals, the ECM

- activates the fuel pump circuit
- determines engine RPM
- determines when to turn the injector groups on
- identifies cylinder TDC from which to start calculating spark advance.

High resolution signals are used to

- identify small changes in crankshaft velocity
- fine tune spark timing
- fine tune injector timing

Opti-Spark Circuit. Notice in Figure A-5 that there is no bypass circuit in this system. When the engine is cranked, the ECM reads the high- and low-resolution signals along with other engine parameters and creates the EST signal as shown in Figure A-4. The leading edge (voltage rise) of the EST signal commands the coil-driver module to

General Motors

FIGURE A–5 Opti-Spark circuit. *Courtesy of Service Technology Group, General Motors Corporation.*

turn the primary circuit on. The E-core coil and the coil-driver module are mounted on the front of the right cylinder head. Current in the coil primary winding will rise to about six amps. The falling edge of the EST signal commands the coil-driver module to open the primary circuit to fire the coil. The ECM calculates how much dwell time is needed and how much spark advance is needed and adjusts the EST signal accordingly.

After the engine starts, the ECM uses the high-resolution signal to fine tune the EST signal, responding to changes in crankshaft speed that might result, for example, from variations in cylinder efficiency or sudden changes in throttle position. This results in better control of the combustion process and more efficient engine performance.

If the ECM sees a high-resolution signal but no low-resolution signal, it will not activate the fuel pump circuit or pulse the injectors and will set a code 16. If the ECM sees the low-resolution signal but not the high-resolution signal, it will start and run, but not with normal efficiency, and will set a code 36. If the EST line is open or shorted to ground, the engine will not run, and the ECM will set a code 41. If the EST line is grounded, the engine will not run and the ECM will set a code 42.

Detonation Sensors. To make the LT1 system more alert to spark knock, two detonation sensors are used, one on each side of the engine block. The sensors are piezoelectric crystals as discussed under Sensing Devices in Chapter 2, but an additional feature has been added to enable the ECM to monitor the sensors and related circuit for faults.

A 3.9K ohm resistor is placed in each sensor, Figure A-6, and the ECM sends a 5-

Appendix: 1992 Automobiles

FIGURE A-6 Detonation sensors with self-diagnostic resistors.

volt signal out to the sensors through a fixed resistor in the ECM. With the two parallel sensor resistors in series with the resistor in the ECM, the available voltage to sensors and the diagnostic circuit should be about 1.5 volts. If one sensor becomes open, the remaining resistance in series with the resistor in the ECM becomes greater, and voltage available at the diagnostic circuit will be between 2.2 and 4.1 volts. If both sensors or the line that serves them becomes open, resistance in series with the in ECM resistor becomes infinite and the voltage available to the diagnostic circuit will be greater than 4.1 volts. If either the sensor or the line becomes grounded, voltage at the diagnostic circuit will be less than .78 volts.

Dual Oxygen Sensors. The LT1 engine has an O_2 sensor in each exhaust manifold. This enables the ECM to monitor and respond to air/fuel mixture problems that might occur in either bank of cylinders.

Acceleration Slip Regulation (ASR)

The 1992 Corvette's Electronic Brake Control Module (EBCM) and ECM work together to provide this feature. The EBCM monitors engine RPM by way of a wire tapped into the tachometer drive circuit, Figure A-7. If one of the drive-wheel speed sensors indicates that wheel slip is occurring during rapid engine-speed increase, the EBCM feeds a voltage signal to the ECM requesting it to retard ignition timing. Retarding ignition timing reduces engine torque so that drive-wheel traction can be restored. Spark retard cannot be retarded beyond TDC and does not last for more than a few seconds per wheel-slip event.

If retarding the spark is not enough to allow the slipping drive-wheel to regain traction, as might occur on a wet or icy surface, the EBCM commands a motor to start moving the throttle valves toward the closed position. This is achieved by a throttle-cable design that routes the throttle cables that link the

General Motors

FIGURE A-7 LT1 ASR circuit.

accelerator pedal to the throttle shaft over a series of pulleys on a shaft. The shaft is connected to a motor through a series of reduction gears. When the throttle is pressed down, it pulls a cable that rotates one of the pulleys. The first pulley drives the pulley beside it by means of a spring. The second pulley pulls a cable that is connected to the throttle shaft.

If the EBCM determines that it needs to reduce throttle opening in order to stop wheel spin, it grounds the motor circuit, thus starting the motor. The energized motor rotates the shaft through a set of reduction gears, and the second pulley is rotated in a reverse direction so that it releases tension on the throttle shaft; the throttle valves start to close. When wheel slip stops, the EBCM deactivates the motor.

The motor's reverse rotation of the second pulley puts additional tension on the spring through which the first pulley drives the second pulley. Meanwhile, the first pulley is held in place by the first cable and by the driver's foot on the accelerator. When the motor stops, the spring starts rotating the second pulley back to its original position, opening the throttle again. The rate of opening is controlled by the resistance of having to turn the motor armature, now being overdriven by the reduction gears.

If needed, the EBCM will also pulse the brakes on the slipping wheel.

Central Port Injection (CPI)

For 1992, the 4.3 liter V/6 features a unique system that combines the simplicity

Appendix: 1992 Automobiles

FIGURE A-8   CPI manifold assembly. *Courtesy of Service Technology Group, General Motors Corporation.*

of throttle-body injection with the advantages of port injection. The CPI system features a two-piece intake manifold, Figure A-8, with an injector solenoid assembly somewhat similar to that used with TBI mounted in the center. The injector solenoid, however, does not open a valve and spray fuel into the plenum of the manifold, rather the injector solenoid lifts a hinged disk to expose six ports to fuel pressure, Figure A-9. Each port is connected to a flexible nylon tube, and each tube leads to a hydraulically opened nozzle. Each nozzle is mounted in an intake port in a way similar to the way an injector would be mounted in a port injection system.

When the solenoid is energized, it lifts the valve disk; fuel pressure forces fuel through the ports and flexible tubes. The fuel pressure causes the nozzles to open, and fuel sprays into each intake port near the intake valve. When the injector solenoid is deenergized, the disk closes, blocking the ports from the fuel pressure in the injector cavity. When the ports are blocked, pressure in the lines drops, and the nozzles close.

The pressure regulator maintains fuel pressure at 370–440 kPa (54–64 PSI). Because of pressure drop across the injector ports and fuel lines, fuel is discharged from the nozzle at 350–420 kPa (51–61 PSI).

Variable-Tuning Control Valve. To obtain the most torque and horsepower at low, mid, and high RPM ranges, a movable, blade-type valve is placed in the intake manifold to vary the shape of the plenum, Figure A-8.

At low and high air-flow rates (low and high RPM), a split plenum puts more air into the cylinder, so when RPM and TPS signals to the ECM indicate low or high air-flow conditions, the ECM positions the valve to divide the plenum into two sections. During mid range air-flow rates, a single plenum puts more air into the cylinders, so the ECM positions the valve to create a single (open) plenum. The ECM controls a reversible electric motor to position the valve.

Linear EGR Valve. The 1992 4.3 liter engine also features an electrically operated EGR valve, Figure A-8. The valve is opened by a large solenoid rather than a vacuum diaphragm. The ECM controls EGR flow by regulating how far it opens the valve. By rapidly turning the solenoid on and off and varying the ratio of on to off time, pulse width modulation, it can control how far the solenoid opens the valve.

A potentiometer is mounted on the top end of the valve assembly with its plunger measuring how far the valve is opened. With this closed-loop control, the ECM can monitor the position of the EGR valve and adjust its position to obtain whatever EGR flow rate its programming calls for. The linear EGR valve provides much faster response time and more precise control of EGR flow.

3800 Engine Supercharger

With the introduction of 1992 models, some vehicle models that use the 3800 engine will be available with a supercharger. The supercharger is a belt-driven, twin rotor design with a bypass valve, Figure A-10. At light to moderate engine loads, the bypass valve is held open by engine vacuum acting on the diaphragm in the bypass actuator. Atmospheric pressure pushes air past the throttle valve and into the rotors. The spinning rotors enable the air to flow by to the low pressure of the intake manifold. The rotors are doing little of the work required to move the air because the difference in pressure between manifold and atmosphere has already set the air moving. Any air velocity increase through the rotors fails and boost pressure in the manifold decreases because the pressure is lost through the open bypass valve.

When the throttle is moved to a wide-open position, the loss of engine vacuum causes the bypass actuator to close the bypass valve. Now the increased air velocity caused by the rotors starts to cause a pressure buildup in the manifold.

Appendix: 1992 Automobiles

FIGURE A-9 CPI injector assembly.

A hose attached to the bottom of the by-pass actuator provides boost pressure below the diaphragm. If boost pressure goes too high, this pressure pushes the diaphragm up, opening the bypass valve. An ECM-controlled solenoid in the boost pressure signal line can vent some of the pressure in the line to atmosphere, lowering pressure applied under

FIGURE A-10 Supercharger.

the diaphragm, thereby allowing boost pressure to go higher if operating conditions are favorable.

FORD

Most of Ford's electronic changes and additions for 1992 such as traction assist; speed-sensitive, electronic variable-orifice (EVO) power steering system; newer antilock brake systems; and electronically controlled suspension systems do not involve the engine control system. Therefore, they are not presented here.

AOD-E Transmission

In 1992, the AOD-E four-speed, automatic, rear-wheel drive transmission joins the AXOD-E and the E4OD in being redesigned for electronic control. It too is controlled by the EEC-IV Module, Figure A-11. Shifts and line pressure are electronically controlled. To enhance shift quality, the AOD's unique dual-input

Appendix: 1992 Automobiles

FIGURE A-11 AOD-E inputs and outputs.

shaft design, used to achieve converter lockup in fourth gear, has been eliminated in favor of a converter clutch.

Inputs

In order to make the AOD-E more responsive to driver demand and operating conditions and still provide the best possible drivability and fuel economy, the ECA uses many of the same inputs it uses to control engine functions plus some from sensors in the transmission.

The ECA uses ACT, MAF, and PIP information to determine how much fuel to inject into the engine. With additional programming, it can use this same information to determine the torque being applied to the transmission and adjust line pressure, shift scheduling, and converter clutch application accordingly. Throttle position information can also be used to adjust shift schedules, converter clutch schedules, and line pressure control.

When the air conditioning clutch is applied, the A/C compressor robs power from the drivetrain. Therefore, when the air conditioning clutch switch signals that the clutch is on, the ECA adjusts line pressure to compensate for the loss of torque applied to the transmission.

The converter clutch puts greater load

FIGURE A-12 Sensors and actuators on the AOD-E. *Courtesy of Ford Motor Company.*

on the engine when it is applied, therefore the ECA will not apply the clutch when the engine is cold to prevent engine stalling. The ECA uses the transmission oil temperature (TOT) sensor to determine if the transmission (and therefore the engine) is warm enough to apply the converter clutch, Figure A-12. Transmission oil temperature information is also used to modify the shift schedule during low-temperature operation, to compensate for increased viscosity of the transmission fluid. If for some reason, the ECA does not receive TOT sensor information, it uses ECT information instead for transmission-control decisions.

When the brake on/off (BOO) switch closes, the ECA knows that the brakes have been applied and that it needs to prepare to take the converter clutch off.

The output shaft speed (OSS) sensor replaces the transmission governor, Figure A-12. It is a magnetic pulse generator; its signals are triggered by the park pawl lugs on the exterior of the output shaft ring gear. The OSS tells the ECA how fast the output shaft is turning, and the ECA uses this information in its decision-making process concerning shift scheduling, line pressure control, and converter clutch control.

The vehicle speed input from the VSS is used to modify upshift scheduling. Keep in mind that both the VSS and the OSS are measuring transmission output-shaft RPM. The ECA decodes VSS signals to determine

Appendix: 1992 Automobiles

FIGURE A-13 AOD-E EPC solenoid. *Courtesy of Ford Motor Company.*

how fast the vehicle is moving and OSS signals to determine how fast the output shaft is turning. Therefore, the OSS data are more relevant to transmission-related decisions.

In order to assume control of transmission functions, the ECA needs to know what position the shift selector is in. The manual lever position sensor (MLPS) provides this information. The MLPS is a tapped resistor similar to those used on the E4OD and AXOD-E transmissions, Figure A-12.

Outputs

Rather than use engine vacuum or some kind of linkage connected to the throttle to move a spool valve in the transmission to control line pressure, the AOD-E (like most other electronically controlled transmissions) uses a variable-force solenoid to move a spool valve that controls line pressure, Figure A-13. Ford refers to it as an electronic pressure control (EPC) solenoid. When the ECA wants line pressure to be at its highest, it turns the EPC solenoid off. An internal spring in the solenoid then forces the armature out (to the right in the illustration). The armature pushes the spool valve back into its bore, allowing oil from the pump's pressure regulator to pass unrestricted through the bore cavity and on to the main regulator valve.

When the ECA wants line pressure to be reduced, it passes current through the EPC solenoid winding. The resulting magnetic field causes the armature to move deeper into the solenoid body. A spring in the spool valve bore causes the spool valve to follow the armature. As the spool valve moves to its left, it begins to restrict the flow of oil around the valve and through the bore cavity. Thus by controlling the amount of current through the EPC solenoid winding, the ECA controls transmission line pressure. With this control, line pressure can be changed much faster and more accurately than it could with previous methods.

Rather than relying on governor pressure

FIGURE A-14 AOD-E shift solenoids. *Courtesy of Ford Motor Company.*

to move shift valves in a staged order as governor pressure rises and balancing governor pressure against T/V pressure to delay shifts at greater throttle openings, the AOD-E uses two ECA-controlled shift solenoids, SS1 and SS2, to control pressure to the shift valves. The ECA monitors relevant information such as engine torque applied, engine speed, throttle position, output shaft speed, vehicle speed, and oil temperature to command shifts with greater discretion to achieve maximum drivability and fuel economy.

Both shift solenoids are alike and work in the same manner. They act as pressure bleeds, Figure A-14. When either is turned off, it opens a ball valve that allows an oil pressure signal directed to the shift valves to bleed off. When turned on, the valve closes, and pressure is developed at the shift valve. The two shift solenoids working in combination can create four different conditions, each condition causing the transmission to be in one of its four different forward speed ratios, as shown in Figure A-15.

The AOD-E is equipped with a torque converter clutch. The clutch operates like those in the A4LD, E40D, and AXOD-E, and its control is similar to that of the AXOD-E. A

	GEAR	SHIFT SOLENOIDS SS1	SS2	MODULATED CONVERTER CLUTCH CONTROL SOLENOID	EPC SOLENOID
Ⓓ RANGE	4	ON	ON	CONTROLLED BY EEC-IV STRATEGY	PRESSURE OUTPUT CONTROLLED BY EEC-IV STRATEGY IN ALL RANGES
	3	OFF	ON		
	2	OFF	OFF		
	1	ON	OFF	CLUTCH HYDRAULICALLY DISABLED	
D RANGE	3	OFF	ON	CONTROLLED BY EEC-IV STRATEGY	
	2	OFF	OFF		
	1	ON	OFF	CLUTCH HYDRAULICALLY DISABLED	
"1" RANGE	2	OFF	OFF	CONTROLLED BY EEC-IV STRATEGY	
	1	ON	OFF	CLUTCH HYDRAULICALLY DISABLED	
REVERSE, PARK, NEUTRAL	—	ON	OFF	CLUTCH HYDRAULICALLY DISABLED	

FIGURE A-15 AOD-E solenoid operation chart. *Courtesy of Ford Motor Company.*

Appendix: 1992 Automobiles

FIGURE A-16 AOD-E modulated converter clutch control solenoid. *Courtesy of Ford Motor Company.*

modulated converter clutch control (MCCC) solenoid, Figure A-16, controls pressure applied to the inboard end of the bypass clutch control valve. With no pressure applied to the end of the bypass clutch control valve, it stays positioned toward the inboard end of its bore. In this position, it directs pressure from the converter pressure limit valve in such a way that the clutch is in the released position. As pressure is applied to the inboard end of the bypass clutch control valve, it moves outward in its bore. This adjusts the release and apply pressures so that the clutch begins to apply. The farther the bypass clutch control valve moves outboard, the more firmly the clutch is applied.

The ECA can duty cycle the MCCC at a rate ranging from zero duty cycle (off, no pressure applied to the bypass clutch control valve, clutch disengaged) to 100% (full on, maximum pressure applied to the bypass clutch control valve, clutch fully applied). This allows the ECA to provide controlled, momentary slippage during shifts to improve shift quality during moderate acceleration rates.

Diagnosis

Transmission-related fault codes are presented as part of the Quick-Test procedure used for engine diagnosis. Vehicles equipped with AOD-E transmissions present three-digit fault codes instead of two-digit codes. After performing service for transmission fault codes, go through the drive cycle test listed in the service manual.

Transmission Tester. A hand-held transmission tester is available, Rotunda 007-00085 or equivalent, that enables the service technician to do voltage tests, check resistance of actuators, and operate the electrical portion of the transmission independent of the ECA. These test features are useful in diagnosing transmission problems, but care must be taken to follow service manual instructions carefully to avoid damage to the vehicle or personal injury.

CHRYSLER

For 1992, Chrysler's 3.0 liter, 3.3 liter and 3.8 liter engines have gone from multipoint injection (MPI) to sequential MPI. The 3.9 liter and 5.2 liter engines have gone from TBI to sequential MPI. The fuel pressure regulator for these engines maintains a pressure differential of 43.5 PSI; pressure in the injector is 43.5 PSI higher than pressure in the intake manifold.

Crankshaft Position Sensor and Sync Sensor. The 3.9 liter and 5.2 liter engines now use a crankshaft position sensor similar to those used on the 3.3 liter and 3.8 liter en-

Chrysler

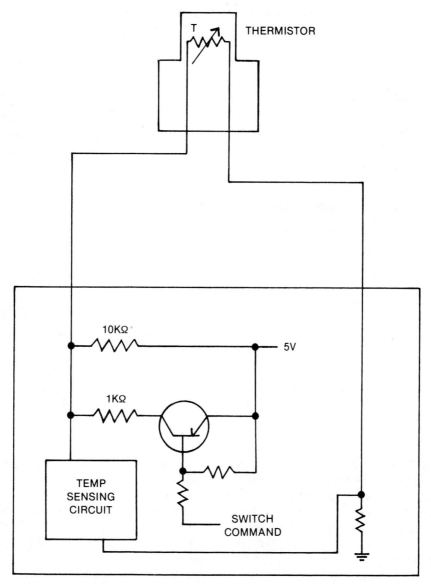

FIGURE A-17 Switching temperature-sensing circuit.

gines. It is a hall-effect device triggered by a sheet metal ring with square windows cut in it that is mounted on the back of the flywheel. The 3.9 liter and 5.2 liter engines, however, continue to use a single-vane hall-effect switch in the distributor as the camshaft or sync sensor.

Temperature Sensors. Chrysler too has gone to switching temperature-sensing circuits for coolant and air temperature sensors, Figure A-17. At temperatures below 125°F, a fixed 10,000 Ω resistor is in series with the thermistor. At about 125°F, the engine controller turns on a transistor that puts a 1,000

Appendix: 1992 Automobiles

Temp. of	Sensor Resistance	Temp. Change	Resistance Change	OHM Change Per Degree	Volt Drop Across Sensor	
−20	156,667 Ω				4.7V	
		10°	50,388Ω	5038.8		
−10	106,279 Ω				4.57V	WITH 10,000 Ω FIXED RESISTANCE
40	25,714 Ω				3.6V	
		10°	6,302Ω	630.2		
50	19,412 Ω				3.3V	
110	4,577 Ω				1.57V	
		10°	1,244	124.4		
120	3,333 □				1.25V	

SENSOR CIRCUIT SHIFT (909 Ω FIXED RESISTANCE RATHER THAN 10,000 Ω)

Temp. of	Sensor Resistance	Temp. Change	Resistance Change	OHM Change Per Degree	Volt Drop Across Sensor	
140	2,338 Ω				3.6V	
		10°	406Ω	40.6		
150	1,932 Ω				3.4V	WITH 909 Ω FIXED RESISTANCE
200	839 Ω				2.4V	
		10°	125Ω	12.5		
210	714 Ω				2.2V	
240	435 Ω				1.62V	
		10°	64.5Ω	6.45		
250	371 Ω				1.45V	

FIGURE A-18 Temperature, resistance, and voltage change table for temperature-sensing circuits.

Ω resistor in parallel with the 10,000 Ω resistor. This reduces the fixed resistance to 909 Ω.

This circuit provides better temperature-change information as temperatures approach the higher end of the operating scale because the resistance change of the thermistor is not linear, Figure A-18. At −20°F, 4.7 of the 5 volts available is dropped across the thermistor. Keep in mind that the engine controller reads what is dropped across the thermistor. At 120°F, with the 10,000 Ω of fixed resistance still in series with the thermistor and the thermistor down to 3333 Ω, only 1.25V is dropped across the thermistor, the rest being dropped across the fixed resistor. At 120°F, the circuit is about halfway through the temperature range it needs to be able to read, but has only about 25% of the voltage range left to work with.

By decreasing the fixed resistance in the engine controller when temperatures increase to 125°F, the fixed resistance is again considerably less than that of the thermistor, so most of the available five volts is again dropped across the thermistor. This is similar to switching an analog voltmeter from the 20-volt scale to the 3-volt scale when low values are being measured but still reading the 20-volt scale.

Shift Indicator. Manual transmission applications are equipped with a shift indicator light to advise the driver when to shift up to the next gear for best fuel economy.

Idle Speed Stepper Motor. Most of

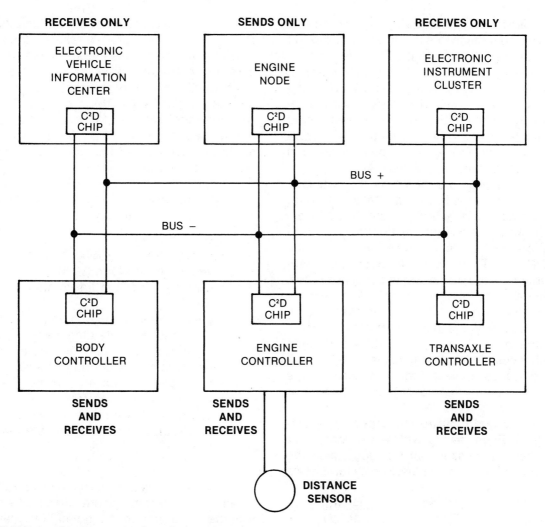

FIGURE A-19 Example data bus network.

Chrysler's engine applications have gone to a stepper motor for idle speed control in place of DC, reversible motor. This allows the engine controller to count the pulses it has applied to move the plunger and therefore keep track of what position the plunger is in.

Vehicle Speed Sensor. Some vehicles are starting to use a hall-effect switch for a vehicle speed sensor in place of the previous sensor. Both sensors produce eight signals per transmission output shaft revolution. The type of sensor used is easily identified, because the earlier type sensor has two wires while the hall-effect type has three wires.

DATA BUS NETWORKS

As the number of controllers used to monitor various functions on a car increases, data bus networks, sometimes called multiplex systems, become more extensively used. Because many of the controllers use some of

Appendix: 1992 Automobiles

the same input information, a method is needed to provide this information to several computers without duplicating sensors or wiring each sensor directly to each controller. This is achieved by wiring each sensor to the controller that is the primary user of that sensor's information and then having that controller broadcast the information to a data bus network where other controllers and devices can receive it, Figure A-19.

Information broadcast on the data bus network takes the form of digital signals that form binary code words, often called bytes, with each byte containing eight bits of information. Each device on the network contains a C^2D chip. The C^2D chip in most controllers is able both to receive and to send messages. Other devices may only receive or may only send messages. Each C^2D chip that can receive scans all messages that are broadcast on the network. Each chip then compares each message to a list in its memory to see if the message is relevant to the device it serves before relaying the message to the device's processor.

Chrysler Collision Detection (C^2D) System. Each device with a C^2D chip that can send messages can broadcast over the network at any time the network is idle. If two devices start to broadcast at the same time, a data collision occurs. To resolve this kind of problem, each message starts with an ID byte that contains coded information about the message's priority, content, and number of bytes, Figure A-20. The message ID byte with the most zeros to the left of the first 1 has highest priority. Binary-code messages are broadcast at a baud rate of 7812.5 (7812.5 bits of information per second).

The devices that may be linked to a data bus network vary from vehicle to vehicle but may include these items:

- body controller
- engine controller (SBEC)
- vehicle theft alarm (VTA)
- electronic vehicle information center (EVIC)
- engine node (a module on the front of the vehicle containing sensors that provide such information as air temperature and direction of travel)
- electronic transaxle controller (EATX)
- air suspension controller
- electronic instrument cluster (EIC)
- mechanical instrument cluster (MIC)
- traveler
- overhead console

The two wires that make up the data bus network and their parallel branches are twisted together to reduce magnetic induction interference. Because the two wires are twisted together, when any magnetic induction is created from other nearby wires that experience rapid changes in current flow, both wires experience the same magnitude of induction at the same time. This tends to cancel the effect of the induction. Otherwise, magnetic induction interference could cause false messages to be broadcast on the network.

FIGURE A-20 Message ID priority.

APPENDIX: 1993 AUTOMOBILES

This appendix is included as a means of informing the reader about significant changes in electronic engine control systems and engine modifications for 1993.

COMMON TERMS

The U.S. government has accepted a list of terms prepared by the Society of Automotive Engineers, entitled J1930, as the terms that all manufacturers selling vehicles in America must use to standardize names for all parts that affect emissions. Use of these terms in service and training publications will become a federal requirement beginning on January 1, 1995. The California Air Regulatory Board mandated that J1930 become law in California on January 1, 1993. Most manufacturers began using these terms in all service and training literature for the 1993 model year. For example, the engine control computer on any manufacturer's vehicle will be called the Powertrain Control Module (PCM). Figure 93-1 lists some examples of these common terms.

GENERAL MOTORS

For 1993, Cadillac introduces a 4.6 liter, dual overhead cam, four valves-per-cylinder engine called the Northstar. Some models such as the Allante feature a 290-horsepower version while others feature a 260-horsepower version. Like the Corvette LT1 engine, the Northstar uses a reverse coolant flow, with coolant from the radiator directed into the heads before going into the block.

Most of the Northstar's engine control system is similar to other GM sequential port injection systems with some additional features described here. This is the first Cadillac-produced engine to use a distributorless ignition system, now referred to as an electronic ignition system (EIS). This system is similar to the earlier GM IDI and DIS (see the Inputs section of Chapter 5), except that the crankshaft position sensor features two magnetic pulse generators, shown in Figure 93-2. The reluctor, mounted on the center of the crankshaft, has 24 notches, spaced every 15 degrees, with 8 additional notches strategically placed between some of the evenly spaced notches to serve as synchronization signals. This puts each sync. notch 7.5 degrees from its neighboring notches. Crankshaft sensors A and B are positioned 27 degrees apart. To identify cylinders so that the Ignition Control Module (ICM) can synchronize its coil firing with crankshaft position, the ICM uses its CKP sensors to count notches, using binary code. Because the extra notches are placed in four groups around the reluctor with one extra notch in the first group, two extra notches in the second and third groups (but spaced differently), and three extra notches in the fourth group, each time a group of extra notches pass the CKP sensors a different binary code is produced. Each binary code identifies a different pair of companion cylinders approaching TDC. Using

Appendix: 1993 Automobiles

Old Acronyms			New Terminology, All Manufacturers	
Chrysler	**Ford**	**GM**	**Acronyms**	**Terms**
SMEC/ SBEC	ECA	ECM	PCM	Powertrain Control Module
Diag. Test Connector	Self-Test Connector	ALDL	DLC	Data Link Connector
DIS	DIS/ EDIS	DIS/IDI/ C^3I	EIS	Electronic Ignition System
CTS	ECT	CTS	ECT	Eng. Coolant Temp. Sensor
EVAP	CANP	CANP	EEC	Evap. Emission Cont. Solenoid
—	SPOUT	EST	IC	Ignition Control
—	DIS Mod.	C^3I Mod.	ICM	Ignition Control Module
CTS	ACT	MAT	IAT	Intake Air Temp. Sensor
KS	KS	DS	KS	Knock Sensor
—	PSPS	PSSw.	PSP	Power Steering Pres. Sw.
PLL/ Check Eng. Light	MIL	Check Eng./ Service Eng.	MIL	Malfunction Indicator Lamp
TPS	TPS	TPS	TP	Throttle Position Sensor
—	BP	BARO	BARO	Barometric Pressure
Brake Sw.	BOO	Brake Sw.	BOO	Brake On/Off Switch
Sync	CID	Sync.	CMP	Camshaft Position Sensor
REF Pickup	CPS	REF	CKP	Crankshaft Position Sensor
—	PFE	—	DPFE	Differential Pres. Feedback EGR
—	TFI-IV	HEI	DI	Distributor Ignition
HO_2	HEGO	HO_2	HO_2S	Heated Oxygen Sensor

FIGURE 93-1 SAE J1930 terminology.

this strategy, the ICM can synchronize coil firing to crankshaft position and fire a sparkplug within the first 180 degrees of crankshaft rotation.

Keep in mind that the ICM uses the signals from crankshaft sensors A and B to decide which coil to fire. A single crankshaft sensor signal sent to the PCM is called the 24X signal. The PCM uses this 24X signal to determine crankshaft speed.

The camshaft sensor produces a signal that identifies cylinder number 1 in the firing order. This signal is passed on to the PCM where it is used for sequential pulsing of the injectors.

Charging System Control. The PCM, along with the BCM, takes on some responsibility for the charging system. The CS144 GEN II alternator has an output capacity of 140 amps at 2000 RPM: CS stands for charging system, and 144 denotes the outside diameter of the stator laminations in millimeters. When the engine is started, the PCM supplies battery voltage to alternator terminal L, Figure 93-3. In response, the alternator's internal voltage regulator applies a duty-cycled volt-

FIGURE 93-2 Northstar electronic ignition system (EIS).

Appendix: 1993 Automobiles

FIGURE 93-3 CS144 GEN II charging circuit.

age to the base of the transistor in the line that supplies power to the rotor from the output terminal. This causes the transistor to be turned on and off to supply a duty-cycled power supply to the rotor. The duty cycle rate is 400 cycles per second, with the percentage of on time during each 1/400 of a second being varied to control alternator voltage. At low engine speed, the transistor will be on for nearly 90% of each cycle; at high engine speed, the transistor will be on for nearly 10% of each cycle. At idle, duty changes are spread over several seconds to help stabilize engine idle speed.

Charging system faults are monitored in two ways. By monitoring voltage values, the regulator can detect such internal problems as a shorted rectifying diode, malfunctioning regulator, open or shorted rotor, open voltage sense circuit at terminal S, or rotor shaft that is not turning. If any of these events occurs, the regulator will ground terminal L. The PCM will see the voltage in this circuit go low and report the information to the BCM by way of the data line. The BCM will then set a code and warn the driver through the Driver Information Center or the volt's telltale light.

The PCM also monitors the circuit from terminal L for voltage; if it detects voltage below 10 or above 16 it sets a code and turns on the *Service Engine Soon* light.

The BCM monitors duty cycle rate through a circuit coming off terminal F. The duty cycle rate is decoded to represent a voltage value. By monitoring this circuit, the BCM can identify and set codes for no charge, voltage below 10 volts, or voltage above 16 volts.

The regulator monitors the charging-system voltage through the circuit at terminal S. This circuit, coming from the positive terminal of the battery, contains a thermistor positioned near enough to the battery so that its temperature is directly related to the battery's. When the thermistor's temperature is low, its resistance is high, therefore the voltage the regulator sees at terminal S is low. The regulator then duty cycles the rotor power to produce a higher charging voltage. If voltage at terminal S is too low, the regulator grounds terminal L and adjusts charging voltage to a default setting.

FORD

The AODE transmission has been renamed; it is now identified as the 4R70W: 4 indicates four speeds forward, R indicates rear wheel drive, 70 represents approximately

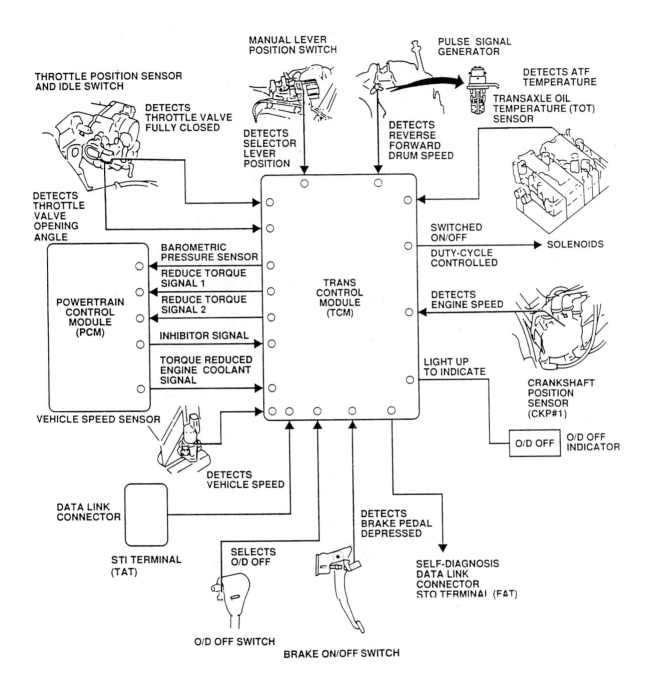

FIGURE 93-4   1993 Probe, 2.5 liter 4EAT-GF transaxle control system. *Courtesy of Ford Motor Company.*

Appendix: 1993 Automobiles

FIGURE 93-5   1993 Probe, 2.0 liter 4EAT-GF transaxle control. *Courtesy of Ford Motor Company.*

Ford

700 ft lbs of torque carrying capacity in stall conditions and W designates a wide ratio transmission. The AXODE has been renamed the AX4S.

For 1993, the Mazda-built (and electronically controlled since 1989) 4EAT (now referred to as the 4EAT-Type GF) transaxle has had some design changes. Because on some 4EAT-GF applications such as the Escort/Tracer and the new 2.0 liter engine for the Probe the Transaxle Control Module has been integrated into EEC-IV system's PCM, it is now being discussed in this text. On other applications, such as the new Probe 2.5 liter engine, the transaxle still has its own control module (TCM), but some engine parameters are shared between the PCM and the TCM to enhance TCM control transaxle functions, as shown in Figure 93-4. Only the PCM-controlled unit is discussed here, but what is said here generally applies to the TCM-controlled units as well.

Figure 93-5 shows the PCM inputs that influence transaxle control and the transaxle control actuators. Figure 93-6 shows the actuators and the operating conditions in

Range	1-2 Sol.	2-3 Sol.	3-4 Sol.	TCCC Sol.	TCC Sol.	Dw. Sh. Sol.	Line Pres Sol.
P	off	off	on	off	1	2	no effect
R < 2.5MPH	off	on	on	off	1	2	no effect
R 3-19MPH	off	off	off	off	1	2	no effect
R > 19MPH	on	off	off	off	1	2	no effect
N < 2.5MPH	off	off	on	off	1	2	no effect
N > 3MPH	on	off	off	off	1	2	no effect
D-1	off	on	on	off	1	2	3
D-2	on	on	on	off	1	2	3
D-2 S/L*	on	on	on	on	< 100% DC*	2	3
D-3	on	off	off	off	1	2	3
D-3 S/L*	on	off	off	on	< 100% DC*	2	3
D-OD	on	off	on	off	1	2	3
D-OD S/L*	on	off	on	on	< 100% DC*	2	3
D-OD L*	on	off	on	on	0% DC*	2	3
2	on	on	off	off	1	2	3
1	off	on	off	off	1	2	3

1. The TCC solenoid has no effect (on or off) if the TCCC solenoid is off.
2. The down shift solenoid is on only during 3-2 downshifts.
3. The line pressure solenoid is duty cycled as a function of throttle pressure: a small throttle angle produces a high duty cycle percentage resulting in a low line pressure.

S/L*. Slip/Lock.—The converter clutch is applied but allowed to slip at a controlled rate.
L*. Locked—The converter clutch is fully applied with no slippage.
DC*. Duty cycled.
< = Less than, > = more than.

FIGURE 93-6 4EAT-GF actuator apply chart.

Appendix: 1993 Automobiles

FIGURE 93-7 1-2 shift valve and solenoid. *Courtesy of Ford Motor Company.*

which they are applied. Each of the actuator solenoids functions as a pressure bleed when energized.

Shift Control. This transaxle has a separate shift solenoid for each shift valve. As seen in Figure 93-7, when the PCM wants the transaxle to shift into second gear, it turns the 1-2 solenoid on. This allows pressurized fluid to bleed from the right side of the 1-2 shift valve; the spring on the left side of the shift valve then drives the valve to the right, causing oil pressure from the low reverse band to exhaust while oil pressure is directed to the 2-4 band. Other shifts are similar, except the 2-3 solenoid turns off to stop a pressure bleed, causing pressure to build up on the right side of the 2-3 shift valve. This moves the valve to the left against a spring, affecting the 2-3 shift.

The addition of the downshift solenoid is one of the changes for 1993 in the 4EAT transaxle, shown in Figure 93-8. During the 3-2 downshift, the downshift solenoid is briefly deenergized to cause pressure to develop on the right side of the 3-2 timing valve. This softens the 3-2 downshift.

Torque Reduction Control. During 1-2 or 2-3 upshifts with more than 50% throttle opening, the PCM momentarily stops pulsing the injectors to reduce engine torque, thus reducing shock on the drivetrain and improving shift feel. If the throttle opening is between 50% and 75%, half the injectors are cut off; if throttle opening is greater than 75%, all the injectors are cut off. During O/D-2, O/D-1, 3-2, 3-1, or 2-1 downshifts, the PCM may momentarily retard ingition timing to reduce engine torque.

Torque Converter Clutch Control. The torque converter clutch control (TCCC) solenoid controls whether or not pressure is allowed to develop on the right side of the TCC shift valve, as shown in Figure 93-9. The torque converter clutch (TCC) shift valve is actually two valves placed end to end in a bore in the valve body, with valve B on the left and valve A on the right. If the TCCC solenoid is off, oil pressure pushes both valves to

FIGURE 93-8 Downshift solenoid and 3-2 timing valve. *Courtesy of Ford Motor Company.*

the left. Valve A blocks oil pressure from the rear (transaxle) side of the torque converter while valve B routes oil pressure to the front converter chamber (engine) side, keeping the converter clutch in the not-applied mode. If the PCM turns the TCCC solenoid on, pressure bleeds off of the right side of TCC shift valves, and both A and B are driven to the right by a spring plus oil pressure on the left side of valve B. Valve A now routes oil pressure to the rear chamber of the converter while valve B exhausts oil from the front chamber, causing the converter to be fully applied.

If the PCM wants the converter clutch applied, but with controlled slippage, it duty cycles the TCC solenoid at a cycle rate of 31.25 Hz. Each on portion of each cycle allows oil pressure to bleed from a passage that feeds the left side of the TCC shift valve plus the right side of the TCCC valve. The TCCC valve starts moving to the right while the TCC shift valve B is moved back to the left by solenoid-reducing pressure. Solenoid-reducing pressure has been there all along, but it has not been strong enough to move valve B until the pressure on its left was reduced; valve A stays to the right. The left movement of TCC shift valve B blocks the exhaust port for the front converter chamber while the right movement of the TCCC valve opens a controlled bleed of front converter chamber pressure. The rate of the bleed depends on the duty cycle rate of the TCC solenoid. The PCM controls the duty cycle rate to achieve the converter clutch slippage rate its program calls for. Slippage rate is determined by comparing engine speed to forward reverse drum speed; forward reverse drum speed is measured by the pulse signal generator a magnetic pulse generator.

Line Pressure Control. When energized, the line pressure solenoid bleeds solenoid-reducing pressure off the left end of the pressure modifier valve, as shown in Figure 93-10. This causes solenoid-reducing pressure that is always applied to the right end of the pressure modifier valve to move the valve to the left. This in turn restricts solenoid-reducing

Appendix: 1993 Automobiles

FIGURE 93-9 TCC shift valve and solenoids. *Courtesy of Ford Motor Company.*

pressure fluid flow between port 20 and port 21, causing a pressure drop at port 21. This pressure is now referred to as modifier pressure. It is applied to the pressure regulator valve in such a way as to affect line pressure: as modifier pressure goes higher, line pressure goes higher.

The PCM duty cycles the line pressure solenoid at 31.25 Hz. When throttle angle is small, a larger percentage of each line pressure solenoid cycle is on time, and modifier pressure is low; when throttle angle is large, a smaller percentage of each cycle is off time, and modifier pressure is high. When the transaxle is in reverse, the manual valve directs oil pressure to the pressure regulator valve to automatically raise line pressure, and modifier pressure is overridden.

Overspeed Protection. If the vehicle is driven to excessive speed in the transaxle range 1 or 2, the PCM will override the manual valve and upshift to protect the transaxle.

Diagnosis. If a fault is detected in any of the main input or output circuits while the vehicle is operating, the PCM will turn on the MIL light (TCM-controlled units will flash the *Overdrive Off* light) and store a fault code in memory. Fault codes can be obtained by grounding the appropriate terminal in the data link connector or by using a Super Stat Tester II with a new MECS (Mazda Engine Control System) adapter, the New Generation Star Scan Tool (NGS) with an adapter, the 4EAT transaxle tester (all available from Motorcraft), or any of several aftermarket test instruments.

Villager. For 1993, Mercury introduced a new mini-van called the Villager. This front-wheel drive vehicle is a joint venture between Ford Motor Company and Nissan Motor Company Ltd. Nissan provides the powertrain, with Ford Motor Co. supplying the rest of the chassis and body. It is assembled in Lorain, Ohio. The engine is a 3.0 liter, 60-degree V6, sequentially injected, featuring a mass air flow sensor and an optic distributor. The distributor is similar to the one discussed in the Inputs section of Chapter 10 for Chrysler's 3.0 liter, Mitsubishi-built engine and the one used by the Corvette, LT1, Opti-Spark distributor. It has a row of 360 slots near the outer circumference of the rotor disc and another row of 6 evenly spaced slots near the center of the disc (see Figure A-3, p. 403). The inner row of slots serves to indicate crankshaft position by causing a voltage pulse each time a piston approaches TDC. One of these slots is slightly

FIGURE 93-10 Line pressure control solenoid.

wider than the others to provide a pulse of longer duration; the PCM uses the longer pulse to identify the number 1 cylinder. The PCM uses the outer row of slots to determine engine RPM; the higher data rate enables it to monitor momentary changes in crankshaft speed.

The 4F20E transaxle has 4 speeds forward and is electronically controlled by a TCM.

CHRYSLER

For 1993 Chrysler introduced a new 3.5 liter, 60 degree V6. This engine is based on and similar to the 3.3 liter engine. The inputs and outputs of its engine control system are shown in Figure 93-11. Most of input and output devices and functions are carryovers from

Appendix: 1993 Automobiles

FIGURE 93-11   3.5 liter engine control system.

other engines and are discussed in the appropriate chapters, but some are new or have significant new features discussed here.

Torque Management. The transaxle is controlled by a transaxle control module (TCM). When the TCM is ready to command a shift (upshift or downshift) it signals the PCM. The PCM determines from engine parameters and programmed instructions if torque reduction is needed. If so, the PCM commands a reduction of fuel delivered by the injectors and spark advance to reduce the harshness of the shift. After the shift, the PCM gradually restores fuel delivery and spark advance to what engine parameters indicate.

On this system, the VSS also provides vehicle speed data to the TCM. The TCM conditions the VSS signal (800 pulses per mile) and sends it to the PCM.

ASD Sense. In addition to powering the injectors, coils, and the heat elements for the heated oxygen sensors, it also provides a 12-volt signal back to the PCM to confirm that the ASD relay has been energized.

Knock Sensors. Piezo-electric knock sensors screw into the engine block, one on each side, in the valley area under the intake manifold.

A/C Select/Evaporator Temp. When the driver turns the air conditioning select switch on, a signal is sent to the BCM. The BCM also receives a signal for a temperature switch in the evaporator. The BCM sends these inputs to the PCM by way of the C²D bus. If evaporator temperature and refrigerant pressure are not too low, the PCM adjusts idle air inlet and turns the compressor clutch relay on. If refrigerant pressure goes too high, the PCM turns the compressor clutch off.

EIS. The 3.5 EIS system (formerly called DIS), like other Chrysler EIS applications, does not use a separate ignition control module; the PCM controls which coil fires and when.

SCI Receive/SCI Transmit. The serial communication interface (SCI) receive and transmit circuits both have terminals in the data link connector through which the diagnostic read-out box II or similar scan tool and the PCM communicate. The SCI receive circuit transmits data from the scan tool to the PCM, and the SCI transmit circuit transmits data from the PCM to the scan tool.

Tachometer. The PCM transmits engine speed information to the body control module (BCM). The BCM then sends signals to drive the tachometer.

EEC Solenoid. The EEC solenoid (new J1930 term, formerly known as the EVAP solenoid) is duty cycled by the PCM. Higher percentages of cycle-on time allow higher purge flow rates.

EGR Solenoid/Transducer. The 3.5 liter engine's EGR valve is controlled by a back-pressure transducer and EGR solenoid. The solenoid is mounted on the transducer. When the solenoid is off, the EGR is open. The PCM turns the EGR solenoid off when MAP values are between minimum and maximum values, TP values are above idle but short of WOT, and ECT value indicates that coolant temperature is above a minimum temperature.

Manifold Tuning Valve (MTV). The 3.5 liter engine features a vacuum-operated, variable, tuning valve in the center of the intake manifold plenum. When the valve is in the closed position, the plenum is divided into separate left and right plenums, one feeding each bank of cylinders. When the valve is in the open position, the left and right plenums are connected as one.

When engine speed is below 3008 RPM or above 4288 RPM, the PCM energizes the MTV solenoid; the solenoid blocks vacuum to the valve's vacuum motor, and the valve closes. At engine speeds between 3008 and 4288 RPM, the PCM deenergizes the MTV solenoid, and vacuum opens the MTV. This

Appendix: 1993 Automobiles

system is similar to the variable-tuning control valve used on GM's 4.3 liter, CPI engine introduced in 1992. Chrysler engineers credit the MTV with a 2% increase in horsepower and a 13% increase in torque.

APPENDIX: 1994 AUTOMOBILES

This appendix is included as a means of informing the reader about significant changes in electronic engine control systems for 1994.

ON-BOARD DIAGNOSTICS II (OBD II)

Electronic engine control systems prior to the 1996 model year, now referred to as OBD I, are programmed to detect components or systems that have failed and then turn on the check engine light or MIL. Research conducted by the California Air Resources Board (ARB) found that emission levels often start to increase as soon as an emissions-related component begins to deteriorate. By the time the component is sufficiently worn to cause the MIL to come on, the vehicle has in most cases been emitting excess emissions for some time. The objective then is to find a way to recognize a component or system in the early stages of failure and alert the driver that service is needed. The California ARB's research also showed that annual or even semiannual tests at emissions inspection stations are not effective enough to identify vehicles with poorly functioning emission systems before excessive emissions have already been released.

In light of these findings, the California ARB developed requirements for electronic engine control systems to monitor not only the performance of key emissions-related systems and components, but also the failure of others. The results of this on-vehicle monitoring must be available to service technicians and emissions inspection personnel without the use of manufacturer-specific test equipment. The Federal Clean Air Act of 1990 mandated that all vehicles sold in the United States by the 1996 model year must adhere to the California ARB requirements. These monitoring requirements are called OBD II. Many manufacturers are introducing one or two engines in the 1994 model year that meet most of the OBD II requirements.

The goal of OBD II in monitoring the efficiency of key emission control systems is both to turn on the MIL and to store a diagnostic trouble code (DTC) whenever efficiency deteriorates to a point where (according to research) the emission level reaches 1.5 times the allowable standard for that vehicle based on federal test procedures.

The additional system hardware and changes, besides those made to the computer required to monitor system efficiency rather than system failures, are limited in most cases to adding another heated oxygen sensor (HO_2S) downstream from the catalytic converter, changing to a standardized 16-cavity data link connector (DLC), and upgrading some connectors and components for federally required 100,000-mile or 10-year reliability. The changes in operating and monitoring strategies in the PCM, however, are enormous.

Appendix: 1994 Automobiles

Standardization

Another impact of OBD II will be standardization designed to eliminate some of the differences in diagnostic features, equipment, and terms between different manufacturers. These standards, prepared by the Society of Automotive Engineers (SAE), are as follows:

- common terms (acronyms) (SAE standard J1930)
- common data link connector (DLC) and its location (SAE standard J1962)
- common diagnostic test modes (SAE standard J2190)
- common scan tools (SAE standard J1979)
- common diagnostic trouble codes (SAE standard J2012)
- standard protocol (SAE standard J1850)

Common terms. All manufacturers have agreed to use the same terms and abbreviations for each part. For example, the device that provides crankshaft position information to the PCM will be called a *crankshaft position sensor* by each manufacturer and abbreviated *CKP* sensor. These new terms first appeared in the manufacturer's literature for the 1993 model year.

Common Data Link Connector (DLC). Each vehicle will have a 16-cavity DLC with the same shape and the same pin assignments and located in the same general area, as shown in Figure 1994-1. It will be located under the instrument panel, somewhere between the left end of the instrument panel and a position 300 millimeters to the right of the vehicle center line.

Common Diagnostic Test Modes. These test modes, common to all OBD II vehicles, can be accessed using an OBD II scan tool. Each mode is described in the following paragraphs.

Cavity	General Assignment
1	Ignition Control
2	BUS (+) SCP
3	Discretionary (Not Used)
4	Chassis Ground
5	Signal Ground (SIG RTN)
6	Discretionary (Not Used)
7	K Line of ISO 9141
8	Discretionary (Not Used)
9	Discretionary (Not Used)
10	BUS (-) SCP
11	Discretionary (Not Used)
12	Discretionary (Not Used)
13	FEPS (Flash EEPROM)
14	Discretionary (Not Used)
15	L Line of ISO 9141
16	Battery Power

FIGURE 1994-1 DLC connector and pin assignments.

Mode 1. Parameter Identification (PID) mode allows access to certain data values, analog and digital inputs and outputs, calculated values, and system status information. Throughout the service literature, there will be references to PID values. Some of the PID references are from a generic OBD II PID list that all scan tools must be able to reference. If a referenced nongeneric PID is not on this list, it can be accessed with the manufacturer-specific scan tool or equivalent. If a generic

On-Board Diagnostics II (OBD II)

scan tool is used for nongeneric OBD II PIDs, a string of characters (a hexadecimal number in some cases) may need to be entered. The necessary characters should be supplied by the scan tool manufacturer.

Mode 2. Freeze Frame Data Access mode permits access to emission-related data values from specific generic PIDs. These values, which represent the operating conditions that existed when the fault was recognized, are stored when the DTC is stored. Once a set of freeze frame data is stored, it will remain in memory even if another emission-related DTC is also stored, except for diagnostic trouble codes (DTCs) indicating misfire of the fuel system. Misfire of fuel system DTCs overwrite any previous data and are not further overwritten. When a DTC is erased, the freeze frame data are erased with it.

Mode 3. The purpose of this mode is to enable all scan tools to obtain stored DTCs. These DTCs are transmitted in response to an OBD II Mode 3 request. Either the DTC, its descriptive text, or both will be displayed. Specific menu access to emission-related DTCs is left up to the scan tool manufacturer, but such data should be relatively simple to access.

Mode 4. The Powertrain Control Module (PCM) reset allows the scan tool to command the PCM to clear all emission-related diagnostic information. When the PCM is reset, an inspection maintenance readiness code is stored in the PCM until all the OBD II system monitors or components have been tested to satisfy an OBD II trip without any other faults occurring.

Mode 5. The Oxygen Sensor Monitoring Test results allows access to the on-board sensor fault limits and actual values during the test cycle. The test cycle contains specific operating conditions that must be met (engine temperature, load, etc.) for completion. This information helps to determine the efficiency of the exhaust catalyst. Here are some of the available tests and test identification numbers.

Test ID	Test Description	Units
01	Rich to lean sensor threshold voltage for test cycle	VOLTS
02	Lean to rich sensor threshold voltage for test cycle	VOLTS
07	Minimum sensor voltage for test cycle	VOLTS
08	Maximum sensor voltage for test cycle	VOLTS

Mode 6. The Output State Mode (OTM) aids in servicing output actuators associated with the PCM. It allows the technician to energize and deenergize most of the system output actuators on command. When the OTM is entered, the outputs can be turned off and on without controlling the cooling fans. The low and high speed cooling fans can be turned on separately without energizing the other outputs.

Common Scan Tools. A scan tool that meets OBD II requirements must assess and interpret emission-related diagnostic trouble codes (DTC) regardless of vehicle make or model.

Common Diagnostic Trouble Codes. SAE J2012 dictates a five-digit alphanumeric code in which each digit has a specific meaning, as shown in Figure 1994-2.

The first digit is the prefix letter indicating function:

    P = Powertrain
    B = Body
    C = Chassis

Appendix: 1994 Automobiles

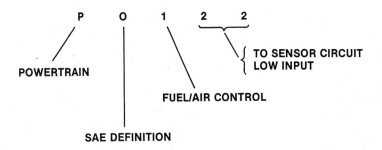

FIGURE 1994-2   DTC example.

The second digit (first number) indicates who was responsible for the DTC definition:

0 = SAE
1 = Manufacturer

A DTC with 0 as the second digit indicates the same kind of problem for the same kind of circuit or component regardless of manufacturer. This may not be true if the second digit is a 1.

The third digit of a powertrain DTC indicates subgroup:

      0 = total system
1 and 2 = fuel/air control
      3 = ignition system/misfire
      4 = auxiliary emission controls
      5 = idle/speed control
      6 = PCM and inputs/outputs
      7 = transmission
      8 = non-EEC powertrain

The fourth and fifth digits relate to the specific fault identified. Some OBD II compliance vehicles will still flash a two-digit code on the MIL. The two digits in the flash-out code will probably not be the same as the last two digits of the J2012 code.

Standard Protocol. Standard protocol compliance refers to the practice of having each manufacturer use the same multiplexing communication language between stand-alone module and devices, including messages and diagnostic data to the DLC.

Monitoring Conditions

OBD II regulations require that detecting faults, setting DTCs, illuminating the MIL, extinguishing the MIL, and erasing DTCs for each monitored system or component be carefully controlled under a specific set of operating conditions. The following definitions help to identify those conditions.

Warm-up Cycle. A warm-up cycle is defined as vehicle operation, after an engine-off period, with engine coolant temperature rising by at least 40° F and reaching at least 160° F. Most OBD II stored DTCs will be erased after 40 warm-up cycles if the concern does not reappear after the MIL is extinguished, as shown in Figure 1994-3.

Drive Cycle. A drive cycle, as opposed to an OBD II drive cycle, consists of an engine start and vehicle operation that at least bring the vehicle into closed loop and include whatever operating conditions are necessary either to initiate and complete a specific OBD II monitor or to verify a symptom or its repair, as shown in Figure 1994-3. A monitor is an operating strategy designed to test the operation of a specific system, function, or component.

OBD II Trip. Usually referred to as a *trip*, an OBD II Trip consists of an engine start following an engine-off period, with subsequent vehicle travel sufficient to allow the following five OBD II monitors to complete their respective tests, as shown in Figure 1994-3:

On-Board Diagnostics II (OBD II)

Monitor	Monitor Type (When it Completes)	Number of Malfunctions on Separate Drive Cycles to Set Pending DTC	Number of Separate Consecutive Drive Cycles to Light MIL and Store DTC	Number and Type of Drive Cycles with No Malfunction to Erase Pending DTC	Number and Type of Drive Cycles with No Malfunction to Turn MIL OFF	Number of Warm-Ups to Erase DTC after MIL is Extinguished
Catalyst Efficiency	Once per OBD II Drive Cycle	1	3	1	3 OBD II Drive Cycle	40
Misfire Type A	Continuous		1		3 (Similar Conditions)	40
Misfire Type B/C	Continuous	1	2	1	3 (Similar Conditions	40
Fuel System	Continuous	1	2	1	3 (Similar Conditions)	40
Oxygen Sensor	Once per Trip	1	2	1 Trip	3 Trips	40
Exhaust Gas Recirculation	Once per Trip	1	2	1 Trip	3 Trips	40
Comprehensive Component	Continuous (When Conditions Allow)	1	2	1 Trip	3 Trips	40

FIGURE 1994-3 OBD II DTC/MIL function chart.

- misfire, fuel system, and comprehensive component. These are checked continuously throughout the trip.
- EGR. This test, which requires a series of idles and accelerations to satisfy conditions necessary for completion, is done only once per trip.
- HO_2S. This test, which requires a steady speed for approximately 20 seconds at 20 to 45 mph after warm-up for completion, is done only once per trip.

All OBD II scan tools feature a readiness function that shows all the monitors on the vehicle and indicates the status of each: complete or not complete. If vehicle travel time, operating conditions, or parameters were not sufficient for a monitor to complete its testing, the scan tool will indicate which monitor has not completed its test.

OBD II Drive Cycle. This specific set of driving conditions will satisfy all requirements for all monitor tests, as shown in Figure 1994-4. The driving conditions are as follows:

1. Start the engine. Once started, the engine must not be turned off at any time during the drive cycle.

Appendix: 1994 Automobiles

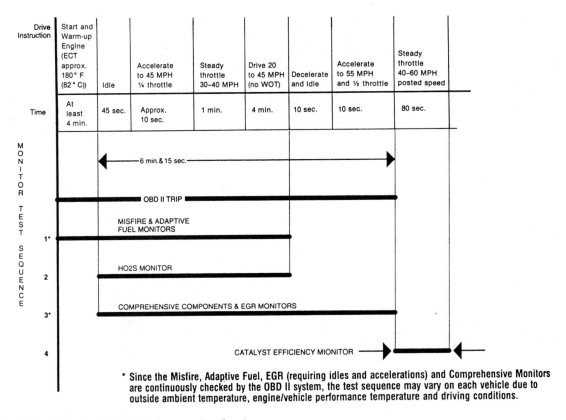

FIGURE 1994-4   OBD II drive cycle chart.

2. Drive the vehicle in any convenient manner for at least four minutes. Any drive mode, including idle, is acceptable for steps 2 and 3.
3. Continue driving until the engine reaches an operating temperature of 180° F or greater.
4. Idle for 45 seconds.
5. Accelerate to 45 mph at ¼ throttle (elapsed time is about 10 seconds).
6. Drive at a speed between 30 and 40 mph, maintaining a steady throttle position for at least one minute.
7. Drive until at least four minutes are spent between 20 and 45 mph. Stop and go conditions causing time to be spent below 20 mph cannot be included in the four minutes; do not operate at WOT.
8. Declerate and idle for at least 10 seconds.
9. Accelerate to 55 mph at ½ throttle (elapsed time is about 10 seconds).
10. Cruise at a speed between 40 and 65 mph, maintaining a steady throttle position for at least 80 seconds. Do not exceed posted speed limit.
11. Bring vehicle down to idle.
12. Check scan tool for on-board system readiness test results and any DTCs.

Similar Conditions. Once the MIL has been illuminated for a misfire condition or a fuel system fault, the vehicle must go through three consecutive drive cycles that include operating conditions similar to those that existed when fault was first detected. Similar conditions are defined as

- engine speed within 375 rpm of same
- engine load within 10% of same
- engine temperature the same (cold or warmed-up)

Note that it is not necessary for a system to fail twice (three times for a catalyst fault) under similar conditions in order to set a DTC and illuminate the MIL. Also note that experiencing operating conditions similar to those that existed when the fault was first detected could take some time. If the problem initially appeared at WOT and high rpm, those conditions may not be revisited until grandmother borrows the car again.

Setting DTCs and Turning the MIL On

When an emission-related fault is detected for the first time, a DTC relative to that fault is stored as a pending or intermittent code. The term for this code varies with manufacturers—for instance Ford calls it a "pending" code, Chrysler calls it a "maturing" code and, General Motors calls it a "failed last time" code. During the next drive cycle, if the monitor that identified the fault originally tests for the same fault but does not find it, the maturing code is erased. If the fault reappears on the second drive cycle, the DTC is stored (as a "history" code in General Motors language) and the MIL is turned on. There are two exceptions to this. In the first, the misfire monitor can store a DTC and start flashing the MIL in response to its first detection of a type A misfire (one that could overheat and damage the three-way catalytic converter [TWC]); and in the second, the catalyst monitor must detect a fault on three OBD II drive cycles before storing a DTC and illuminating the MIL.

The criteria for turning the MIL off vary depending on the fault. If the MIL was turned on by the HO_2S, EGR, or comprehensive component monitor, it will be turned off after three consecutive OBD II trips during which the concern does not reappear. If the MIL was illuminated by either the misfire or fuel system monitor, it will be turned off after three consecutive drive cycles under similar operating conditions without the concern reappearing. If the MIL was illuminated by the catalyst efficiency monitor, it will be turned off after three consecutive OBD II drive cycles without the concern reappearing.

Erasing a DTC requires 40 warm-up cycles without the concern reappearing after the MIL has been turned off. DTCs can also be erased by a technician using a scan tool.

Diagnostic Managing Software

The PCM contains diagnostic managing software to orchestrate the complex testing process. The term for this managing software varies without manufacturers. Ford calls it "diagnostic executive," while Chrysler calls it "task manager." It is called the managing software here.

Each monitor must perform its tests under a unique set of operating conditions relative to engine termperature, rpm, MAF/MAP values, throttle position, and time from start-up in order to be accurate. The managing software determines the order in which tests are done, when and if the correct operating conditions exist for each test, and whether the correct operating conditions lasted long enough for the test to be completed. If not, the test is absorbed and to trip or drive cycle is counted relative to that test; the managing software waits for the next opportunity to have the appropriate monitor run the test.

As an example, let's say that an EGR monitor has detected a fault on the previous trip, as shown in Figure 1994-3. A maturing

DTC has been set and the managing software is waiting for the next trip to either confirm the fault (store the DTC and turn on the MIL) or erase the maturing code. In the next period of vehicle operation, it sees the engine at the correct temperature and idling, with over four minutes elapsed since start-up, so it commands the EGR monitor to begin the EGR test. Idle time proves to be sufficient and the correct amount of acceleration occurs, followed by a steady throttle position at 34 mph for well over a minute. Then there is about five minutes of driving with speeds ranging from 23 mph to 44 mph with no WOT operation, followed by deceleration and idle for 15 seconds. Then the ignition is turned off.

A trip (the minimum for an EGR monitor test) has not occurred because there was no ½-throttle acceleration to 55 mph. The EGR system has neither passed nor failed its second test; the managing software is still waiting for the next "trip" so the test can be completed. There was, however, a sufficient drive cycle to enable the misfire and fuel system monitors to complete their tests.

The managing software may elect to not conduct a test due to a related problem. For example, if the MIL is turned on for an HO_2S concern, a catalyst monitor test would obviously result in a failure. Therefore, the managing software does not ask for the catalyst monitor test (see Figure 1994-5).

A test may not be run because it conflicts with a test in progress. For example, if the EGR monitor test is being run, the managing software will not allow the catalyst monitor test to be run. The EGR monitor's manipulation of the EGR temporarily causes conditions in the catalyst that would keep the catalyst monitor test data from being representative of normal catalyst operation.

The managing software may run a test and then suspend its results until another monitor has run its test and the results are evaluated. For example, the results from the

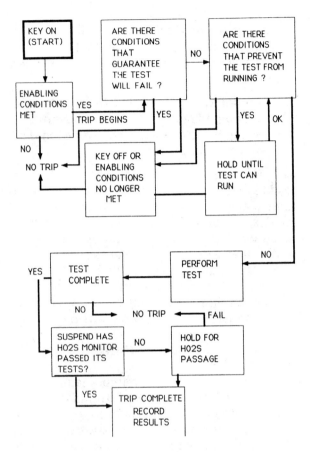

FIGURE 1994-5 Managing software decision chart.

catalyst monitor may be held until the HO_2S test has been successfully completed.

Freeze Frame Data. In addition to storing DTCs, the managing software stores a "snapshot" of information about relevant engine parameters that existed when the fault was first detected. These data are used by the managing software for comparison when similar operating conditions must be identified. They are also useful to the technician for diagnostic purposes and can be retrieved with a scan tool. Information in a freeze frame typically includes:

- the related DTC
- engine rpm
- engine load
- fuel trim (short term and long term)
- engine coolant temperature
- MAP or MAF values
- operating mode (open or closed loop)
- vehicle speed

The freeze frame data storage area will only accommodate the data relative to one DTC at a time. If two DTCs are stored, freeze frame data are stored for the one that occurred first. If, however, a DTC is recorded relative to a misfire or fuel system fault, the managing software replaces any stored data relative to a lower priority DTC with data relative to the misfire or fuel system DTC.

Monitors

As noted earlier, a monitor is an operating strategy designed to test the operation of a specific system, function, or component. The specific monitors are described in this section.

Catalyst Efficiency Monitor. During lean engine operation, an efficient TWC stores oxygen. During rich engine operation, the oxygen is used to oxidize hydrocarbon and carbon monoxide molecules. Because of this behavior, the oxygen discharged from an efficient catalytic converter tends to be constant even though the exhaust stream going into the converter has a rapid rise and fall in oxygen content. A TWC that is not efficient does not store oxygen well, so its oxygen discharge rises and falls with oxygen content of the exhaust stream coming into the converter.

Catalyst efficiency, tested once per OBD II drive cycle, can be determined by comparing the readings of the second HO_2S placed downstream from the TWC with those of the first, HO_2S upstream of the converter. The second HO_2S is called a catalyst monitor sensor (CMS). As the PCM rapidly commands the air fuel mixtures rich and lean across stoichiometry, the first HO_2S should produce a rapidly rising and falling voltage signal while the CMS produces a signal with a much lower frequency and amplitude, as shown in Figure 1994-6.

To prevent cracking of the ceramic body in the CMS, some manufacturers delay turning on the heater in the CMS until the ECT signal indicates a warmed-up engine. This allows condensation in the exhaust system

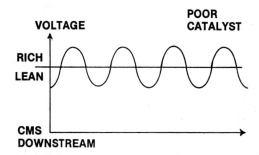

FIGURE 1994-6 Catalyst efficiency detection.

Appendix: 1994 Automobiles

to evaporate. The CMS has a different harness connector from that of the HO₂S even though it is essentially the same part.

Misfire Monitor. When a cylinder misfires, raw hydrocarbon and oxygen are pumped into the TWC. When this much hydrocarbon is burned in the TWC, the temperature goes too high. The catalytic honeycomb begins to melt into a solid mass. Emissions go up dramatically, and the exhaust is severely restricted.

Misfires are monitored continuously by measuring the contribution each cylinder makes to engine performance. The crankshaft should accelerate each time a cylinder fires. Using a high-data crankshaft position sensor (CKP-HDR), shown in Figure 1994-7, the misfire monitor can calculate the time lapse between CKP-HDR signals and determine the acceleration produced by each cylinder firing.

The misfire monitor contains an adaptive feature that compensates for variations in engine characteristics caused by manufacturing tolerances and component wear. It also has adaptive ability to allow for vibration, even vibration variations that occur at different engine speeds and loads. When an individual cylinder's acceleration deviates beyond a set point, the misfire monitor calculates vibration and tolerance allowances and determines whether a misfire has occurred.

There are two different misfire monitor thresholds: misfire type A and misfire type B/C.

FIGURE 1994-7 High data rate crankshaft position sensor.

Type A Misfire. During a 200-revolution period, if the misfire monitor detects a misfire rate that (according to a stored speed/load table) would cause the catalyst temperature to reach 1600° F or more, the MIL begins to flash and the system defaults to open loop in order to prevent the fuel control system from commanding a greater pulse width. Once the vehicle is out of the operating range where the high misfire rate is occurring, the MIL stops flashing and stays on steadily. A DTC is immediately set. Some manufacturers have even programmed into the PCM the ability to turn off one or even two injectors in offending cylinders. As a safety exception, if the engine is under load, as in passing or climbing a hill, the PCM does not deny fuel to the misfiring cylinder or cylinders.

Type B/C Misfire. During a 1000-revolution period, if the misfire monitor detects a misfire rate of 2 to 3 percent, a pending DTC is recorded, and the operating conditions at the time the fault was detected are stored on the freeze frame. The managing software will turn the MIL on if the fault occurs again on the next drive cycle.

Fuel System Monitor. Although this is one of the highest-priority monitors, it is also one of the simplest. While the system is operating in closed loop, the fuel system monitor continuously watches short-term fuel trim and long-term fuel trim (see "Adaptive Fuel Control" in Chapter 7).

If problems such as a vacuum leak, air flow restriction, or incorrect fuel pressure cause the adaptive fuel control to make changes exceeding a predetermined limit in the short- or long-term fuel trim tables, the fuel system monitor reports a failure, and a maturing DTC is set. On the next drive cycle, if the failure does not reappear, the maturing code is erased; if it does appear again, a DTC is set and the MIL is turned on.

Heated Oxygen Sensor (HO₂S) Monitor. Separate tests are run for upstream

On-Board Diagnostics II (OBD II)

FIGURE 1994-8 HO$_2$S switching frequency response rate.

(HO$_2$S) and downstream (CMS) sensors, and each test is completed once per drive cycle. When the managing software identifies operating conditions that meet HO$_2$S testing criteria, the injectors are pulsed at a fixed-duty cycle rate. The HO$_2$S monitor checks the frequency response of the HO$_2$S to see that it produces a signal proportionate to the cycle rate of the injectors. The frequency is high enough that a slow responding HO$_2$S will not be able to keep up and will exhibit a reduced amplitude signal, as shown in Figure 1994-8. In order to pass the HO$_2$S monitor test, the HO$_2$S must generate a voltage output greater than .67, switch across .45 volts a minimum number of times over a 120-second period, and demonstrate a rapid voltage rise and fall.

Because of the behavior of TWC, the rear HO$_2$S or catalyst monitor sensor (CMS) should produce a low amplitude and fairly low frequency signal. During CMS testing, the PCM forces the air-fuel ratio from rich to lean to force the CMS to produce higher amplitudes. If a rich condition is momentarily sustained, available oxygen in the TWC is consumed and the TWC discharge contains less oxygen, so the CMS signal goes high. If a lean condition is momentarily sustained, the TWC is saturated with oxygen and the CMS signal goes low.

Some manufacturers use gold-plated pins and sockets for the HO$_2$S harness connector to obtain increased reliability.

EGR System Monitor. Because of the variety in methods manufacturers use to get EGR feedback, there are a variety of methods to monitor EGR efficiency. The simplest is probably the method used by Chrysler Motors.

In normal operation, when the EGR valve is opened, the pulse width to the injectors is reduced to compensate for the oxygen displaced by exhaust gases. Chrysler's strategy is to select an operating condition that meets test criteria, including the criterion that the EGR valve be open. Then, without any change in the pulse width command to the injectors, the EGR valve is disabled. This should cause the air-fuel ratio to go lean. Monitoring the HO$_2$S reveals whether it does.

Comprehensive Component Monitor (CCM). the remaining inputs and outputs that affect emissions and are not specifically tested by another monitor are checked by the CCM. In many cases, the monitoring of these circuits is not much different from the testing that occurs in OBD I systems.

Analog inputs are checked for opens, shorts, and out-of-range values by monitoring the analog to digital converter input voltages. Typical examples are

- intake air temperature (IAT)
- engine coolant temperature (ECT)
- throttle position (TP)
- manual level position (MLP)
- mass air flow (MAF)/manifold absolute pressure (MAP)

Digital and frequency signal inputs are checked by rationality checks. This is done by using other sensor value and calfulations to see if a sensor's reading is appropriate for existing conditions. For example, the managing software compares the PIP signal to the CID signal. Some examples of these are

- crankshaft position (CKP)
- cylinder identification (CID)
- ignition diagnostic monitor (IDM)
- vehicle speed (VSS)
- output shaft speed (OSS)

Some inputs are checked continuously; others can be checked only when they are actuated so that a change of state can be observed. All PCM outputs are tested while the vehicle is operating. Some are checked continuously, others require system actuation to observe changes of state.

A malfunctioning idle air control (IAC) system can cause an incorrect idle speed or a rough idle, either of which can produce increased or unstable emissions. The IAC is a closed-loop system with the PCM adjusting commands to the IAC solenoid if the idle speed reported by the CKP is not correct. The CCM checks to see if needed corrections exceed predetermined limits.

Other outputs are checked for opens and shorts by observing the voltage in the actuator's driver circuit. The actuator is turned

FIGURE 1994-9 Output state monitor.

on by switching its driver circuit to ground, which pulls the voltage to nearly zero, as shown in Figure 1994-9. When the actuator is not energized, the voltage should be near battery voltage. The following is a list of actuators typically monitored in OBD II systems:

- fan control (high)
- heated oxygen sensor (HO_2S) heater
- catalyst monitor sensor (CMS)
- wide open throttle A/C cutout (WAC)
- electronic pressure control (EPC) solenoid*
- shift solenoid 1 (SS1)*
- shift solenoid 2 (SS2)*
- torque converter clutch (TCC)*

Control of the HO_2S and CMS heaters varies between systems. Ford, for example, uses the PCM to actuate the heater circuits. Therefore it is easy to monitor whether the circuit is working by monitoring circuit voltage (see Figure 1994-9).

General Motors feeds a bias voltage of approximately .45 volts to the HO_2S signal terminal, as shown in Figure 1994-10. When the HO_2S reaches operating temperature and starts to generate a signal, the bias voltage

---

*Faults in these actuator circuits will in most cases cause the transmission control indicator light (TCIL), rather than the MIL, to be illuminated.

On-Board Diagnostics II (OBD II)

FIGURE 1994-10 General Motors HO$_2$S and heater circuit.

is removed and the system operates normally. When the ignition is turned on, it feeds battery voltage to the HO$_2$S heater circuit, which has a fixed ground. Following a cold start, the PCM measures how long it takes for the HO$_2$S to start generating signals. The HO$_2$S reaches operating temperature faster if the heater is working. If the PCM decides that it took too long for the HO$_2$S to start producing signals, a "failed last time" DTC is set. The amount of time allowed for the HO$_2$S to start producing signals depends on ECT and IAT values.

Chrysler Motors' HO$_2$S heater monitoring strategy is also different. The HO$_2$S heater is powered by the automatic shutdown (ASD) relay which is controlled by the PCM. After the ignition is shut off, the PCM uses the battery temperature sensor to sense ambient temperature. It then waits for an amount of time based on sensed ambient temperature for the HO$_2$S to cool down far enough to stop generating signals. After the time has elapsed, the PCM energizes the ASD relay. If the heater brings the HO$_2$S back to operating temperature, it generates a signal, even though the engine is not running. If the HO$_2$S does not produce a signal within a predetermined length of time, a maturing DTC is set.

Evap-Fuel System Integrity. To further guard against the possibility of hydrocarbons leaking from the evaporative emission system, California ARB OBD II requirements call for a detection system that can recognize a .040-inch opening in the fuel evaporative system. Most manufacturers are still developing that system; some expect to implement it by 1995.

Secondary Air (AIR) Monitor. Little information is available yet relative to AIR systems and OBD II compliance. Most vehicles equipped with OBD II systems for 1994 do not use AIR. OBD II systems that do use AIR will probably monitor the HO$_2$S or the CMS while commanding air upstream and downstream in the exhaust system as a functional test.

Chlorofluorocarbon (CFC) Monitoring. By federal mandate, refrigerants containing CFCs will no longer be produced after 1995. Therefore, while CFC monitoring is an OBD II requirement, it is not likely that any vehicles will be equipped with CFC monitoring capability.

INDEX

A/C and cooling fan electronic module, 296-98
A/C cut-out relay, 352
Air charge temperature sensor, 263-64
Air-conditioning clutch signal/switch, 272, 347
Air-conditioning control, 196
Air-conditioning cut-out, 223-26
Air-conditioning relay, 148-49
Air-conditioning switch, 92, 140, 174-76, 346
Airflow meter, 374
Air/fuel ratio, 319-20
Air management valve, 99-104
Air switching solenoid, 364-35
Air temperature sensor, 374
ALDL request, 213
Altitude sensor, 377
Amperage, 10
Analog computers, 39
Atomic structure, 3-4
Automatic idle speed motor, 351

Barometric pressure (BP) sensor, 87, 256-57, 312, 318
Battery switch, 347
Battery temperature sensor, 346
Battery voltage, 210, 377
Baud rate, 43
Block learn, 197-98
Boost control, 351
Bosch (Robert), Motronic system, 369-85
Brake on/off switch, 271-72
Brake switch, 346

Cadillac Allanté, 205-7, 216
Cadillac, digital fuel injection system, 205-43
Canister purge, 151, 289, 333
Canister purge solenoid, 320
Catalytic converter, 43-47, 149
Central fuel injection, 274-75
CFI canister purge, 323
CFI injectors, 323
Charcoal canister purge, 191-93, 218-20

Charge temperature sensor, 342, 360
Charging circuit control, 352
Charging circuit voltage, 346
Check engine light, 115, 151
Chrysler, computerized engine controls, 337-66
   oxygen feedback system, 357-66
   single-point and multipoint fuel injection, 337-55
Circuits, 12-17, 18-19. *See also* Electricity
   excessive resistance, 18
   opens, 18
   parallel, 15-16
   series, 14-15
   series-parallel, 16-17
   shorts, 18-19
Cold-start injector, 180-81
Components, common to all systems, 55-79
   actuators, 73-76
   computers, 55-57
   diagnosis and service, 76-79
   five-volt reference, 57
   fuel injection, 57-60
   sensing devices, 60-73
Computer controlled dwell, 278
Computers, automobile, 37-51
   binary code, 40-41
   clock pulses, 43
   communication signals, 40
   data links, 43-44
   development, from 1963, 37-40
   emission control, 44-48
   fuel economy, 48-50
   interface, 41
   memories, 41-43
   technician's approach, 50-51
Computer command control, General Motors, 84-131
   electronic control module, 84
   inputs, 86-93
   operating modes, 84-86
   outputs, 93-116
   system diagnosis and service, 116-31

Index

Computer-controlled coil ignition (C³I), 166, 171-72
Conductors, 12
Constant spray injection, 57
Controlled canister purge, 115
Coolant temperature sensor, 86-87, 168, 342, 358-60, 374
Cooling fan control, 148
CP sensor, 312-13, 318
Cranking signal, 176
Cruise control, 176, 354
Cruise control brake switch, 213
Cruise control enable, 213
Cruise control vacuum solenoid and power valve, 223
Cruise resume/accelerate switch, 213
Cruise set/coast switch, 213
Cylinder identification sensor, 261-63

DC motor idle speed control, 285-86
Detonation control, 349
Detonation (knock) sensor, 71, 91, 140, 174, 346, 362
Digital computer, 39-40
Digital EGR valve, 189-91
Digital fuel injection, Cadillac, 205-43
    Allanté, 205-7
    body computer module, 226-35
    electronic control module, 207-9
    inputs, 210-13
    operating modes, 209-10
    outputs, 213-26
    system diagnosis and service, 235-43
Diodes, 30-32
Distributorless ignition, 279-85
Distributor pickups, 360
Distributor reference pickups, 342
Distributor reference pulse, 89-90, 139, 210
Distributor sync pickup, 342-43
Dwell control, 380

Early fuel evaporation, 220
E-cells, 72-73, 115
E core coil, 278-79
EGR control and vent solenoids, 289, 351
EGR diagnostic switch, 174
EGR flow, 320
EGR pressure feedback sensor, 265
EGR shut-off solenoid, 289-90
EGR solenoid, 364
EGR vacuum diagnostic control switch, 93
EGR vacuum regulator solenoid, 291-93
EGR valve control, 382
EGR valve position sensor, 264-65
EGR valve temperature sensor, 351
Electric backlite switch, 346
Electricity, 1-36
    amperage, 10
    circuits, 12-17, 18-19
        faults, 18-19
        types of, 12-17
    conductors, 12
    electron theory, 2-4
        atomic structure, 3-4
        elements, 2-3
        molecules, 2
        potential, 4
    insulators, 12
    integrated circuits, 36
    magnetism, 4-10
    Ohm's law, 11-12
    polarity, 17
    resistance, 10-11
    semiconductors, 25-35
    testing, 19-25
        ammeter, 22-23
        available volts, 19-20
        resistance, 23-25
        self-powered light, 25
        voltage drop, 20-22
    voltage, 10, 11
Electric motors, as actuators, 75-76
Electromagnets, 5-7
Electronic engine control I, Ford, 311-18
    description, 311
    electronic control assembly, 311-12
    inputs, 312-13
    outputs, 313-17
    system diagnosis and service, 317-18
Electronic engine control II, Ford, 318-20
    description, 318
    inputs, 318
    outputs, 319-20
    system diagnosis and service, 320
Electronic engine control III, Ford, 321-22
    description, 321
    electronic control assembly, 321
    inputs, 321
    outputs, 321-23
    system diagnosis and service, 323
Electronic engine control IV, Ford, 247-308
    electronic control assembly, 247-54
    inputs, 254-72
    outputs, 272-300
    system diagnosis and service, 301-8
    vehicle speed control, 300-301
Electronic fuel injection, General Motors, 135-56
    electronic control module, 136
    inputs, 138-41
    operating modes, 137-38
    outputs, 141-51
    system diagnosis and service, 151-56
Electronic spark control, 98-99, 146, 185
Electronic spark timing, 97-98, 146, 181-85, 214
Electronic transmission control, 383
Emission controls, 44-48
    air injection system, 48
    catalytic converter, 45-47
    EGR valve, 47-48
    fuel evaporation control, 48
    heated air inlet, 48
    intake charge heating, 48
    PCV system, 48
    three major emissions, 44-45

Index

Engine calibration, 38
Engine coolant temperature sensor, 138, 254, 318
Engine speed sensor, 374
Evaporative emissions control, 382-83
EVP sensor, 318
Exhaust gas oxygen sensor, 60-61, 263, 318, 327
Exhaust gas recirculation valve, 111-13, 149-51, 188-91, 216-18

Feedback control carburetors, 272-74
Ford, computerized engine controls, 247-335
    EEC I, 311-18
    EEC II, 318-20
    EEC III, 321-22
    EEC IV, 247-308
    microprocessor control unit, 325-35
Fuel control devices, 331-33
Fuel economy, computerized, 48-50
    air-conditioning control, 49
    air/fuel mixture, 48
    closed loop, 49-50
    EGR valve control, 49
    electronic ignition timing, 48-49
    idle speed control, 49
    open loop, 50
    torque conversion clutch control, 49
    turbocharger boost control, 49
Fuel injection, 57-60
    fuel pressure regulators, 58-60
    multipoint, 58
    single-point, 57-58
Fuel pressure pulsation damper, 379
Fuel pressure regulator, 58-60, 142, 214, 275, 348, 378-79
Fuel pump, 275
Fuel pump control, 144-46
Fuel pump operation, 214
Fuel pump relay, 275-77, 378
Fuel pump voltage signal, 141

General Motors, computerized engine controls, 84-200
    computer command control, 84-131
    electronic fuel injection, 135-56
    port fuel injection, 149-200
Generator monitor, 210-13

Hall-effect switch, 69-71, 258
Heated oxygen sensor, 60-61
Heated rear window switch, 346
Hood louver control, 148

Idle air bypass valve solenoid, 287-88
Idle air control, 142-44, 185-87
Idle speed control, 92, 104-7, 214-15, 320
Idle tracking switch, 267
Ignition crank position, 141
Ignition diagnostic monitor, 263
Ignition switch, 92, 140, 272
Ignition timing, 313-15, 320, 348-49, 363-64, 379-80
Inferred mileage sensor, 271

Injector assembly, 141
Injectors, 177-81, 213-15, 347-48, 377-78
Inlet air solenoid, 298
Insulators, 12
Integrated circuits, 36
Integrator, 197-98

Keep-alive memory, 84, 160, 209
Knock sensor. *See also* Detonation sensor, 271, 331, 376
Knock sensor response, 285

Lambda sensor, Bosch, 374
Lock-up converter, 353
Low coolant temperature switch, 328

Magnetism, 4-10. *See also* Electricity
    electromagnets, 5-7
    induction, 9-10
    lines of force, 4-5
    motors, 7-9
Malfunction indicator light, 301-2
Manifold absolute pressure sensor, 87, 138-39, 210, 255-56, 312, 318, 349
Manifold air temperature sensor, 168, 210
Mass airflow sensor, 168, 267
Microprocessor control unit, Ford, 325-35
    description, 325
    inputs, 327-31
    MCU module, 325-27
    outputs, 331-34
    system diagnosis and service, 334-35
Mid- and dual-temperature switches, 328-29
Mixture control solenoid, 93-97
Motronic, Robert Bosch, 369-85
    control unit, 369-70
    injectors, 374
    inputs, 374-77
    operating modes, 370-74
    outputs, 377-84
    system diagnosis and service, 384-85
Multec injector, 178-80
Multipoint injection system, 58, 275

Neutral switch, see Park/neutral switch

Octane adjustment rod, 279
Ohm's law, 11-12
Open circuit, 10
Optical distributor, 343-46
Oxygen feedback system, Chrysler, 357-66
    combustion control computer, 358
    inputs, 358-62
    outputs, 362-65
    system diagnosis and service, 365-66
Oxygen sensor, 87-88, 139, 174, 342, 361

Park/neutral switch, 92, 140, 174, 210, 346
Peak coil-current cut-off, 380-81
Polarity, 17
Port fuel injection, General Motors, 149-200

Index

electronic control module, 160
fuel supply system, 162-63
injectors, 163-65
inputs, 168-76
non-ECM emission control, 166
operational modes. 161-62
outputs, 177-96
system diagnosis and service, 197-200
throttle body, 165-66
Potentiometers, 63-64
Power loss lamp, 353-54
Power relay, 253-54
Power steering anticipate switch, 213
Power steering pressure switch, 272
Power steering switch, 176
Pressure differential (VAC) sensor, 87
Pressure sensors, 64-65
Profile ignition pickup sensor, 258-61

Radiator fan relay, 352, 365
Reference mark sensor, 374
Reference pulse, 171-74
Relays, 75
Resistance, 10-11
Rotary idle adjustor, 381-82

Scanner, 117, 151, 197
Second-generation distributor, 323
Semi-closed loop, 138
Semiconductors, 25-35. *See also* Electricity
    diode, 30-32
        how it works, 30-31
        symbols, 31-32
    doping, 25-28
    materials, 25
    PN junction, 28-30
    transistor, 33-35
        bipolar, 33-34
        power, 34-35
        switching, 35
        symbols, 35
Service and coolant over-temperature lights, 226
Shift light indicator, 300, 365
Short circuit, 18-19
Single-module engine control, 338-39
Single-point and multipoint fuel injection, Chrysler, 337-55
    inputs, 341-47
    logic module/power module, 338-41
    outputs, 347-54
    system diagnosis and service, 354-55
Solenoids, 73-75
    duty-cycled, 74
    pulse width, 74-75
Spark knock control, 380
Spark output, 277-78
Spark retard, 333
Speed density formula, 267
Speed sensor, 362
Starter signal, 377
Start stop control, 383-84

Stoichiometric air/fuel ratios, 47
Switches, as sensing devices, 71-72
System diagnosis and service, 50-51, 76-79, 116-31, 151-56, 197-200, 235-43, 301-8, 317-18, 320, 323, 334-35, 354-55, 365-66, 384-85
    Bosch (Robert) Motronic, 384-85
    Cadillac digital fuel injection, 235-43
    Chrysler, 354-55, 365-66
    for common components, 76-79
    Ford, 301-8, 317-18, 320, 323, 334-35
    General Motors, 116-31, 151-56, 197-200
    technician's approach, 50-51
System voltage/fuel pump voltage, 176

Tach drive, 365
Tach signal, 328
Temperature-compensated accelerator pump solenoid, 298-99
Thermactor air bypass and Thermactor air divert, 287-89
    solenoids, 288-89
    valves, 288
Thermactor air control, 315-16, 320, 333
Thermistor, 62-63
Third- and fourth-gear switches, 210
Throttle body injection unit, 136, 141
Throttle kicker, 285
Throttle position sensor, 88-89, 139, 168-71, 257-58, 318, 341
Throttle position switches, 329-31
Throttle stop switch, 361
Throttle switch, 210
Throttle valve switch, 374
Timed spray injection, 57
Torque converter clutch, 107-11, 146-48, 215-16, 293-95
Transistors, 33-35
Transmission converter clutch, 194-95
Transmission switches, 93, 140, 176, 267-71
Troubleshooting, see System diagnosis and service
Turbocharger, 193-94
Turbocharger boost control, 49, 382
Turbocharger boost control solenoid, 295-96

VAC sensor, 87
Vacuum secondary, 365
Vacuum transducer, 360-61
Vane airflow sensor, 266-67
Vane air temperature sensor, 267
Vane meter, 265-66
Variable voltage choke, 298
Vehicle speed control, 300-301
    programmed ride, 300-301
    top speed limiting, 300
Vehicle speed sensor, 91-92, 140, 174, 271, 346
Viscous converter clutch temperature sensor, 210
Voltage, 10, 11, 20-22
Voltmeter, 5

Wastegate control solenoid, 349-51
Wide-open throttle A/C cut-off relay, 298

Zener diode, 31